Revolutionizing Solar Energy Harvesting

This book provides insights into emerging semiconductor device technology, challenges, and solutions for harnessing solar power to produce sustainable energy and meet the escalating demand for electricity generation.

Revolutionizing Solar Energy Harvesting provides desired exposure to the ever-growing field of semiconductor electronic devices and technologies to produce power by harnessing solar energy. The authors highlight the role of semiconductors and process technologies in meeting global energy demand. They also explore international policies and standards for harnessing solar power. The authors then discuss the impact of semiconductor materials and architecture designs on photovoltaic performance. Finally, the authors then discuss the manufacturing and selection of materials using artificial intelligence (AI)–machine learning (ML) techniques and emphasize enhancing the production of defect-free semiconductor materials by employing AI–ML techniques.

The book is intended for researcher professionals in the field of nanomaterials and semiconductor devices for harnessing solar power codesign issues, as well as undergraduate/postgraduate students within Electronics or Electrical Engineering programs.

Revolutionizing Solar Energy Harvesting

Advanced Semiconductor Devices and Technology with Artificial Intelligence and Machine Learning Integration

Edited by
Piyush Kuchhal, Deepak Kumar,
Rupendra Kumar Pachauri,
and Vijay Kumar Sharma

CRC Press
Taylor & Francis Group
Boca Raton London New York

CRC Press is an imprint of the
Taylor & Francis Group, an **informa** business

Designed cover image: © Shutterstock

First edition published 2026
by CRC Press
2385 NW Executive Center Drive, Suite 320, Boca Raton FL 33431

and by CRC Press
4 Park Square, Milton Park, Abingdon, Oxon, OX14 4RN

CRC Press is an imprint of Taylor & Francis Group, LLC

ISBN: 978-1-032-85714-5 (hbk)
ISBN: 978-1-032-84957-7 (pbk)
ISBN: 978-1-003-51578-4 (ebk)

DOI: 10.1201/9781003515784

Typeset in Times
by Apex CoVantage, LLC

Contents

About the Editors

Dr. Piyush Kuchhal has extensive experience in education, research, and administration. He holds a Ph.D. in Physics from the IIT Roorkee and an M.S. in Physics with a specialization in Engineering Physics from the same institution. He has overseen operations, curriculum design, and academic quality as Cluster-Head of Electrical and Electronics Engineering at University of Petroleum and Energy Studies (UPES). He has also held the positions of Academic Coordinator, Associate Dean of Applied Sciences, and Department Head of Physics.

Dr. Deepak Kumar is currently a research associate at the Technische Universität Dresden Germany and the chair for Electron Devices and Integrated Circuits. He has continued his research with the DFG (Deutsche Forschungsgemeinschaft)-funded project "Experimental characterization and compact modeling of high-field effects in CNTFET channels." He has worked as an IEEE student branch counsellor at UPES and faculty coordinator for spoken tutorial-IITB. He is also serving as an active member of the Germany Chip Design program. He is working in solar cell technology, semiconductor devices, and compact modeling for self-heating in CNTFETs. He has also completed the virtual teacher final project from Coursera in collaboration with the University of California Irvine (UCI).

Dr. Rupendra Kumar Pachauri is an expert in renewable energy technology and electrical hybrid power generation systems (micro-grid advancements). His research is based on hydrogen fuel-cell power generation, energy management, advanced strategies for photovoltaic systems (stand-alone: ground-mounted and floating photovoltaic systems) for power enhancement during adverse environmental conditions (dust aerosol, partial shading conditions), and power generation forecasting using artificial intelligence and machine learning techniques. Dr. Pachauri is a distinguished academic and an associate professor holding an esteemed position within the Department of Electrical and Electronics Engineering at UPES in Dehradun, India.

Dr. Vijay Kumar Sharma brings over 15 years of research expertise in nanomaterials, optoelectronics, and photonic devices. He is a Senior Scientist and Program Manager at the LUMINOUS! Centre of Excellence for Semiconductor Lighting and Displays, Nanyang Technological University, Singapore. His work has led to impactful innovations, including a mercury-free UV disinfection technology now in commercial use. With a Ph.D. in Physics from IIT Roorkee and postdoctoral experience in Turkey and Singapore, Dr. Sharma is deeply committed to translating scientific research into practical, real-world solutions.

Contributors

Tayab Abu
School of Mechanical Engineering
Yanshan University
Qinhuangdao, China

S. Aditya
CSE
Vellore Institute of Technology
Chennai, India

Syed Ahmad
School of Mechanical Engineering
Yanshan University
Qinhuangdao, China

Sengupta Anindita
Department of Electrical Engineering
Indian Institute of Engineering Science and Technology (IIEST)
Shibpur, Howrah, India

Banu Anushiya
CSE
Vellore Institute of Technology
Chennai, India

Pangas Arijit
School of Advanced Materials
Green Energy and Sensor Systems (SAMGESS)
Indian Institute of Engineering Science and Technology (IIEST)
Shibpur, Howrah, India

Nazar Asif
Department of Sustainable Engineering
TERI School of Advanced Studies
New Delhi, India

K. Deepa
Department of Electronics and Communication Engineering
Manav Rachna University
Faridabad, India

S. Dhaneshver
SMEC
Vellore Institute of Technology
Chennai, India

Das Gourab
RCT Solutions
GmbH, Konstanz, Germany
and School of Advanced Materials, Green Energy and Sensor Systems (SAMGESS)
Indian Institute of Engineering Science and Technology (IIEST)
Shibpur, Howrah, Germany, India

Rahaman Hafizur
School of VLSI Technology, Indian Institute of Engineering Science & Technology
Shibpur, Howrah, India

Ghosh Kasturi
Department of Computer Science & Engineering-Artificial Intelligence
Brainware University
Barasat, Kolkata, India

Ghosh Dibyendu Kumar
School of Advanced Materials, Green Energy and Sensor Systems (SAMGESS)
Indian Institute of Engineering Science and Technology (IIEST)
Shibpur, Howrah, India

Gill Vijay Kumar
Department of Electronics and Communication Engineering
Manav Rachna University
Faridabad, India

Devi Velpuri Leela
Electrical Cluster
UPES
Dehradun, Uttarakhand, India

Gopal Lakshay Madan
CSE
Vellore Institute of Technology
Chennai, India

M. Maghutesan
SMEC
Vellore Institute of Technology
Chennai, India

Anwer Naqui
Department of Sustainable Engineering
TERI School of Advanced Studies
New Delhi, India

Sankaran Venkateswaran Pedinti
CADFEM India Pvt. Ltd
Hyderabad, India

Chaudhary Bhanu Pratap
Department of Electronics and Communication Engineering
Manav Rachna University
Faridabad, India

Acharyya Shiladitya
School of Advanced Materials, Green Energy and Sensor Systems (SAMGESS)
Indian Institute of Engineering Science and Technology (IIEST)
Shibpur, Howrah, India

S. Sridevi
SCOPE
Vellore Institute of Technology
Chennai, India

Bose Sukanta
Department of Electronics and Communication Engineering
Academy of Technology (AOT)
Adisaptagram, Hooghly, India

Mukhopadhyay Sumita
School of Advanced Materials, Green Energy and Sensor Systems (SAMGESS)
Indian Institute of Engineering Science and Technology (IIEST)
Shibpur, Howrah, India

Chaudhari Sushmita
Department of Computer Science & Engineering
Brainware University
Barasat, Kolkata, India

Li Yanwen
School of Mechanical Engineering Yanshan University
Qinhuangdao, China

1 An Overview of AI–ML Powered Systems with Computer Vision for Wafer Inspections for Solar Cells Applications

Abu Tayab, Yanwen Li, and Ahmad Syed

1.1 INTRODUCTION

The responsibility to safeguard the planet for future generations has never been more urgent, especially as the threat of global warming intensifies. Over the past century, greenhouse gas emissions have risen dramatically, primarily due to industrialization and the extensive use of fossil fuels. This increase in greenhouse gases, such as carbon dioxide (CO_2) and methane, has significantly increased the world's average surface temperature, triggering a series of climatic changes [1]. These changes destabilize the planet's delicate climate systems, leading to more repeated and unadorned climate events, including extremely high-temperature waves, hurricanes, and unpredictable rainfall patterns [2]. This process, known as environmental change, is not just a natural change in the Earth's atmosphere but is heavily influenced by human activities, particularly since the Industrial Revolution.

The significance of environmental change is comprehensive and poses important possibilities to both the natural environment and human societies. Ecosystems and biodiversity are under severe threat, with habitats like coral reefs, rainforests, and polar ice caps experiencing unprecedented stress because of increasing temperatures and ocean acidification. The warming and rising of oceans, increased frequency of droughts, and the likelihood of mass extinction are all direct outcomes of this climatic shift [3]. Moreover, climate change is exacerbating socio-economic issues, particularly in vulnerable regions. Increased poverty, displacement of populations due to uninhabitable conditions, heightened health risks from vector-borne diseases, and widespread food insecurity are becoming more common, disproportionately affecting developing nations [4]. The scientific consensus is clear: climate change is primarily human-induced, and, without immediate and drastic action, humanity is headed toward irreversible environmental damage. The global scientific community

DOI: 10.1201/9781003515784-1

agrees that only by making bold and coordinated efforts can we redirect the current trajectory and prevent the worst outcomes of climate change [5]. We are standing at a critical juncture, not only facing the precipice of environmental collapse but also holding the power to steer toward a sustainable future. The International Panel on Climate Conversion (IPCC), the leading global authority on climate science, has reported that nearly three-quarters of all greenhouse gas emissions stem from human energy consumption, predominantly from burning fossil fuels like coal, oil, and natural gas [6]. Addressing climate change therefore necessitates a comprehensive overhaul of the global energy sector, moving away from carbon-intensive sources of energy toward renewable, carbon-free alternatives. Surrounded by several renewable energy sources, photovoltaics (PV) stands out as one of the most promising solutions. Solar energy is abundant, with the Earth receiving more solar energy in a single hour than humanity uses in an entire year [7]. This makes solar power a natural candidate for addressing both global energy demands and the need to mitigate greenhouse gas emissions.

The IPCC has projected that, by 2030, nearly one-third of the global energy should be generated by PV power to prevent the most catastrophic effects of the weather revolution[8]. The key advantage of PV power is its cost efficiency. Over the past decade, the costs associated with producing and installing solar panels have plummeted, making PV the most affordable source of electricity in many parts of the world [9]. However, to accelerate the widespread adoption of solar energy and ensure it plays a pivotal role in global energy systems, several challenges must be addressed. The International Technology Roadmap for PV ensures outlined three critical schemes: budget optimization throughout the PV fabrication value chain, the development of specialized PV components for various market sectors, and, most essentially, improvements in the efficacy and dependability of PV devices [10]. Parallel to the global effort to combat climate change is the rapid advancement of artificial intelligence (AI) technologies. In recent years, AI has revolutionized various sectors, offering new tools that are transforming industries from healthcare [11] to engineering and even education [12]. The development of machine learning (ML) and deep learning, subsets of AI, has created opportunities for innovation in areas previously limited by human capabilities or traditional computation methods. In the situation of weather change, AI holds enormous possibilities to support humanity's efforts to stabilize the climate by improving our consideration of complex weather models, fast-tracking the discovery of new materials, and optimizing existing resolutions for better efficiency and sustainability [13].

Simultaneously, the semiconductor industry, which underpins much of the technological advancements in AI, Internet of things [14], automotive industry, and 5G networks, is undergoing rapid growth. Integrated circuits (ICs), the foundation of modern electronics, are densely packed electronic circuits etched onto silicon wafers. As the demand for advanced technologies continues to rise, semiconductor fabrication companies must improve their manufacturing processes to meet the growing need for high-performance ICs. The fabrication of high-grade semiconductors is a complex and delicate development, and even minor defects during the wafer production stage can lead to critical flaws [15]. Reducing faults and improving yield is therefore critical for semiconductor manufacturers. Traditional methods of defect detection, which rely on manual inspection, are increasingly seen as insufficient due to their high cost and limited accuracy. In recent years, ML techniques, particularly in image processing, have emerged as more proficient and commercial solutions for defect detection. One of the critical tools in power semiconductor engineering is the wafer bin map (WBM), which visually represents the spreading of imperfect defects on a wafer. WBMs use color coding to indicate the status of each die, helping engineers diagnose defects, identify patterns, uncover root causes, and monitor production quality [16]. Defects on semiconductor wafers often form spatial patterns, and, by studying these patterns, engineers can make targeted improvements in the fabrication process. This, in turn, leads to better product quality and higher yields of defect-free chips.

ML algorithms are principally suitable for this task, as they can process large datasets and detect subtle patterns that may not be obvious to human inspectors. These algorithms use powerful computational tools to analyze data from wafer inspections, identifying defects that are critical for quality control in the semiconductor industry. One of the most promising approaches in this field is deep learning (DL), a subset of ML that is particularly adept at recognizing complex patterns in data [17]. DL models offer several advantages over traditional methods, including the ability to learn openly from raw data without the need for manual feature abstraction. This allows for more autonomous, accurate, and efficient defect detection processes. Furthermore, DL models are highly adaptable and scalable, meaning they can be easily updated to recognize new defect varieties as the difficulty of semiconductor manufacturing increases [18]. Leading semiconductor companies such as Intel, Micron, and TSMC have been at the forefront of integrating AI–ML technologies into wafer inspection and defect detection. These companies leverage advanced ML algorithms and computer vision (CV) systems to significantly enhance the quality control of their semiconductor products. The successful implementation of these technologies in semiconductor fabs provides a promising avenue for similar innovations in the solar cell manufacturing industry, where AI–ML can similarly improve efficiency and defect detection.

In the realm of PV and semiconductor manufacturing, AI technologies, especially ML, significantly enhance wafer inspections for solar cell applications. By integrating CV with ML algorithms, researchers can develop vast quantities of data to detect patterns and defects in wafers with remarkable accuracy. This capability not only optimizes production processes but also improves the total efficacy and reliability of solar power, playing a vital role in the evolution to carbon-free power systems. Despite the advancements in ML for wafer defect detection, comprehensive reviews of the algorithms employed in this area remain scarce. This chapter aims to bridge that break by offering a comprehensive analysis and categorization of various ML classification algorithms used for wafer inspections, evaluating their methodologies, and analyzing their strengths and weaknesses. Numerous application areas of ML that have been extensively studied on both sides are depicted in Figure 1.1. Through this analysis, we suggest a basis for prospect research intended at refining ML techniques in both the PV and semiconductor manufacturing sectors.

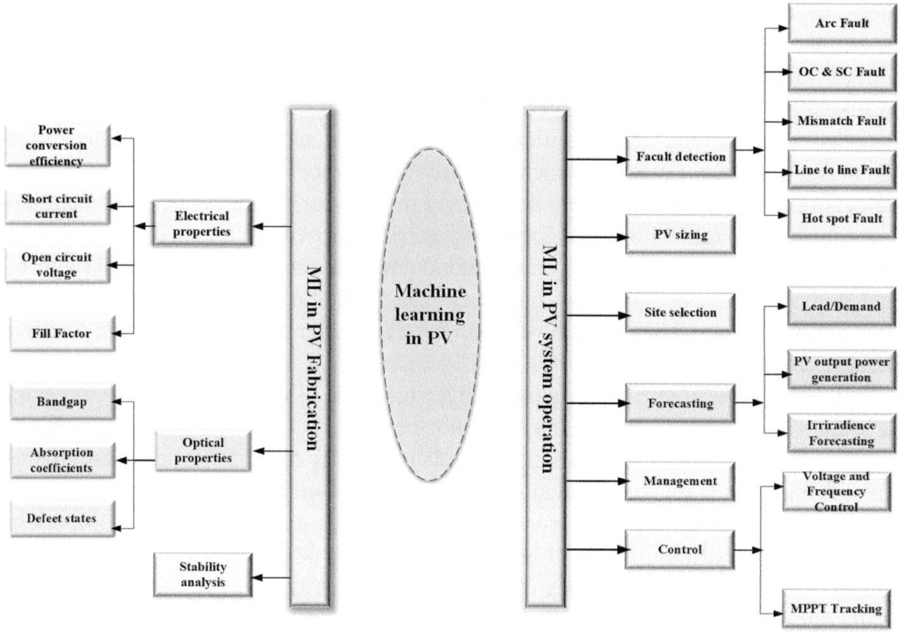

FIGURE 1.1 Applications of machine learning in photovoltaic fabrication and operation.

1.2 TYPICAL ML TECHNIQUES IN SOLAR CELLS

1.2.1 Single-Label Wafer Defect Scheme Using Convolutional Neural Network

This method utilizes convolutional films to extract and learn hierarchical structures from wafer imageries. This design enables the convolutional neural network (CNN) to effectively recognize complex patterns and attributes associated with various wafer defects, resulting in precise classification based on the acquired features. In CNNs tailored for wafer defect classification, the layered architecture is crucial. By employing a mix of convolutional and pooling layers, the network skillfully detects intricate features that signify defects, including scratches, irregular patterns, and surface anomalies on the wafer. These layers collaborate to extract and emphasize features essential for maintaining the quality and reliability of wafers.

In the realm of detecting defects in wafers, convolutional layers employ unique filters to examine wafer images for key attributes such as edges and textures. This initial feature extraction stage is crucial as it enables the CNN to identify regions that may have defects by recognizing patterns that differ from the standard, thus allowing for timely intervention during the manufacturing process. Given the extensive datasets commonly found in semiconductor production, pooling layers in CNNs play an essential role in reducing data size. By downsampling the feature maps, these layers significantly decrease both the computational burden and memory demands. After convolutional and combining layers, CNNs combine fully associated layers to categorize the features detected in the wafers into distinct defect types. The softmax layer in CNNs assigns probabilities to each defect type for classification purposes. Figure 1.2 shows the CNN methodology for identifying wafer defects. Highlighting how this technique is particularly effective in recognizing specific patterns and spatial relationships within small areas of images is vital for accurately detecting wafer defects. In automatic feature learning, CNNs can independently recognize features in wafer images, such as fundamental components like edges and textures, as well as more complex shapes. This capability assists in identifying wafer defects. Hierarchical modeling is utilized, where CNNs have multiple layers that learn features at different levels of complexity, allowing them to detect detailed patterns that signify wafer defects. Training on extensive datasets empowers CNNs to comprehend a range of wafer image structures, building them versatile for responsibilities such as defect detection, when labeled data are scarce.

For optimal performance of the technique, utilize pretrained CNN models, for instance, individuals developed on ImageNet, and to perfect them for specialized datasets like wafer defect data. Select a CNN architecture, such as ResNet, that effectively captures the complexities of the data while maintaining computational efficiency. Ensure that the CNN's depth is appropriately balanced to recognize intricate patterns without risking overfitting the data. Additionally, select optimizer algorithms and refine hyperparameters such as learning rates, batch sizes, and regularization techniques. Research papers applying this technique include a study [19], which presents a predefined CNN model paired with transferral learning that utilizes pretrained parameters to help the system identify fundamental patterns in wafer map defects. Yu et al. [20] developed a deep transfer learning model named JFLAN that

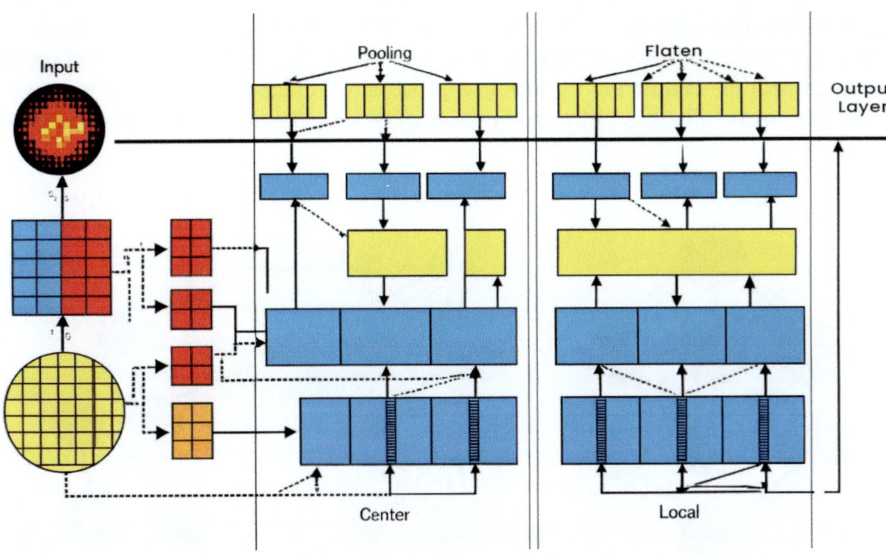

FIGURE 1.2 Diagram illustrating the model for detecting wafer defect patterns utilizing CNN.

employs CNNs to excerpt exchangeable features from wafer maps, introducing a distinctive feature learning methodology through transfer learning, which incorporates multilayer domain adaptation via adversarial training. Chithra et al. [21] established an automatic system for classifying wafer defects that employ a CNN paired with a memristor crossbar erection. This crossbar configuration uses pretrained neural network weights, with classification performed via a softmax layer. In a related study, engineers from GlobalFoundries evaluated a traditional support vector machine (SVM) method, which is usually used in CV, against a four-layer deep CNN for classifying wafer test maps. They aimed to improve the precision of detecting low-yield defects. While the SVM model required innovative engineering for its training, the CNN utilized the remaining image datasets and underwent training for 120 epochs. All models were created by 300 to 500 automatically labeled imageries corresponding to each of the 12 different wafer map patterns. The CNN significantly outstripped the SVM in terms of precision, achieving an average of 90% across the signatures and showing a high sensitivity to the shapes of the patterns.

1.2.2 AI-Based Neural Network (AI–NN)

An AI–NN is a computation ideal created on a directed acyclic graph, designed for learning intricate rules from large datasets. This approach allows for various tasks such as pattern recognition, data arrangement, and performance expectation, [22] to be executed primarily by complete learning from models rather than requiring the manual programming of specific rules. An artificial neural network (ANN) is structured with several layers of interconnected nodes known as artificial neurons. These artificial neurons are capable of receiving, processing, and transmitting signals based on different algorithms. A standard ANN model embraces three layers: the input layer, the unknown layer, and the output layer given in Figure 1.3. The neurons in the input layer take in the raw data provided to the network, while the hidden and output layer neurons handle the information processing. The output from the unknown layer neurons is influenced by the activities of the input layer neurons along with the

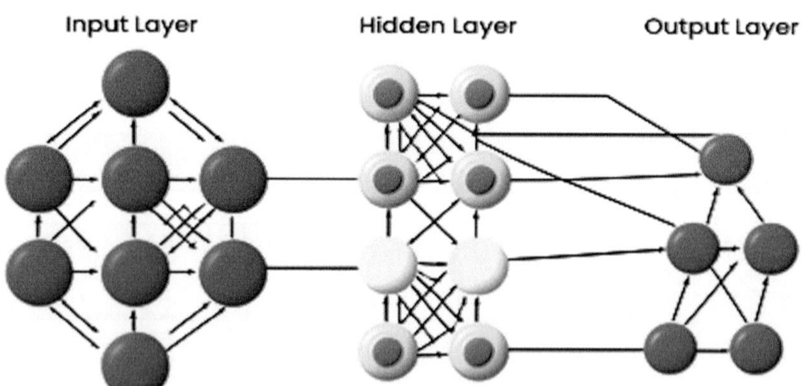

FIGURE 1.3 Configuration of an artificial neural network.

connection weights between these two layers. Conversely, the outputs of the output layer neurons are determined by the hidden layer neuron activities and the loads connecting the hidden neurons to the output neurons.

1.3 GENETIC ALGORITHM STUDIES

The genetic algorithm (GA) is a technique that emulates regular assortment and can address equally constrained and unconstrained optimization challenges. This algorithm begins with an initial population of genes. Through selection rules, it identifies certain individuals referred to as "parents," which will contribute to the next generation known as "children" (see Figure 1.4). The algorithm employs crossover and mutation rules to generate children that exhibit optimal fitness function values. The population of individual solutions is iteratively adjusted within the GA, progressing toward an optimal outcome. GA is suitable for tackling optimization problems involving objective functions that are discontinuous, stochastic, non-differentiable, or highly nonlinear. It can effectively handle difficult optimization responsibilities in areas such as pattern acknowledgment, image improvement, and separation [23]; transportation signal coordination [24]; software analysis; intrusion detection; and unmanned aerial vehicles (UAVs) [25].

1.4 PARTICLE SWARM-BASED OPTIMIZATION

Particle swarm-based optimization (PSBO) is a powerful optimization technique encouraged through a flock of birds foraging for sustenance [26]. Individually, birds in the flock navigate based on their own experiences and the movements of their companions until they collectively settle on a spot where food is available. In PSO, a group of potential solutions, referred to as particles, mimics the birds, and the optimal explanation to a problem is achieved by iteratively adjusting the situations of these particles within the search space using straightforward scientific formulas [27]. PSBO algorithms are smeared in numerous areas, for instance, path planning for UAVs, digital image watermarking, groundwater resource management, and optimizing radio frequency identification (RFID) networks by eliminating redundant readers.

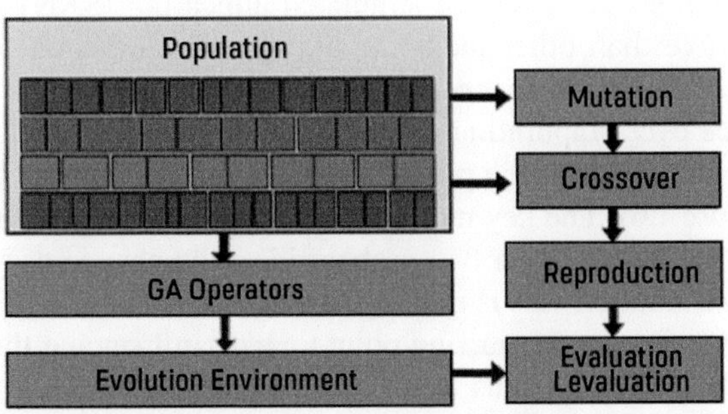

FIGURE 1.4 Configuration of genetic algorithm.

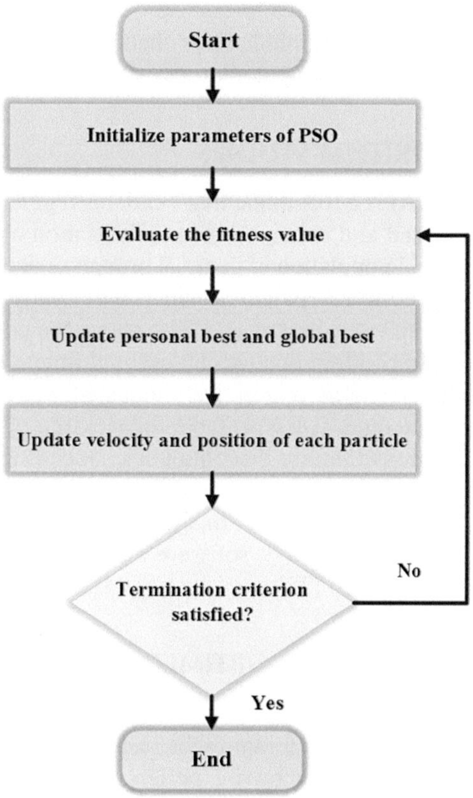

FIGURE 1.5 Configuration of particle swarm optimization algorithm.

1.4.1 Simulated Annealing

The system replicates the galvanizing progression used in materials science, and the temperature is intentionally lowered to allow for larger crystal formation and to minimize defects. It is commonly utilized to identify a global optimum within an extensive search space [28]. The simulated annealing (SAN) algorithm proves to be more effective than other methods, like gradient descent, particularly when dealing with discrete search spaces and when approximating the global optimum, taking precedence over pinpointing a precise local optimum. Typically, in the SAN algorithm, a clarification near the present is selected at a relatively high temperature (as shown in Figure 1.6). The key methods can be substituted with the casually chosen option or retained, based on two probabilities: the chance of progressing to an original, improved solution versus moving to a new, inferior one. The temperature is progressively lowered from its starting point to zero, influencing these probabilities. SAN is applicable across various domains, including very large-scale integration [29], manufacturing development, control engineering [30], ML, signal dispensation, and image renovation [31].

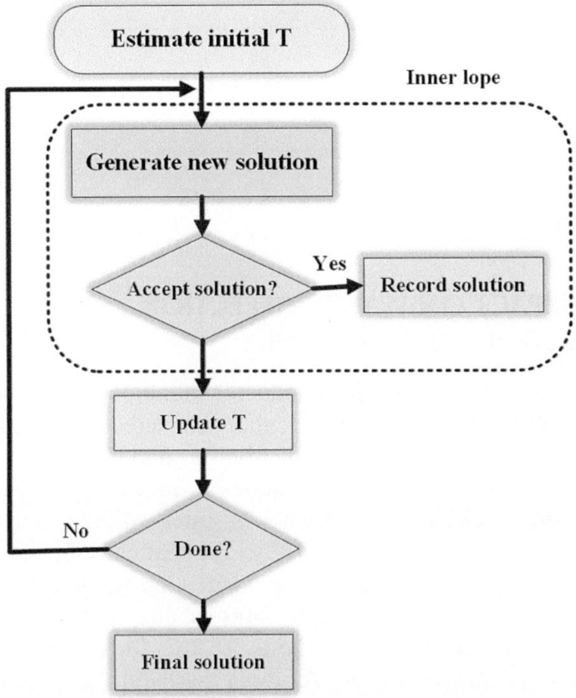

FIGURE 1.6 Structure of simulated annealing algorithm.

1.4.2 Decision Tree

It is among the most effective and straightforward ML systems. While primarily associated with classification tasks, they are also widely used for regression purposes [32]. A decision tree (DET) contains origins, divisions, and trees. The individual division evaluates a true or false statement, and the final decision is derivative from the trees. The key challenge in DETs is determining the optimal tree structure, which involves selecting an excruciating principle at the origin and divisions that most effectively divide the input models. The process is recognized as the feature collection extent (FCE). The most ordinarily used FCE techniques include statistics improvement and the Gini key, deliberate according to equations (1) and (3), respectively. Intended for any given nodule, the entropy, previously a split, is represented as entropy (S), where S denotes the overall sum of examples at that knob, and S_v represents the total of models belonging to class v. The entropy for N structures is defined by equation (2), with P_j indicating the possibility of casually choosing a component from class j. When using info improvement, DETs aim to capitalize on this assessment, while minimization is applied for Gini scores [33].

$$\text{Information gain}(S) = \text{Entropy}(S) - \sum_{v} \left(\frac{|S_v|}{|S|} \cdot \text{Entropy}(S_v) \right) \tag{1.1}$$

where

$$\text{Entropy}(s) = -\sum_{j=1}^{N} P_j \log_2 P_j \tag{1.2}$$

$$\text{Gini} = 1 - \Sigma_{j=1}^{N} p_j^2 \tag{1.3}$$

1.5 AI–ML IMPACT ON WAFER INSPECTION AND PRODUCTION EFFICIENCY

The adoption of AI and ML in semiconductor manufacturing [34], particularly in wafer inspections, has significantly enhanced defect detection and process optimization. Industry leaders like Intel, Micron, and TSMC have successfully integrated AI-driven tools and CV techniques to improve wafer quality, reduce defects, and optimize yields (see Figure 1.7). Intel, Micron, and TSMC utilize deep learning algorithms, especially CNN, to automatically detect defects such as micro-cracks, scratches, and surface irregularities on wafers. These AI systems can learn hierarchical patterns from large datasets, enabling high-precision detection far beyond traditional manual methods. TSMC, for instance, has integrated real-time process optimization using AI tools that adjust fabrication parameters based on inspection data, further minimizing waste and improving yield. The implementation of AI has led to significant improvements in defect detection accuracy and production efficiency. Intel's AI-driven system has increased defect detection rates by 30%, while process optimization has resulted in up to 20% improvements in yield. These advancements not only reduce manufacturing costs

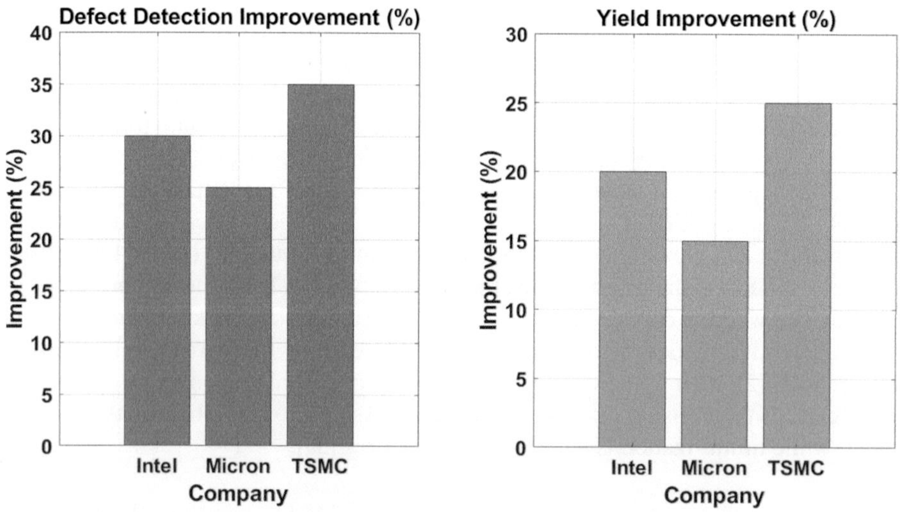

FIGURE 1.7 AI–ML-based defect detection and process optimization in wafer inspections.

but also ensure a more reliable and consistent product, keeping these companies competitive in the fast-paced semiconductor market. The success of these AI–ML techniques in semiconductor manufacturing provides valuable insights into solar cell production. By adopting similar AI-driven defect detection and process optimization methods, solar cell manufacturers can improve the efficiency, reliability, and longevity of their products. DL and CV technologies [35] could help detect defects such as micro-cracks in solar cells, while AI-powered real-time adjustments to fabrication parameters could improve yield and reduce waste. While these technologies offer substantial benefits, challenges such as high implementation costs, data requirements, and system integration remain. However, as AI–ML technologies continue to mature, these obstacles are likely to diminish. For solar cell manufacturers, adopting AI–ML technologies could lead to similar transformative benefits, enhancing both production efficiency and product quality in the future.

1.6 ML-BASED PEROVSKITE PV CELLS

Over a few years ago, P-PVCs have garnered considerable interest due to their affordability [36], high absorption coefficients [37], adjustable bandgaps, extended carrier lifetimes, and diffusion lengths. They also exhibit high defect easiness, low exciting energy, and excellent carrier flexibility [38]. Figure 1.8 shows the organization of P-PVCs featuring a layer designed to absorb energy, commonly known as the perovskite layer, situated among the whole transport layer (TL) and the electron transport layer (ETL). Significant advancements have been made in enhancing the power conversion efficiency (PCE) of P-PVCs. Rising significantly from 3.8% to 25.7%, this growth is attributed to extensive study and the favorable properties of these cells [39]. This swift development situates P-PVCs as a viable and interesting substitute for conventional solar technologies [40].

Despite their rapid efficiency improvements, perovskite solar cells (PSCs) still encounter several challenges before they can be considered commercially viable. The perovskite material typically follows the molecular formula ABX3, Here, A stands for a monovalent organic or inorganic cation, B represents a

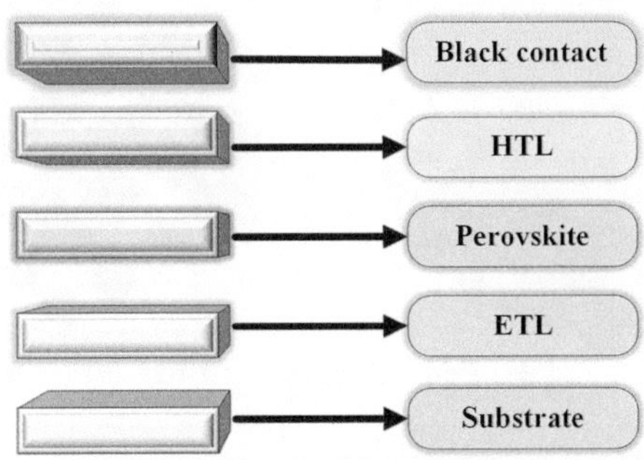

FIGURE 1.8 Structure of perovskite solar cells (PSC).

divalent metal ion, and X denotes a halide anion [41]. Figure 1.9 shows the mineral arrangement of perovskite materials. Common choices for the A site include cesium (Cs^+) and rubidium (Rb^+), while lead (Pb_2^+) and tin (Sn_2^+) are typically found at the B site. Here X is engaged by iodide (I^-), bromide (Br^-), or chloride (Cl^-). This versatility enables countless perovskite combinations and configurations for applications in solar cell technology. One notable representative of perovskite materials is their adjustable band gap, consenting to engage a wide spectrum of photons. Adjusting perovskite structure can fine-tune the band gap to optimize the material's absorption properties [42]. To maximize PSC efficiency, classifying a perovskite material with a perfect band gap aligned with desired performance parameters is crucial.

The structural instability of P-PVCs poses a major obstacle to their commercialization, as they can play a crucial role in the degradation of cells over time. Various issues, like high-temperature fluctuations, dampness levels, and bright coverage, subsidize this variability. Degradation of perovskite materials depressingly powers both the efficacy and the duration of PV cells. Moreover, the presence of lead [43], a toxic element, raises environmental concerns, as it can leach from the cells and the soil in the surrounding area. To mitigate these stability topics, academics are focused on increasing different perovskite materials that offer enhanced stability and reduced toxicity [44]. This includes the creation of lead-free alternatives, the incorporation of stabilizing materials, and the application of encapsulation techniques via environmental solar cell factors.

Recent years have witnessed the emergence of ML techniques as powerful tools for accurately predicting PSCs' band gap, PCE, and immovability, thereby accelerating their development and optimization. Figure 1.10 illustrates the integration of ML into the PSC fabrication process. Numerous investigations have established the effectiveness of ML techniques in approximating key parameters such as band gap, PCE, and constancy. This segment summarizes the role of ML methods in PSC production.

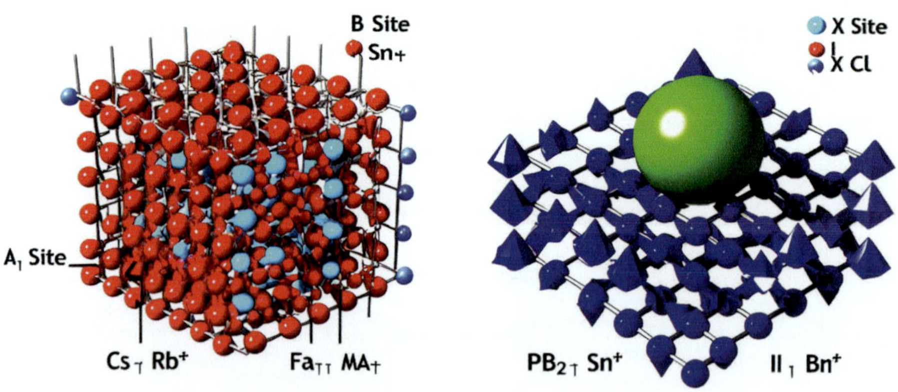

FIGURE 1.9 Crystal structure of perovskite material.

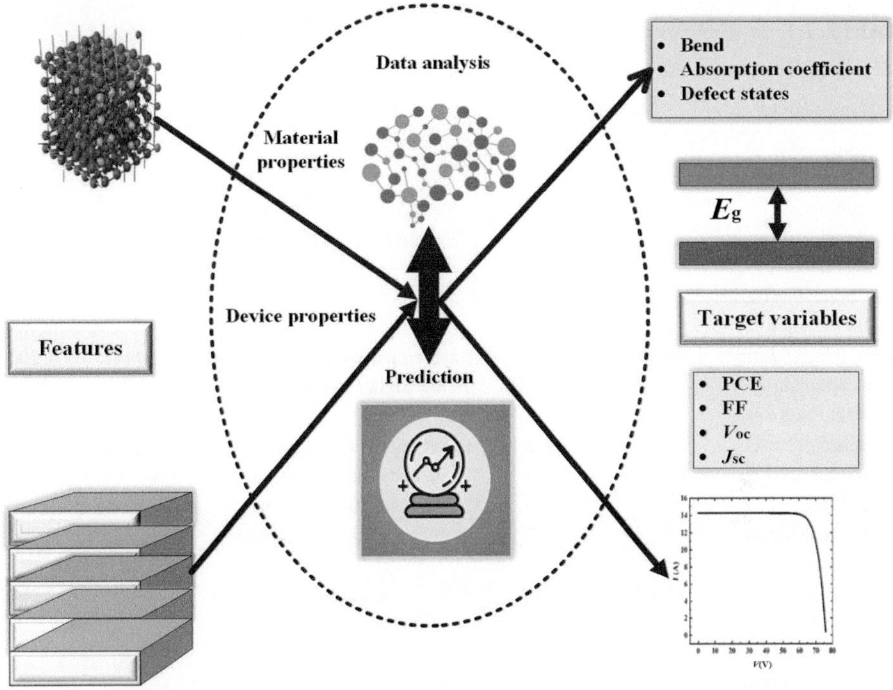

FIGURE 1.10 ML application in PV fabrication.

1.6.1 ML in the Prediction of Stability

The researchers explored how storage conditions, materials used in cell development, and statement methods affect individually the constancy [44] of the policies and compiled their findings in Table 1.1 They also examined how environmental factors influence stability, concluding that high humidity, elevated temperatures, and intense lighting negatively impact the stability of solar cells. Hu et al. [45] devised an ARM-based ML approach to identify the primary factors that affect the stability of 2D/3D PSCs. Their results designate that encrusted perovskites exhibit greater stability than passivized or varied assemblies. Among common 2D cations, butylammonium (BA) proved more stable than PEA, although it achieved advanced ordinary control adaptation efficacy. Varied 3D camion perovskites, especially those comprising formamidinium (FA), demonstrated enhanced constancy in 2D/3D configurations associated with single-cation structures. Hu et al. [45] applied an extremely randomized tree technique to predict perovskite solar cell stability, revealing the effect of individual levels on constancy. The model was trained on 1,050 perovskite device models varying in materials, installation systems, and storing settings. Results showed that incorporating hydrophobic materials in PSC layers improves device stability. In addition, perovskite covers with multiple cations and

TABLE 1.1
Existing Research Analysis on PV Stability by ML

Research Focus	Methodology	Key Findings	Reference	Year
Properties of manufacturing materials, deposition, storage, and ambient environments on cell constancy	ARM, decision trees	Low humidity enhances stability; mixed-cation perovskites and multi-spin coating improve stability.	[43]	2020
Identifying stability descriptors in 2D/3D PSCs	ARM	Layered perovskite structures show higher stability.	[44]	2020
Influence of materials and individual perovskite layers on device stability	Extra trees classifier	Hydrophobic PSC layers enhance stability; 2D–3D gemstone configurations improve long-term constancy.	[36]	2023
Influence of grain size, defect compactness, band gap, and external coarseness on constancy	PLSR, ridge regression, SVR, GPR	Surface roughness and grain size are key factors for long-term stability.	[45]	2022
Optimization of coating layer for $MAPbI_3$ stability	RF	Phenyltriethylammonium iodide increases MAPbI3 stability by approximately 4 ± 2 times.	[46]	2020
Relationship between reproducibility, hysteresis, and stability in PSCs	ARM, pooled variance	Favorable deposition techniques enhance reproducibility, low hysteresis, and long-term stability.	[47]	2020

2D/3D crystal erections contribute to ongoing constancy. For standard cells, the combination of TiO_2/m-TiO_2 as the ETL, 2D/3D perovskite as the absorber layer, P_3HT as the hole transport-TL, $LiTFSi^+$ 4-tert-butylpyridine (TBP) as the secondary HTL layer, and carbon as the back contact, along with DMF^+ DMSO as the pioneer key and chlorobenzene as the antisolvent, enhanced device stability. For reversed cells, constancy is enhanced by bathocuproine (BCP), Phenyl-C_{61}-butyric acid methyl ester (PCBM), $MAPBI_{3-x}Cl_x$, NiO, DEA, and aluminum as the back communication.

Hartono et al. [46] utilized ML to develop a highly efficient PSC with the configuration of glass/TiO_2/$FA_{0.85}MA_{0.1}Cs_{0.05}Pb(I_{0.97}Br_{0.03})_3$/spiro-OMeTAD/Au, achieving a

PCE of 23.4% along with excellent long-term ambient constancy. In this study, ML was implemented to formulate models that connect stability and efficacy, specifically open-circuit voltage; it correlated five distinct aspects: band gap, defect density, particle size, fluorescence lifetime, and surface coarseness, with Jsc and F. Those ML systems (partial least squares regression [PLSR], ridge regression, support vector regression [SVR], and Gaussian process regression [GPR]) were employed to determine the optimal correlation concerning structures and object assets. SVR yielded the best results. Their research indicated that bandgap significantly influences both PCE and Jsc, while surface coarseness and grain scope are major factors affecting device stability. They tested various galvanizing temperatures (60°C, 90°C, 110°C, 120°C) to modify perovskite thin-film grain size and employed different organic compounds (ortho-, meta-, and para-methoxyphenylethylamine iodized salts) to adjust surface roughness. The highest scheme constancy was observed with the major particle extent and minimal seeming coarseness. Additionally, low-dimensional perovskite, used as a capping coating atop the perovskite absorber, enhances solar cell ecological constancy.

Hartono et al. [47] introduced an ML method aimed at enhancing the coating layer for $MAPbI_3$. The authors evaluated 21 organic salts for use as coating-layer materials, which included iodine- and bromine-based salts with diverse sizes, structures, and chemical characteristics. Random forest models and SHapley Additive exPlanations (SHAP) principles indicated that minimal hydrogen-bond benefactors and small topological polar surface area (TPSA) are key features influencing device constancy. Phenyltriethylammonium iodide (PTEAI) was identified as the most stable capping layer due to its low hydrogen-bond donors and small TPSA. PTEAI increased $MAPbI_3$ constancy period by 4 ± 2 epochs compared to untreated $MAPbI_3$ and by 1.3 ± 0.3 times over octylammonium bromide (OABr). Table 1.1 summarizes our review of recent research papers (past three years) using ML methods to explore factors affecting PSC stability.

1.6.2 Prediction of Material Properties

Identifying appropriate materials for solar cells is an important contest due to the vast array of potential candidates. Traditionally, the development of new PV materials with the desired characteristics has relied on a basic trial-and-error method, which is costly and time-concentrated. Conversely, with the swift progress of quick-speed calculating and the growing volume of available data, it's now possible to use algorithms to expect the assets of PV-based materials on specific examples. This is an exciting development, as these techniques enable initial-stage predictions of key substantial properties and earlier synthesis of different materials, potentially facilitating more targeted synthetic approaches. Tailored experiments can be designed and executed to create materials as predicted. Table 1.2 summarizes the properties predicted using ML techniques.

It has been shown that ML methods, like ANN and GA, effectively forecast the assets of PV materials. The majority of published estimates focus on "dye-sensitized solar cells (DSSC)" and "organic solar cells (OSC)."

TABLE 1.2
Summary of Properties Projected by ML Techniques

References	Property	ML Methods
[48]	Absorption maxima	AN-NN, GA
[49]	Band gap	RF, GA, SVM, KRR
[50]	Conductance	AN-NN
[50]	Charge transport	AN-NN
[50]	Particle concentration	AN-NN
[50]	Resistivity	AN-NN
[51]	Solubility	AN-NN
[52]	Constancy	ERT + KRR, GA
[53]	I-V enactment	KNN, RF, GA, GB
[54]	Physical separation	GA
[54]	Reformation energy	GA
[55]	Concentration plane	AN-NN
[55]	Tension stickiness	AN-NN

1.6.3 Optimization of Fabrication Processes

Determining the precise connection between the preferred characteristics of PV resources and the adjustable fabrication situations is highly challenging due to the extensive collection of handling parameters and the trade-offs that exist between various assets. ML systems can generate models that can be well-trained with existing data to accurately forecast material properties within uncharted chemical spaces, allowing for their specific design and synthesis. These ML techniques have been utilized to enhance fabrication processes for components such as Si_3N_4 thin films, metal contacts, texturing, and emitters in silicon solar cells; TiO_2 photoanodes in DSSC; laser "scribing in thin-film" astral cells; statement of thin-film layers of an absorber; and the processing of silicon wafers. This approach has significantly decreased both the costs and time involved in experimental designs. Table 1.3 summarizes the fabrication processes that have been optimized using ML techniques.

In reference [56], experimental findings from the PECVD of Si_3N_4 thin films were utilized to develop the ANN system, which effectively demonstrated how deposition conditions impact film characteristics. The study revealed that parameters such as temperature and the flow rates of SiH_4 and NH_3 are essential for improving hydrogenation, thereby enhancing carrier lifetime in polysilicon solar cells. ANNs were employed to explore the impact of process parameter interactions on the refractive index. SiH_4 flow rate was identified as the most influential factor. Numerous previously unrecognized interaction effects, particularly related to the SiH/NH bond ratio, were visualized and interpreted. Consequently, the ANN model of refractive key not only enhanced the appreciation of statement

TABLE 1.3
Reviews of the Fabrication Processes Optimized by ML Methods

References	Route	Material	ML Methods
[56], [57]	Plasma-enhanced chemical vapor deposition (PECVD)	Si_3N_4	ANN, GA
[58]	Co-firing	Ag grid	ANN + GA, AN-NN + PSO, AN-NN
[58]	Surface patterning	Si wafer	AN-NN + GA, AN-NN + PSBO
[59]	Diffusion	Emitter	GA, AN-NN + GA, AN-NN + PSO
[60]	Deposition	TiO_2	GA, AN-NN
[61]	Deposition	Perovskite	GA
[62]	Filling	Dye	AN-NN + GA
[63]	Imperfect zone Subtraction	Silicon wafer (SCW)	FL
[63]	Cutting	SIW	AN-NN
[63]	Lapping	SIW	AN-NN

mechanisms but also facilitated the identification of valuable trade-offs among variables. An ANN was utilized to enhance a room temperature "PECVD") [21] method, resulting in a comparatively high statement rate through the control of duty ratio and bias power. It was found that the insufficient availability of nitrogen atoms concerning silicon on the film surface was the main factor that reduced the statement amount at minor duty ratios. Akiki [57] combined ANN and GA to develop a technique that enhances development factors for PECVD Si_3N_4 thin films, including power, density, SiH_4 current ratio, NH_3 current ratio, N_2 flow rate, and substrate temperature. This approach resulted in improvements of 73% and 81% in carrier lifetime, compared to traditional and statistical regression methods, respectively.

In reference [58], ANN and GA have been utilized together to enhance the co-sacking procedure settings for a screen-published silver grid on the obverse side and aluminum on the back side of monocrystalline silicon solar modules. The effectiveness of this technique was validated through further investigation, resulting in an average efficacy of 15.44%. This represents a 3.9% improvement over previous experiments using the design of experiments approach. ANN was integrated with GA and PSBO, correspondingly, to improve co-firing route parameters, counting the temperature of the individual belt furnace zone and belt speed. Results indicated that PSBO-optimized efficiency was slightly higher than GA-optimized efficiency. Repeated PSBO applications produced method parameters with reduced standard deviations, suggesting greater consistency in parameter generation. AN-NN was used to reduce interaction confrontation between the obverse anode and silicon. The optimization route focused on parameters like paste arrangement, silicon substrate morphology, co-firing temperature, laser beam power, and laser beam perusing rapidity during co-firing and selective laser sintering.

1.7 ML IN THE PRODUCTION OF CONVENTIONAL SOLAR MODULES

Abadi et al. [64] introduced an innovative approach for forecasting the performance of devices and elucidating the correlations between structure and properties in organic PV by integrating ML with Theory of Descriptors (TODE). They constructed a comprehensive database containing 240 small-particle gradual solar modules, incorporating 220 unique contributor resources paired with two fullerene-centered acceptors (PC61BM and PC71BM). The predictive example employed ten important chemical topographies related to energy adaptation as descriptors. By combining TODE and AN-NN systems, they derived valuable plan insights. The results indicate that decreasing the visual band gap improves organic solar cell enactment by growing photon immersion [65]. Moreover, the model identifies ideal values for ΔL and ΔH for each optical band gap. It has been perceived that organic solar module materials with a low optical band gap exhibit higher PCE when ΔL and ΔH are minimized. The prediction model suggests that achieving near-zero ΔL and ΔH values is favorable for high PCE outcomes.

Zhu et al.'s paper [66] applied ML methods to explore the key factors that mark the enactment of Cu $(In_{1-x}Ga_x)Se_2$ (CIGS) solar cells, as well as the connections between these factors. They built upon previous research and experiments to identify 15 input features for the ML models. These features included the Cu/Ga/In ratio, grain size, CIGS level thickness, maximum substrate temperature in manufacturing, production technique, alkali conduct, substrate style, and other factors. Between the various algorithms evaluated, the RF ideal completed the maximum precision in predicting the efficiency of CIGS solar cells. Their findings indicated that the most significant factors influencing device performance were alkali PDT, CIGS thickness, barrier level thickness, Cu/Ga ratio, and Ga/In incline. Therefore, the authors recommended using a co-vaporization production technique for CIGS solar cells with a CdS barrier level. They also recommended that optimal CIGS thickness should be between 2 and 2.3 micrometers, with the substrate temperature set between 540°C and 600°C, while maintaining Cu/Ga and Ga/In ratios of 0.87 to 0.95 and 0.37 to 0.40, respectively, to optimize efficiency in CIGS solar cells.

Salman et al. [67] developed an ML model based on ANN and utilized to identify the definitive copper nobbling profiles that arise from the dispersal galvanizing and cooling phases in the fabrication of CdTe solar modules. The photovoltaic reverse design (PVRD)–fast adaptive solver for poisson (FASP) solver was used to generate a substantial dataset. The goal was to replicate the copper dispersal process executed by the PVRD–FASP solver through an ANN model. The main aim was to simulate the copper diffusion process performed by the PVRD–FASP solver using the ANN ML model. The results from the simulation of the PVRD–FASP solver closely matched the predictions from the ANN algorithm. Furthermore, it was observed that creating Cu-doping profiles with the PVRD–FASP solver is lengthy, whereas the ANN algorithm can accomplish this almost immediately. This efficiency renders the ANN algorithm an effective tool for producing optimal doping profiles for CdTe solar cells.

Mohammad et al. [68] employed ANN-based ML models to forecast the PCE of three types of silicon solar cells: monocrystalline, multi-crystalline, and amorphous crystalline. They gathered training data by testing investigational solar module samples under different light intensity and ambient temperature scenarios. The predictive outcomes were assessed using the correlation coefficient and mean squared error (MSE), aligning well with established physics-based solar cell studies. Moreover, the authors identified the ideal number of hidden layers in the ANN [69] architecture for each solar cell type, highlighting how crucial the hidden layer count is for the accuracy of predictions. Their analysis showed that having three or four hidden layer units resulted in high correlation coefficients and low MSE for multi-crystalline and amorphous crystalline cells. In comparison, the best results for monocrystalline cells occurred with eight hidden layer units.

Anti-reflective coatings (ARCs) are utilized on the surfaces of solar modules to decrease light reflection and enhance its absorption, ultimately boosting power generation. In [70] the authors suggested employing an AN-NN to predict the most effective depth of ARCs based on the voltage and current of the solar cells. The team used PC1D, a one-dimensional numerical simulation tool, to compile the training dataset for their ANN model. Findings indicated that a thicker ARC could increase solar cells' power output.

Kaya and Hajimirza [71] created an NN alternate model to estimate the visual characteristics of thin-film multilayer amorphous-silicon solar cells utilizing NN techniques. They generated a training dataset through finite-difference time-domain (FDTD) simulations. Their NN model showed both accuracy and notable speed advantages over FDTD simulations when predicting optical absorption with the wavelength of incident light and the geometry of the cell. This study investigates the factors that enable an ML-based surrogate model to effectively replicate the outcomes of device physics-based simulations. The optimization findings reveal significant enhancements in exterior substantial efficiency related to traditional silicon and randomly designed systems.

The researchers [72] suggested using an ensemble of decision stumps trees (DST) that incorporates contained sub-sampling as the ML model to study the PCE of mass-produced Q. ANTUM solar cells made from p-type czochralski silicon (Cz–Si) wafers. Their main objectives were to comprehend the hourly global trends in solar cell efficiency over a week's worth of data and to explore how different production factors affected these changes in performance over time. They collected 50k training data from PV cells via inline extents conducted throughout the seven days. Utilizing hierarchical clustering, they grouped 329 features based on their similarity in affecting cell performance. The developed ML model predicted solar cell efficiency over time, treating time as a feature. The authors proposed that this highly accurate predictive model could be implemented in real-world production settings.

1.8 DEVELOPMENTS OF ML-BASED SOLAR CELL METHODS

To analyze the recent improvements in ML design methods in solar cell production, a quantitative review of literature from the past 10 years has been conducted. The findings presented in this article do not provide a precise count of the publications related to ML techniques in solar cells, but they do reveal observable trends. As illustrated in Figure 1.11, ANN and GA are the normally utilized ML schemes. AN-NN models are advantageous due to their less statistical training for development. The selection of AN-NN due to their nonlinear relationships between independent and dependent variables can be implicitly identified. Furthermore, various well-trained systems can choose to tailor the AN-NN system to diverse circumstances. Correspondingly, GA is user-friendly and does not require complex mathematical methods, making it accessible to researchers. GA primarily employs stochastic search strategies inspired by natural genetics and selection. Moreover, GA's evolutionary operators are well-suited for global searches. GA is highly adaptable and can combine with other systems to create effective hybrid systems for specific scenarios. Figure 1.12 illustrates the distribution of ML techniques in solar cell design and fabrication. ML methods have been widely used for optimizing device configurations and fabrication processes, indicating their suitability for these applications.

Here the whole ML schemes and their solar cell production are depicted in Figure 1.13. Before 2006, the use of ML studies in this area was quite uncommon, largely due to a general lack of interest in PV technologies. However, since 2006, substantial advancements in PV technologies have occurred, driven by increased financial backing from governments, leading to a global surge in PV system installations and the development of various PV technologies. As AI has rapidly evolved, its potential has become widely acknowledged. Implementing ML design techniques and solar cell manufacturing started gradually but has seen more remarkable growth in recent decades (Figure 1.13).

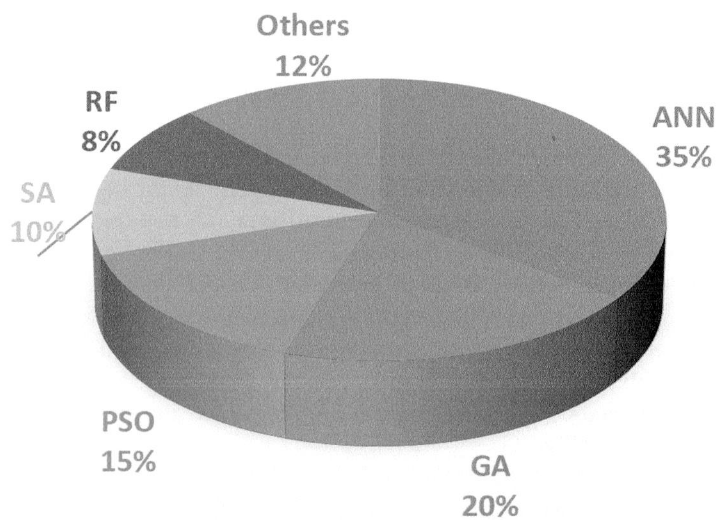

FIGURE 1.11 Assessment of ML studies on solar cell fabrication.

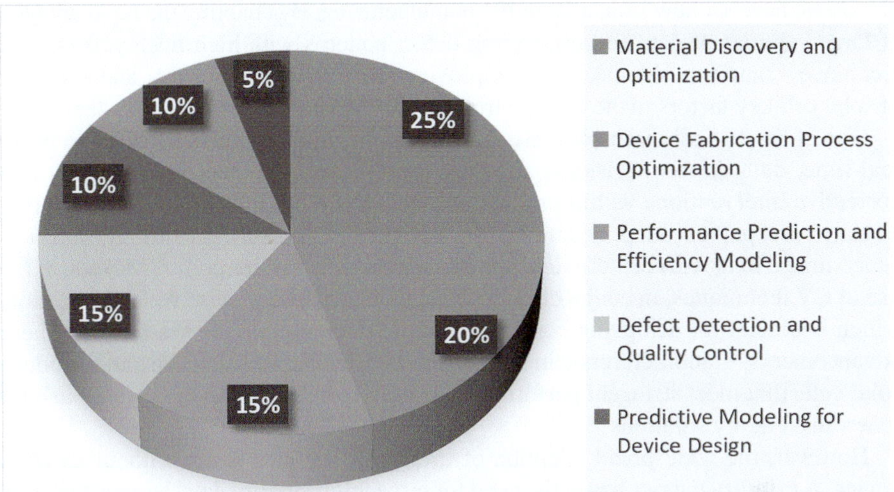

FIGURE 1.12 Various solar cell fabrication ML schemes.

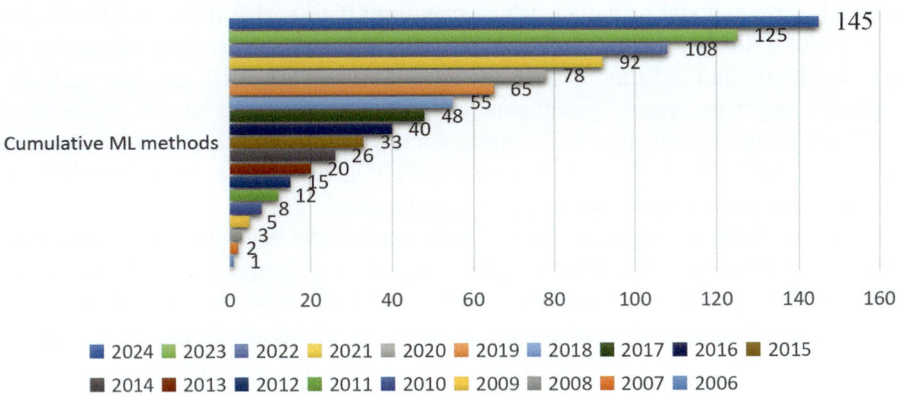

FIGURE 1.13 Summary of cumulative solar cell fabrication ML schemes.

1.9 CONCLUSION

This chapter highlights the transformative impact of the integration of AI, ML, and CV technologies into solar cell fabrication and wafer inspection processes, which represents a pivotal advancement in PV production. Through these technologies, the ability to detect, classify, and address defects in semiconductor wafers has become both faster and more precise, surpassing the limitations of traditional, labor-intensive inspection methods. This review underscores how AI-driven methodologies, such as CNNs for defect localization, AN-NNs for property prediction, and GAs for optimizing production processes, have set new standards in PV manufacturing. By enabling the rapid analysis of large volumes of data and uncovering defect patterns with high fidelity, these ML techniques contribute significantly to improving the reliability, efficiency, and longevity of solar cell key factors in the quest to transition to carbon-neutral energy systems.

One of the critical merits of these AI–ML applications is their capability to make real-time, data-driven decisions, allowing for early-stage defect identification and corrective interventions within production lines. Deep learning frameworks, in particular, have proven adept at automating complex tasks, such as linking luminescence image data with cell efficiency metrics to predict failure points. Moreover, the use of CV techniques, in conjunction with AI tools, enables precise wafer inspection, which is crucial for achieving consistent quality and minimizing waste. With these advancements, manufacturers can streamline operations, reduce costs, and produce solar cells that meet stringent performance criteria, aligning with the global push for renewable energy solutions.

However, the widespread adoption of these technologies is not without its challenges. A primary barrier lies in the need for large, high-quality datasets to effectively train and validate ML models. Data availability and quality can vary significantly across production environments, often necessitating the development of custom data-collection and preprocessing protocols. Furthermore, these models require substantial computational resources, which may limit their scalability in smaller manufacturing facilities. Another challenge is the complexity of real-time integration of AI algorithms into existing manufacturing systems, which often requires multidisciplinary expertise spanning semiconductor physics, AI, and systems engineering. Addressing these challenges will require continued research collaboration between academia and industry, along with investments in infrastructure that can support the computational demands of advanced ML applications.

Looking forward, future research should explore the development of more generalized and adaptive ML models capable of handling diverse materials and defect types across varying environmental conditions. This could be achieved by incorporating techniques such as alteration learning, which assists models in leveraging information from interrelated fields to progress enactment in different tasks, thereby reducing the dependency on large labeled datasets. Additionally, advancements in unsupervised and semisupervised learning could mitigate data scarcity issues by allowing models to extract meaningful patterns from unlabeled or partially labeled data. The integration of reinforcement learning, which adapts in real time to dynamic environments, also holds promise for enhancing the autonomy and resilience of PV production lines, particularly in the context of fluctuating manufacturing conditions.

Ultimately, the continued refinement of AI and ML methodologies in PV fabrication and inspection offers deals toward ecological energy solutions. As PV technology evolves, AI-powered systems can serve as a foundation for more intelligent, automated production processes that meet the increasing global demand for clean energy. By bridging the gap between semiconductor manufacturing and AI-driven optimization, the PV industry is well-positioned to achieve new levels of efficiency, reliability, and scalability. This examination serves as a foundational source for dynamic researchers and industry well-qualified professionals and gives a comprehensive roadmap for the next generation of ML-enhanced PV systems that have a decisive role in the energy landscape of tomorrow.

1.9.1 Future Directions

Here, AI and ML have made impressive strides in PV wafer inspection and solar cell manufacturing; several key areas of future research could address current challenges and expand the applications of these technologies:

1. **Development of Transfer Learning and Few-Shot Learning Techniques**: Most AI–ML models in PV systems require substantially considered datasets for active training, which can be a barrier in environments where data collection is limited. Future research should explore transfer learning and few-shot learning approaches, which enable models to adapt pre-trained knowledge from related domains to new applications with minimal labeled data. This direction could make AI–ML techniques more accessible and practical for smaller or data-constrained production environments.
2. **Integration of Reinforcement Learning for Real-Time Process Control**: RL algorithms offer significant potential for dynamic optimization in PV production. By training models to learn from real-time feedback, RL can enable systems to adjust production parameters on the fly, improving efficiency and yield under varying manufacturing conditions. This could lead to a new generation of self-optimizing manufacturing lines capable of responding autonomously to environmental changes and process fluctuations.
3. **Advanced Defect Prediction and Prevention Models**: Although current AI models focus on defect detection, future research could explore predictive models that identify potential defect formation during the early stages of fabrication. By predicting and preventing defects before they manifest, manufacturers can proactively address quality issues, reducing waste and enhancing production efficiency.
4. **Hybrid AI Models for Complex Data Fusion**: PV manufacturing generates data from diverse sources, such as luminescence imaging, environmental monitoring, and electrical testing. Hybrid AI models that combine techniques from deep learning, CV, and traditional statistical methods could be developed to integrate data from multiple sources, providing a more holistic understanding of production conditions. Such models could enable more accurate and comprehensive defect characterization and process optimization.
5. **Scalable, Energy-Efficient AI Algorithms**: As the petition for computationally intensive models is produced, it will become necessary to develop more energy-efficient algorithms that maintain accuracy without excessive resource consumption. Innovations in hardware acceleration, lightweight model architectures, and efficient algorithms could enable scalable AI solutions that are viable for larger production lines and reduce the environmental footprint of AI-driven PV manufacturing.
6. **Standardization of AI Practices in PV Manufacturing**: Establishing standard AI practices for PV manufacturing could benefit the industry by creating consistent benchmarks, protocols, and performance metrics. Standardization could facilitate easier comparisons of model performance, improve data sharing across organizations, and foster collaborative research, accelerating AI adoption in the PV sector.

REFERENCES

[1] L. F. Cook, "The latest IPCC synthesis report: What we learned about climate science, and what we learned about ourselves," Environ. Sci. Policy Sustain. Dev., vol. 65, no. 5, pp. 19–25, 2023.
[2] A. E. Dessler, Introduction to modern climate change. Cambridge University Press, 2021.
[3] Y. Malhi et al., "Climate change and ecosystems: Threats, opportunities and solutions," Philos. Trans. R. Soc. B, vol. 375, no. 1794. The Royal Society, p. 20190104, 2020.
[4] T. Dietz, R. L. Shwom, and C. T. Whitley, "Climate change and society," Annu. Rev. Sociol., vol. 46, no. 1, pp. 135–158, 2020.
[5] K. F. Myers, P. T. Doran, J. Cook, J. E. Kotcher, and T. A. Myers, "Consensus revisited: Quantifying scientific agreement on climate change and climate expertise among Earth scientists 10 years later," Environ. Res. Lett., vol. 16, no. 10, p. 104030, 2021.
[6] V. Masson-Delmotte, P. Zhai, H.-O. Pörtner, D. Roberts, J. Skea, and P. R. Shukla, Global warming of 1.5 C: IPCC special report on impacts of global warming of 1.5 C above pre-industrial levels in context of strengthening response to climate change, sustainable development, and efforts to eradicate poverty. Cambridge University Press, 2022.
[7] M. Perez and R. Perez, "Update 2022—A fundamental look at supply side energy reserves for the planet," Sol. Energy Adv., vol. 2, p. 100014, 2022.
[8] V. P. Masson-Delmotte et al., Ipcc, 2021: Summary for policymakers in: Climate change 2021: The physical science basis contribution of working group i to the sixth assessment report of the intergovernmental panel on climate change, 2021.
[9] S. Bouckaert et al., Net zero by 2050: A roadmap for the global energy sector, 2021.
[10] A. Jäger-Waldau, "Snapshot of photovoltaics—May 2023," EPJ Photovoltaics, vol. 14, p. 23, 2023.
[11] K. Guo, Z. Yang, C.-H. Yu, and M. J. Buehler, "Artificial intelligence and machine learning in design of mechanical materials," Mater. Horizons, vol. 8, no. 4, pp. 1153–1172, 2021.
[12] T. Ilkka, "The impact of artificial intelligence on learning, teaching, and education," European Union, 2018.
[13] J. Cowls, A. Tsamados, M. Taddeo, and L. Floridi, "The AI gambit: Leveraging artificial intelligence to combat climate change—Opportunities, challenges, and recommendations," Ai Soc., pp. 1–25, 2023.
[14] M. S. K. Pheng and L. G. David, "Artificial intelligence in back-end semiconductor manufacturing: A case study," in 2022 IEEE International Conference on Distributed Computing and Electrical Circuits and Electronics (ICDCECE), IEEE, 2022, pp. 1–4.
[15] J. W. Lambrechts, S. Sinha, K. Sengupta, A. Bimana, S. Kadam, S. Bhandari, J. Du Preez, Z. Shao, X. Huang, Z. Liu, and E. A. Karahan, "Intelligent integrated circuits and systems for 5G/6G telecommunications," IEEE Access, vol. 12, no. 2, pp. 21402–21419, 2024.
[16] R. Busch, M. Wahl, P. Czerner, and B. Choubey, "Yield prediction with machine learning and parameter limits in semiconductor production," in 2022 International Symposium on Semiconductor Manufacturing (ISSM), IEEE, 2022, pp. 1–4.
[17] R. Wang, S. Wang, and B. Niu, "Shape prior guided defect pattern classification and segmentation in wafer bin maps," J. Intell. Manuf., pp. 1–12, 2023.
[18] K. Taha, K. Salah, and P. D. Yoo, "Clustering the dominant defective patterns in semiconductor wafer maps," IEEE Trans. Semicond. Manuf., vol. 31, no. 1, pp. 156–165, 2017.
[19] J. Yu, Z. Shen, and S. Wang, "Wafer map defect recognition based on deep transfer learning-based densely connected convolutional network and deep forest," Eng. Appl. Artif. Intell., vol. 105, p. 104387, 2021.

[20] J. Yu, Z. Shen, and X. Zheng, "Joint feature and label adversarial network for wafer map defect recognition," IEEE Trans. Autom. Sci. Eng., vol. 18, no. 3, pp. 1341–1353, 2020.
[21] R. Chithra, A. R. Aswani, and A. P. James, "Memristive cnn for wafer defect detection," in 2022 IEEE International Symposium on Circuits and Systems (ISCAS), IEEE, 2022, pp. 3284–3288.
[22] A. Majumdar, "Comparative analysis of regression and artificial neural network models for the prediction of yarn hairiness," Res. J. Text. Appar., vol. 14, no. 3, pp. 85–93, 2010.
[23] C. Shanmugam and E. C. Sekaran, "IRT image segmentation and enhancement using FCM-MALO approach," Infrared Phys. Technol., vol. 97, pp. 187–196, 2019.
[24] Y. Zhang, Y. Zhou, H. Lu, and H. Fujita, "Spark cloud-based parallel computing for traffic network flow predictive control using non-analytical predictive model," IEEE Trans. Intell. Transp. Syst., vol. 23, no. 7, pp. 7708–7720, 2021.
[25] X. Gao et al., "Conditional probability based multi-objective cooperative task assignment for heterogeneous UAVs," Eng. Appl. Artif. Intell., vol. 123, p. 106404, 2023.
[26] H. Jiang, C. K. Kwong, W. Y. Park, and K. M. Yu, "A multi-objective PSO approach of mining association rules for affective design based on online customer reviews," J. Eng. Des., vol. 29, no. 7, pp. 381–403, 2018.
[27] A. P. Piotrowski, J. J. Napiorkowski, and A. E. Piotrowska, "Particle swarm optimization or differential evolution—A comparison," Eng. Appl. Artif. Intell., vol. 121, p. 106008, 2023.
[28] A. H. C. Correia, D. E. Worrall, and R. Bondesan, "Neural simulated annealing," in International Conference on Artificial Intelligence and Statistics, PMLR, 2023, pp. 4946–4962.
[29] S. Dilek, S. Tosun, and A. Cakin, "Simulated annealing-based high-level synthesis methodology for reliable and energy-aware application specific integrated circuit designs with multiple supply voltages," Int. J. Circuit Theory Appl., vol. 51, no. 10, pp. 4897–4938, 2023.
[30] R. Sitzenfrei, Q. Wang, Z. Kapelan, and D. Savić, "Using complex network analysis for optimization of water distribution networks," Water Resour. Res., vol. 56, no. 8, p. e2020WR027929, 2020.
[31] S. A. Qureshi, A. U. Rehman, A. A. Mir, M. Rafique, and W. Muhammad, "Simulated annealing-based image reconstruction for patients with covid-19 as a model for ultralow-dose computed tomography," Front. Physiol., vol. 12, p. 737233, 2022.
[32] M. Bansal, A. Goyal, and A. Choudhary, "A comparative analysis of K-nearest neighbor, genetic, support vector machine, decision tree, and long short term memory algorithms in machine learning," Decis. Anal. J., vol. 3, p. 100071, 2022.
[33] E. Demirović et al., "Murtree: Optimal decision trees via dynamic programming and search," J. Mach. Learn. Res., vol. 23, no. 26, pp. 1–47, 2022.
[34] S. Kim, Y. Jeong, D. W. Han, and C. B. Mo, "Machine learning-assisted defect analysis and optimization for PIN-structured perovskite solar cells." J. Electron. Mater., vol. 52, no. 9, pp. 5861–5871, 2023.
[35] J. Li, H. Li, Y. Wu, H. Zhou, L. Manfredi, P. Li, and H. Zhang, "Deep learning based defect detection algorithm for solar panels," in 2023 WRC Symposium on Advanced Robotics and Automation (WRC SARA), IEEE, 2023, pp. 438–443.
[36] M. Mammeri, L. Dehimi, H. Bencherif, and F. Pezzimenti, "Paths towards high perovskite solar cells stability using machine learning techniques," Sol. Energy, vol. 249, pp. 651–660, 2023.
[37] M. Abbas, L. Zeng, F. Guo, M. Rauf, X.-C. Yuan, and B. Cai, "A critical review on crystal growth techniques for scalable deposition of photovoltaic perovskite thin films," Materials (Basel), vol. 13, no. 21, p. 4851, 2020.
[38] Q. Jiang et al., "Surface passivation of perovskite film for efficient solar cells," Nat. Photonics, vol. 13, no. 7, pp. 460–466, 2019.

[39] W. Yan et al., "Machine learning enabled development of unexplored perovskite solar cells with high efficiency," Nano Energy, vol. 99, p. 107394, 2022.
[40] D. Lin et al., "Stable and scalable 3D-2D planar heterojunction perovskite solar cells via vapor deposition," Nano Energy, vol. 59, pp. 619–625, 2019.
[41] E. C. Gok et al., "Predicting perovskite bandgap and solar cell performance with machine learning," Sol. RRL, vol. 6, no. 2, p. 2100927, 2022.
[42] F. Gao et al., "Enhanced performance of tin halide perovskite solar cell by addition of lead thiocyanate," RSC Adv., vol. 8, no. 25, pp. 14025–14030, 2018.
[43] Ç. Odabaşı and R. Yıldırım, "Machine learning analysis on stability of perovskite solar cells," Sol. Energy Mater. Sol. Cells, vol. 205, p. 110284, 2020.
[44] B. Yılmaz, Ç. Odabaşı, and R. Yıldırım, "Efficiency and stability analysis of 2D/3D perovskite solar cells using machine learning," Energy Technol., vol. 10, no. 3, p. 2100948, 2022.
[45] Y. Hu, X. Hu, L. Zhang, T. Zheng, J. You, B. Jia, Y. Ma, X. Du, L. Zhang, J. Wang, and B. Che, "Machine-learning modeling for ultra-stable high-efficiency perovskite solar cells," Adv. Energy Mater., vol. 12, no. 41, p. 2201463, 2022.
[46] N. T. P. Hartono et al., "How machine learning can help select capping layers to suppress perovskite degradation," Nat. Commun., vol. 11, no. 1, p. 4172, 2020.
[47] Ç. Odabaşı and R. Yıldırım, "Assessment of reproducibility, hysteresis, and stability relations in perovskite solar cells using machine learning," Energy Technol., vol. 8, no. 12, p. 1901449, 2020.
[48] N. Tomar et al., "Molecular structure-based prediction of absorption maxima of dyes using ann model," Big Data Cogn. Comput., vol. 7, no. 2, p. 115, 2023.
[49] C.-Z. Zhang and X.-Q. Fu, "Applications and potentials of machine learning in optoelectronic materials research: An overview and perspectives," Chinese Phys. B, vol. 32, no. 12, p. 126103, 2023.
[50] C. Meijuan, "Application of ANN technique to predict the thermal conductivity of nanofluids: A review," J. Therm. Anal. Calorim., vol. 145, no. 4, pp. 2021–2032, 2021.
[51] M. Abdi-Khanghah, A. Bemani, Z. Naserzadeh, and Z. Zhang, "Prediction of solubility of N-alkanes in supercritical CO2 using RBF-ANN and MLP-ANN," J. CO2 Util., vol. 25, pp. 108–119, 2018.
[52] F. Sharmin and M. A. Basith, "Simple low temperature technique to synthesize sillenite bismuth ferrite with promising photocatalytic performance," ACS Omega, vol. 7, no. 39, pp. 34901–34911, 2022.
[53] M. Kaya and S. Hajimirza, "Rapid optimization of external quantum efficiency of thin film solar cells using surrogate modeling of absorptivity," Sci. Rep., vol. 8, no. 1, pp. 1–9, 2018.
[54] K. Huwig, C. Fan, and M. Springborg, "From properties to materials: An efficient and simple approach," J. Chem. Phys., vol. 147, no. 23, 2017.
[55] M. A. S. Abourehab, A. M. Shawky, K. Venkatesan, S. Yasmin, A. M. Alsubaiyel, and K. M. AboRas, "Efficiency development of surface tension for different ionic liquids through novel model of machine learning technique: Application of in-thermal engineering," J. Mol. Liq., vol. 367, p. 120391, 2022.
[56] T. B. Asafa, N. Tabet, and S.-A. M. Said, "Taguchi method–ANN integration for predictive model of intrinsic stress in hydrogenated amorphous silicon film deposited by plasma enhanced chemical vapour deposition," Neurocomputing, vol. 106, pp. 86–94, 2013.
[57] G. Akiki, Area selective deposition of microcrystalline silicon by PECVD: Physical origin, challenges and solutions. Institut Polytechnique de Paris, 2021.
[58] F. Li et al., "Machine Learning (ML)-assisted design and fabrication for solar cells," Energy Environ. Mater., vol. 2, no. 4, pp. 280–291, 2019.

[59] R. Ding et al., "Application of machine learning in optimizing proton exchange membrane fuel cells: A review," Energy AI, vol. 9, p. 100170, 2022.
[60] J. Watté, M. Van Zele, K. De Buysser, and I. Van Driessche, "Recent advances in low-temperature deposition methods of transparent, photocatalytic TiO2 coatings on polymers," Coatings, vol. 8, no. 4, p. 131, 2018.
[61] M. Powalla, S. Paetel, E. Ahlswede, R. Wuerz, C. D. Wessendorf, and T. Magorian Friedlmeier, "Thin-film solar cells exceeding 22% solar cell efficiency: An overview on CdTe-, Cu (In, Ga) Se2-, and perovskite-based materials," Appl. Phys. Rev., vol. 5, no. 4, 2018.
[62] A. M. Ghaedi and A. Vafaei, "Applications of artificial neural networks for adsorption removal of dyes from aqueous solution: A review," Adv. Colloid Interface Sci., vol. 245, pp. 20–39, 2017.
[63] Z.-W. Zhong, "Advanced polishing, grinding and finishing processes for various manufacturing applications: A review," Mater. Manuf. Process., vol. 35, no. 12, pp. 1279–1303, 2020.
[64] E. A. J. Abadi, H. Sahu, S. M. Javadpour, and M. Goharimanesh, "Interpretable machine learning for developing high-performance organic solar cells," Mater. Today Energy, vol. 25, p. 100969, 2022.
[65] Z. W. Zhao, Y. Geng, A. Troisi, & H. Ma, "Performance prediction and experimental optimization assisted by machine learning for organic photovoltaics." Adv. Intell. Syst., vol. 4, no. 6, p. 2100261, 2011.
[66] C. Zhu et al., "Key factors governing the device performance of CIGS solar cells: Insights from machine learning," Solar Energy, vol. 228, pp. 45–52, 2021.
[67] G. Salman et al., "Machine learning for optimal copper doping profile design in cdte solar cells," 2021 IEEE 48th Photovoltaic Specialists Conference (PVSC), IEEE, 2021.
[68] Abdulrahman Th Mohammad, Hasanen M. Hussen, and Hussein J. Akeiber, "Prediction of the output power of photovoltaic module using artificial neural networks model with optimizing the neurons number," International Journal of Renewable Energy Development, vol. 12, no. 3, 2023.
[69] M. A. Khan, H. Ali, B. Ashraf, S. Khan, A. Wadood, and T. Khurshaid, "Output power prediction of a photovoltaic module through artificial neural network," IEEE Access, vol. 10, pp. 116160–116166, 2022.
[70] Shivam Kumar, Jong-Chyuan Tzou, and Shang-Chen Wu, "A multi-objective predictive energy management strategy for residential grid-connected PV-battery hybrid systems based on machine learning technique," Energy Convers. Manag., vol. 237, pp. 114103, 2021.
[71] Mine Kaya and Shima Hajimirza, "Rapid optimization of external quantum efficiency of thin film solar cells using surrogate modeling of absorptivity, Sci. Rep., vol. 8, no. 1, pp. 1–9, 2018.
[72] Sven Wasmer, Kati Hübener, and Bernhard Klöter, "Explaining the efficiencies of mass-produced p-type Cz-Si solar cells by interpretable machine learning," Solar RRL, vol. 6, no. 5, p. 2100477, 2022.

2 Introduction to Characterization and Fabrication Techniques Employed for Advanced Solar Cell Structures

Dibyendu Kumar Ghosh, Arijit Pangas, Shiladitya Acharyya, Sukanta Bose, Gourab Das, Sumita Mukhopadhyay, and Anindita Sengupta

2.1 FABRICATION TECHNIQUES

2.1.1 WET CHEMICAL TREATMENT OF WAFERS

Wafer treatment is the first and foremost step for any wafer-based Si solar cells. It is considered one of the most effective tools in deciding the photovoltaic (PV) performance of all the advanced c-Si-based solar cells. Proper wafer treatment is undoubtedly a challenging but essential task. All the processes are carried out in a chemical bath.

- **Saw damage removal:** In this step, the wafers are merged into the 10% KOH solution with deionized (DI) water for anisotropic etching [1]. The process temperature is generally kept at 80°C to 85°C, whereas the usual process time is 2–3 minutes. This results (due to the presence of OH ions in the solution) in the anisotropic etching of the Si wafers. The loss of thickness from both wafer surfaces is around 10 μm; the corresponding chemical reaction is given in the following equation [2]:

$$Si + H_2O + 2KOH \rightarrow K_2SiO_3 + 2H_2O$$

The wafers are then washed in DI water for 4 to 5 minutes to remove the chemical species and byproducts from the surface.

 DOI: 10.1201/9781003515784-2

- **Texturization:** Front-surface texturization is an inevitable requirement to enhance the optical confinement of the incident sunlight. The texturization of the saw damage, which removed Si wafers, is generally performed in a mixture of low-concentration (1.8%) KOH solution, DI water, and isopropyl alcohol solution. The process is carried out at 75–80°C for 35–40 minutes. The textured wafers are then washed in DI water. The textured surface is realized due to different etching rates of the planes of the wafers. To be specific, the (100) plane gets faster etched compared to the (111) plane, whereas the etching rate of the (110) plane is the highest among all three planes. The different etching rate of the planes forms a pyramidal shape on the (100) plane [3]. This pyramidal structure is responsible for realizing a lower front-surface reflection of the incident beam. The pyramid shape is largely affected by the KOH concentration, process temperature, and process time.
- **Wafer Cleaning:** The textured wafers are cleaned using state-of-the-art RCA-1 and RCA-2 wet chemical procedures. Subsequently, the wafers are merged into 1–2% HF solution for a few seconds, mainly due to removing the formed native oxide during the RCA1 and RCA2 process. Subsequently, the wafers are washed with DI water before undergoing a drying process with a hot air dryer.

2.1.2 Diffusion

The doping of the p-type/n-type c-Si wafers is realized through the diffusion process. Optimization of the penetration of the dopant atoms, doping concentration, and overall doping profile is essential for realizing advanced c-Si based high-efficiency solar cells. Doping is the second step in fabricating these type of solar cells after the wafer treatment. A diffusion furnace is also used to fabricate the SiO_x layer on the Si wafers [4]. High-temperature annealing may also be conducted using the diffusion furnace. Therefore, the diffusion furnace is widely recognized as an important tool for its multidimensional activities.

2.1.2.1 Working mechanism

The whole diffusion mechanism is governed by Fick's law of diffusion process. The process is also termed thermal diffusion process as the dopant atoms migrate from the dopant source to the silicon wafer due to thermal excitation. A thermal diffusion furnace comprises heaters, quartz tubes, a gas flow system, and temperature controllers. Diffusion is generally accomplished at higher temperatures to activate the dopants; therefore, the proper heating arrangement is made within the furnace so that the maximum temperature may attain as high as 1,200°C. A standard diffusion furnace typically consists of three chambers. Two chambers are used for diffusion, whereas oxidation/annealing is carried out in the remaining chamber. A quartz tube is placed in each chamber where the samples, that is, the wafers, are kept throughout the process. The quartz tube is used to avoid contamination on the wafer surface; hence, high-quality material may be realized. Out of the two diffusion chambers, one is used for phosphorous diffusion (using a $POCl_3$ bubbler), and the other is used for the diffusion of boron (using a BBr_3 or BCl_3 bubbler). The dopant source is positioned on one end of the quartz tube. The same end is connected to the ultra-high pure (UHP) gas cylinders through mass-flow controllers (MFCs), whereas the other end is connected to the exhaust through a wet scrubber to maintain the gas pressure inside the tubes and simultaneously remove the toxic mixture from the tubes.

The doping profile and the penetration depth mainly depend on three parameters, that is, the dopant concentration, diffusion temperature, and diffusion time. The microcontroller-based control system regulates the diffusion chamber's temperature precisely to ensure uniformity in the diffusion process and oxidation (if required) throughout the wafer surface. In modern state-of-the-art solar cell architectures, a shallow emitter is sometimes required, which may be realized by controlling the diffusion time and operating temperature.

On the other hand, high-temperature annealing of chemically grown SiO_x layer and crystallization and dopant activation of plasma-enhanced chemical vapor deposition (PECVD)-grown a-Si:H layer are perceived in the oxidation/annealing chamber. In tunnel oxide passivated contact (TOPCon) solar cells, the SiO_x tunneling layer is grown in this chamber; the same may also be superior to the chemically grown SiO_x layer and plasma-assisted SiO_x layer in terms of the passivation properties and thermal stability. UHP N_2 is used during the heating and cooling period, whereas UHP O_2 is used for oxidation. The layer properties are largely dependent on the diffusion time and diffusion temperature.

2.1.3 Plasma-Enhanced Chemical Vapor Deposition

PECVD, an advanced form of conventional chemical vapor deposition method, also known as glow discharge, is quite a popular thin-film deposition technique due to featuring several advantages such as the ability to deposit high-quality thin films by maintaining the film integrity, operating at low temperature, and maintaining productivity and uniform deposition over large area [5]. In the field of solar photovoltaics (SPV), the technology is widely recognized for its capability of producing high-quality a-Si:H, μc-Si:H, SiN_x, and SiO_x layers and, hence, is used for fabricating thin-film Si solar cells/modules as well as advanced c-Si-based solar cells/modules [6–9]. Figure 2.1 illustrates one multi-chamber PECVD unit installed at DST-IIEST Solar PV Hub.

2.1.3.1 Working mechanism

The whole PECVD process can be segregated into different stages: sample loading, attaining the required vacuum and temperature, the exposure of the source gas inside the chamber, excitation of the source gas, electron-induced dissociation of the source gas, gas-phase chemical reactions, diffusion of radicals, the deposition process, and the sample unloading. After sample loading and achieving the required vacuum and the substrate temperature, the source gas (precursor) is exposed to the chamber. The gas is then ionized with the help of the excitation source. The excitation source such as direct current (DC), radio frequency (RF)

FIGURE 2.1 Photograph of multi-chamber PECVD unit installed at DST-IIEST Solar PV Hub.

or microwave results in the formation of an electric field between two parallel electrodes. In the PECVD process, RF (13.56 or 27.12 MHz) is the most commonly used excitation source for creating and sustaining high-density plasma, even at low pressure and various load conditions. In the microwave glow discharge technique, plasma is generated using a microwave frequency (2.45 GHz). In this technique, a high deposition rate can prevail due to its high-energy and efficient ionization capabilities. Electron cyclotron resonance is a specialized version of this microwave discharge method, also referred to as a remote plasma technique, where energetic ions are generated by combining microwaves with a magnetic field in a separate chamber/in one zone of the plasma chamber and then transferred to the deposition chamber/another zone of the plasma chamber (where deposition takes place) [10]. In this way, the possible damage on the substrate surface can be minimized as well as precise control of the deposition process can also be realized. Besides, the remote plasma deposition technique offers uniformity in the thickness and deposition temperature, thus enabling the formation of superior films. On the other hand, in the DC glow discharge technique, plasma is established between two metal electrodes at a low chamber pressure. This is a relatively less-used excitation source compared to the other two sources. The common purpose of different excitation sources is to establish a sufficient electric field, which not only ignites and sustains the glow discharge mechanism (plasma) but also facilitates the acceleration process of the electrons so that they (the accelerated electrons) can acquire sufficient energy to fragment the precursor gas molecules into neutral radicals and ions. These neutral radicals, ions, and molecules participate in the complicated gas-phase reactions within the bulk plasma. The reaction products adsorb on the substrate surface, and the deposition process is

TABLE 2.1
Different Layers Fabricated Using PECVD Technique for Applications in SPV

Layer	Applications	Ref.
SiN_x	Anti-reflecting coating	[11, 12]
AlO_x	Surface passivation	[13, 14, 15]
SiO_x	Surface passivation	[9, 15]
Intrinsic and doped a-Si:H and μc-Si:H layer	Absorbing/surface passivation (undoped) and field formation layer (doped)	[6, 15, 16]

eventually completed. The reaction products, referred to as growth precursors, are therefore considered a deciding factor in enabling the fabrication of high-quality films featuring required structural, compositional, and optoelectronic characteristics. The optimized condition of any film deposition includes various attributes such as the base vacuum, power density, deposition temperature, process pressure, and gas mixture ratio. Any alteration of any of these parameters can change the film properties in a significant way. The deposition chamber is rarely exposed to the ambient atmosphere (only for maintenance and cleaning purposes) so that the vacuum can be maintained and repeatability can be realized. Consequently, one separate chamber is used for sample loading and unloading purposes. Most of the PECVD units feature magnetic arms to (i) enhance the ionization possibility of the source gas(es), (ii) stabilize the plasma by confining the charged particles, (iii) deposit uniform films, (iv) control the deposition parameters and, hence, the film properties. Table 2.1 highlights the different layers fabricated through the PECVD technique for the applications in SPV.

2.1.4 Low-Pressure Chemical Vapor Deposition

Low-pressure chemical vapor deposition (LPCVD) is another tool primarily used to fabricate different layers in various solar cell architectures. Transparent conducting oxides (TCOs) and a-Si:H layers can be fabricated using this technique [15, 17]. In TOPCon solar cells, LPCVD is an alternative tool for fabricating the TOPCon structure [15]. Besides, SiN_x used as anti-reflecting coating (ARC) layer, can be deposited using this advanced fabrication tool [18].

2.1.4.1 Working mechanism

As the name suggests, the deposition occurs at low pressure. First, the substrates are placed inside the quartz tube. The quartz tube is connected to the vacuum pump to reduce the tube pressure. On the contrary, the other end of the tube is connected to the gas cylinders through MFCs. The heater is kept underneath the quartz tube and, hence, the tube wall is heated up, and, eventually, the substrate is heated up. As the quartz tube wall is heated, it is termed a hot-wall LPCVD reactor [19]. The gases are introduced inside the chamber upon attaining the required pressure and temperature. The exposure of gases in the deposition chamber leads to chemical reactions. The reaction products get deposited on the substrate surface; thus, the film is deposited using this technique. The byproducts and the unused gases are removed from the chamber using the pumps attached to the chamber. In this way, the low pressure is maintained inside the chamber. The film quality with uniform deposition is largely dependent on the deposition conditions, especially the chamber pressure.

2.1.5 Atomic Layer Deposition

Atomic layer deposition (ALD), a sophisticated technique is specially designed for fabricating high-quality, low-thick films. ALD is an integral part of the fabrication of passivated emitter and rear contact (PERC) solar cells as the rear-side surface passivation layer; that is, the Al_2O_3 layer in PERC solar cell architecture is generally deposited through the ALD process [20]. High precision, uniformity, precise controllability at the atomic level, and self-limiting deposition are considered the significant advantages of this technique [21]. Depending on the excitation sources, ALD can be classified into various categories, namely, thermal ALD, photo-assisted ALD, plasma ALD, laser ALD, and so on.

2.1.5.1 Working mechanism

The film growth mechanism in the ALD process is fundamentally based on the sequential gas-phase chemical process; that is, the substrate is exposed to the alternate gaseous species during the deposition time [22]. These gaseous species are termed precursors. The gaseous species are introduced in the deposition chamber one by one; that is, the deposition is carried out in non-overlapping mode. Assume there are two precursors required for the deposition of a thin film. At first, precursor A (first reactant) is introduced in the deposition chamber; hence, it chemically reacts with the active sites on the substrate surface. The reaction continues until all the active sites on the surface are saturated, and no reaction will occur beyond saturation; that is, self-limiting deposition will occur. After that, an inert gas, usually N_2, is introduced into the chamber to bring out the remaining precursor A (if any) and the unwanted byproducts. Precursor B (second reactant) is then exposed to the deposition chamber, which reacts with the substrate's surface. Again, the inert gas is introduced into the chamber to extract the unused precursor B and the by-products. Each introduction of precursor in the deposition chamber is known as the pulse, whereas each introduction of the inert gas to the chamber is known as the purge steps. This process is continued cyclically. Hence, one cycle in ALD consists of several pulses (depending on the number of precursors), that is, a single exposure of all the precursors. The chemical reaction between the precursor and the substrate's surface determines the uniformity and thickness of the developed material. Simultaneously, the number of cycles also determines the characteristics of the fabricated materials. One of the key challenges of this technique is overcoming the slow growth rate, that is, to enhance scalability. Table 2.2 shows the different layers developed using the ALD technique for SPV applications.

TABLE 2.2
Different Layers Fabricated Using ALD Technique for Applications in SPV

Layer	Applications	Ref.
SiO_x	Surface passivation	[15]
AlO_x	Surface passivation	[20]
MoO_x	Carrier selective layer	[23]

2.1.6 Sputtering

Sputtering, a physical vapor deposition technique, is well known for its ability to develop thin, high-quality films [24]. TCO films, such as indium tin oxide (ITO), aluminum-doped zinc oxide (AZO), and boron-doped zinc oxide (BZO), are generally deposited through sputtering due to several advantages, such as high deposition rate, uniformity, repeatability, and scalability [24, 25]. In 1852, Grove first discovered the sputtering phenomenon from the cathode during electrical gas discharge [26].

2.1.6.1 Working mechanism

In the sputtering process, the material to be deposited is known as the target material, and it is placed inside the deposition chamber. Upon the loading of the substrates, the chamber pressure is decreased using the rotary pump, and the turbo molecular pump (TMP) is attached to the system. The deposition process starts after achieving an adequate vacuum ($>10^{-6}$ bar) and the required temperature inside the chamber. The target material is mounted on the cathode, whereas the substrate is placed on the anode, and an electric field is applied between these two. A noble gas, usually argon, also known as the sputtering gas, is introduced into the chamber during the deposition. The applied electric field ionizes the portion of the applied argon gas into plasma. The ionized high-energy particles collide with the target material and eject the atoms from the target. The sputtering process is also known as the momentum transfer process. The dislodged atoms condense on the rotating substrate. The substrate rotation is carried out to ensure uniformity in the substrate temperature and the film deposition. The movement of high-energy electrons can ionize the neutral gas molecules through collisions. The secondary electrons can also be ejected when the argon gas ions strike the cathode. These secondary electrons also participate in the ionizing process of the neutral gas molecules through collisions. To triumph over the challenge of low deposition rate, magnetron sputtering is employed in which the improvement of the mean free path of the sputtered particles and ionization efficiency is perceived [27]. The magnetic field ensures the magnetic confinement of the plasma to the target side of the chamber apart from forming Lorentz force, causing electrons to move in a drifting closed loop, whereas the exerted force on argon ions is negligible due to their greater mass. Figure 2.2 depicts the photograph of a magnetron sputtering

FIGURE 2.2 Photograph of a magnetron sputtering system installed at DST-IIEST Solar PV Hub.

system, whereas Table 2.3 shows the different layers fabricated through the sputtering technique for SPV applications.

2.1.7 Thermal Evaporation

The metallic contacts of different thin-film solar cells, a-Si:H solar cells, μc-Si:H solar cells, and micromorph (combination of a-Si:H and μc-Si:H cells) solar cells, are realized by a thermal evaporation system. Apart from metal deposition, metal oxide or any other material may be deposited using this technique. Figure 2.3 shows the photograph of an installed thermal evaporation system.

TABLE 2.3
Different Layers Deposited through Sputtering Technique for Applications in SPV

Layer	Applications	Ref.
ITO	TCO and backside reflection layer (BRL)	[28, 29]
AZO	TCO	[30]
BZO	TCO	[31]
ITO/AZO	TCO	[6]

2.1.7.1 Working mechanism

The metal wire (or metal oxide powder in pellet form) to be deposited is placed in the boat. At first, the substrates are clipped onto a circular substrate holder plate, which is later attached to a rotating mechanism facing down above a tungsten boat. Concurrently, the heater is also switched ON if the deposition needs to be carried out at an elevated temperature. The deposition process is initiated when the chamber reaches the required vacuum (~ 5×10^{-6} mbar) and the temperature. The low chamber pressure ensures a high deposition rate by increasing the mean free path of the vapors, reduces the possibility of contamination by ensuring the minimal impact of the vapors from other sources inside the chamber, facilitates heat transfer control inside the chamber, and avoids the oxidation reaction of the metals. First, a high current is provided to the boat to heat the metal wire through the Joule effect. The evaporated metal wire vapors condense on the surface of the sample, and, thus, the coating is accomplished. The control of the heating process is achieved by adjusting the applied voltage and current to the boat. The steps for the deposition of any other material are also identical. A quartz crystal oscillator microbalance is used to observe the thickness of the film during the deposition. In most thermal evaporation systems, an electron beam evaporation facility is also attached for the deposition of the material, which requires a higher vacuum level and a higher deposition temperature. It offers a higher deposition rate and can deposit a wide range of materials compared to the thermal evaporation system.

2.1.8 Screen Printing

The metallic contacts in any advanced c-Si based solar cells are carried out by the screen-printing and firing technique [32]. It is an industrially adopted, well-established, simple, cost-effective technique to realize high-quality metallic contacts on both sides of the solar cell architectures. High throughput and easy automation are the other two salient features of the technique.

2.1.8.1 Working mechanism

In general, the whole screen printing and firing process can be classified into different stages, such as printing of the front-side metallic contacts, drying of the

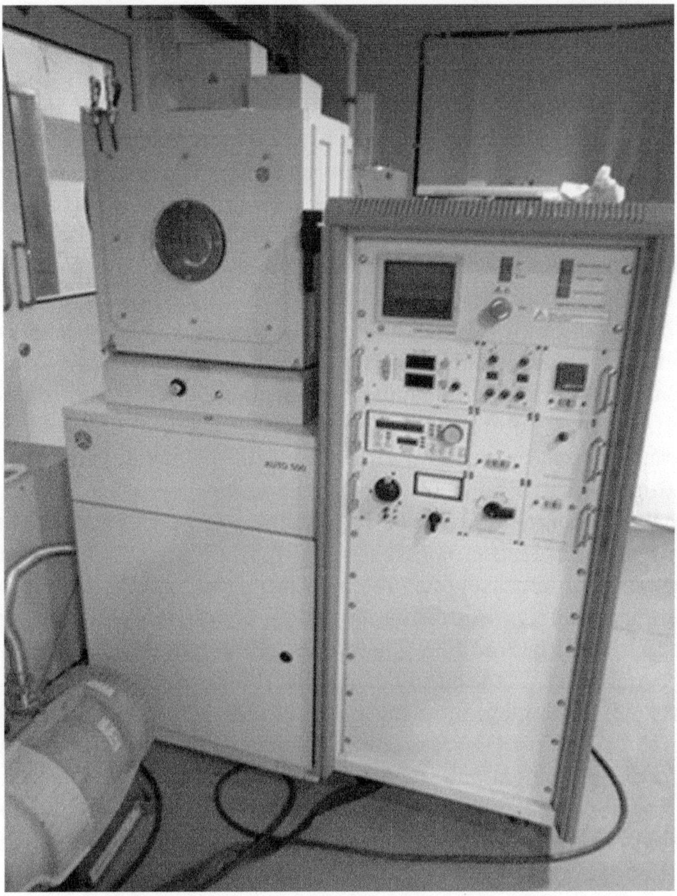

FIGURE 2.3 Photograph of thermal evaporation system installed at DST-IIEST Solar PV Hub.

front-side metallic contacts, printing of the rear-side metallic contacts, drying of the rear-side metallic contacts, co-firing, and, finally, cooling down process. At first, the back side of the device is screen-printed using a conducting paste (in general, Al or Ag paste, depending on the type of the cell architecture) through a mesh patterned screen (made using stainless steel). A squeegee is used to transfer the paste to the wafers. The quality of metallic contacts is influenced by various attributes: solvent quality of the paste, squeegee speed, squeegee pressure, fiducial alignment accuracy, and the spatial distance between the screen and the wafer. The next step, that is, the drying process, is accomplished through a belt furnace at around 225°C to 275°C for ~2 minutes. However, the operating temperature and time of the drying process depend on the type of paste used. In the next part, the front metallic contact is also realized similarly; however, the mesh patterned screen features the finger-busbar arrangement in this case. Again, the drying process is carried out at 225°C to 275°C for ~2 minutes. Two different screen printers and belt furnaces are utilized if the paste used for front- and back-side metallization is different, so the possibility of contamination between the pastes can be avoided. After printing, the wafers are put in a firing furnace. One of the most significant attributes of a firing furnace is the number of temperature zones. The number of temperature zones is a deciding factor in replicating the temperature profile provided by the paste manufacturers, which is vital for realizing the ohmic contacts. The firing time is significantly less (a few seconds). In a three-zone temperature furnace, in temperature zone I, the burn-out process is executed at 350°C to 510°C to detach the organic binders from the paste, whereas, in temperature zone II (between 750°C and 850°C), the Al–Si alloy gets formed. In temperature zone III (between 880°C and 940°C), Al-BSF formation occurs in the state-of-the-art Al-BSF solar cells [33]. Simultaneously, the contact between the front-side metallic contact (Ag for n^+ emitter) and the emitter (n^+ layer) is also accomplished in this temperature zone by penetrating through the front-side SiN_x anti-reflecting layer utilizing glass frit particles of Ag paste. However, fast movement of the belt causes the set temperature to be higher than the temperature realized by the wafers. Process optimization is essential for realizing higher power output from the fabricated cells because poor metallization may result in maximizing the series resistance value and minimizing the shunt resistance value—this will, in turn, provide low fill factor (FF). Therefore, the temperature profile of the firing process is a deciding factor in realizing the PV performance of any solar cell architecture. After completion of the firing process, the wafers are cooled down by passing the same through the cooling zone. A thickness measurement profilometer checks the height and width of the metallic grid. An optical microscope is also used to observe the obtained pattern. In recent times, possible uses of low-temperature pastes have been explored to address the issue of thermal budget [34]. Using low-temperature pastes also enables the cost-effectiveness of the fabrication process of solar cells.

2.2 CHARACTERIZATION TECHNIQUES

2.2.1 X-Ray Diffraction

X-ray diffraction (XRD) is a sophisticated technique through which the structural characteristics of a material can be attained [35]. Precisely, various structural information, such as the presence of crystallites within the samples, the orientation of the crystal planes, crystalline nature (single crystal or polycrystalline), lattice parameters, the size of the crystallites, the phase composition, dislocation energy, microstress, and microstrain, can be realized by analyzing the obtained XRD pattern of the films [30, 35, 36].

2.2.1.1 Working mechanism

The interaction between the incident X-ray beam (generated through bombarding a metal target with high-energy electrons) with a specific wavelength and the test sample will effectively cause the formation of scattered beams in all possible directions [35]. However, in some specific orientations of scattered beams, the condition for constructive interference is satisfied, forming a diffraction pattern. The direction and intensity of diffracted beams are determined by the crystal structure of the material and the angle of the incident X-ray beam. Detectors placed in the XRD setup detect the intensities of the diffracted beams (as illustrated in Figure 2.4). The recorded response is represented as a pattern of peaks obtained by plotting the intensity of the diffracted beam with respect to 2θ (where θ is the incident angle of the X-ray beam termed Bragg angle). Eventually, the obtained peaks are analyzed to reveal the structural attributes of the test material such as the nature of crystallinity, crystallite size, lattice spacing, lattice parameter (*c*-axis), stress, strain, and dislocation density. Figure 2.5 shows the variation of 2θ peak position, full width at half maxima (FWHM) of (002) plane, and grain size concerning B_2H_6 gas concentration ratio of sputter-deposited BZO thin films. Table 2.4 summarizes different structural attributes of 900 nm of BZO thin films. By observing Figure 2.5 and Table 2.4 one may realize the importance of gas mixture (here, Ar:B_2H_6) on the structural attributes of the BZO thin films. This may facilitate the path of choosing the appropriate deposition condition of the film so that the same may find its applications in the field of SPV.

(i) **Calculation of crystallite size:** The crystallite size (D) of any test sample may be evaluated as [30, 36]:

$$D = \frac{K\lambda}{\beta\cos\theta} \tag{2.1}$$

where K = shape factor (usually taken as 0.89–0.9) and λ = wavelength of the incident X-ray.

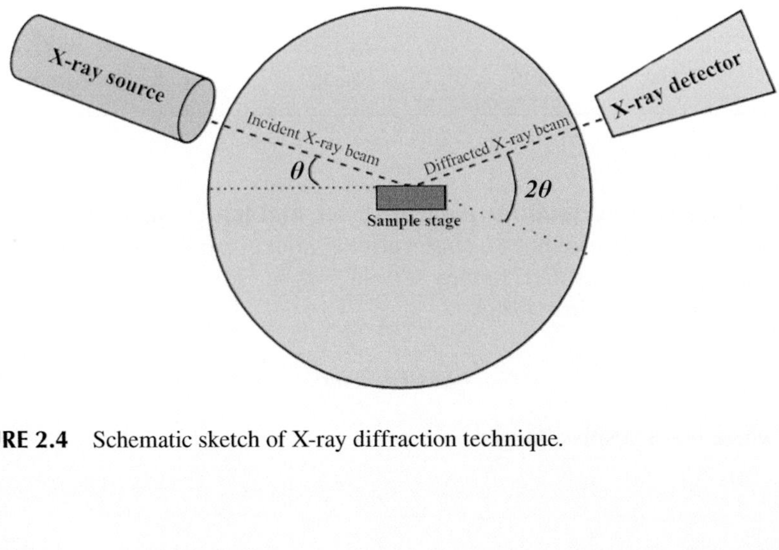

FIGURE 2.4 Schematic sketch of X-ray diffraction technique.

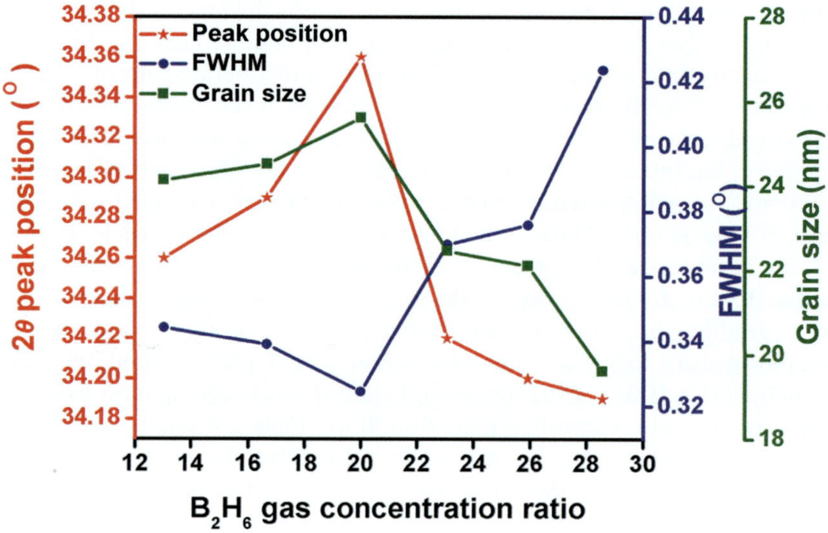

FIGURE 2.5 Impact of B_2H_6 gas concentration ratio on the 2θ peak position, FWHM of (002) reflections, and grain size of BZO films [31].

TABLE 2.4
Different Structural Attributes of the Sputtered BZO Thin Films [31]

Sample ID	$Ar:B_2H_6$	Lattice Spacing (d) in Å	Lattice Parameter (c) in Å	Strain ($\times 10^{-3}$)	Stress (GPa)
BZO-1	5:0.75	2.614	5.228	4.41	−1.02
BZO-2	5:1.00	2.612	5.224	3.65	−0.85
BZO-3	5:1.25	2.606	5.212	1.34	−0.31
BZO-4	5:1.50	2.617	5.234	5.57	−1.30
BZO-5	5:1.75	2.618	5.236	5.95	−1.39
BZO-6	5:2.00	2.619	5.238	6.34	−1.48

(ii) **Calculation of crystalline plane distance and lattice constant:** The distance (d_{hkl}) between the two consecutive similar crystalline planes or lattice spacing for indices (hkl) for any sample may be estimated by using Bragg's formula as stated here [35]:

$$n\lambda = 2d_{hkl}sin\theta \tag{2.2}$$

where n is a positive integer.

(iii) **Calculation of microstrain:** The induced microstrain (ε) in sample material may be realized by using the following formula [31, 36]:

$$\varepsilon = \frac{\beta}{4tan\theta} \tag{2.3}$$

(iv) **Calculation of dislocation density:** The dislocation density (δ) of any sample can be revealed using the equation [36]:

$$\delta = \frac{1}{D^2} \tag{2.4}$$

2.2.2 Four-Point Probe Method

The four-point probe method, a better alternative to the two-point probe method, is a beneficial and popular technique to measure the lateral conductance of any thin film [37, 38]. Contrary to the two-point probe method, the sheet resistance of any thin-film coating can be realized with more accuracy by eliminating the impact of contact resistance.

2.2.2.1 Working mechanism

As the name suggests, four equidistant probes are made in contact with the sample surface while measuring the sheet resistance (as depicted in Figure 2.6). The sheet resistance (R_{sh}) of the sample is measured per the following formula [39]:

$$R_{sh} = \frac{\pi}{ln2}\frac{V}{I} \approx 4.53\frac{V}{I} \tag{2.5}$$

where 4.53 is the geometry rectification factor generally used for the state-of-the-art dual configuration of the instrument, V is the measured voltage, and I represents the magnitude of the provided current. Table 2.5 shows the measured sheet resistance value of different TCO layers using the four-point probe method.

2.2.3 Hall Effect

The hall measurement technique is employed to explore in-depth information regarding the electrical properties of a thin-film semiconducting material. First, the type of the semiconducting material, that is, n-type or p-type, can be realized. Most importantly, detailed information regarding the carrier concentration (n), mobility (μ), and, hence, overall conductivity (σ) may be revealed through the following formula [6]:

$$\sigma = ne\mu \tag{2.6}$$

2.2.3.1 Working mechanism

In this technique, an electric current (I) is flown through the test sample in the x-direction. Concurrently, a magnetic field (B) is applied in the z-direction.

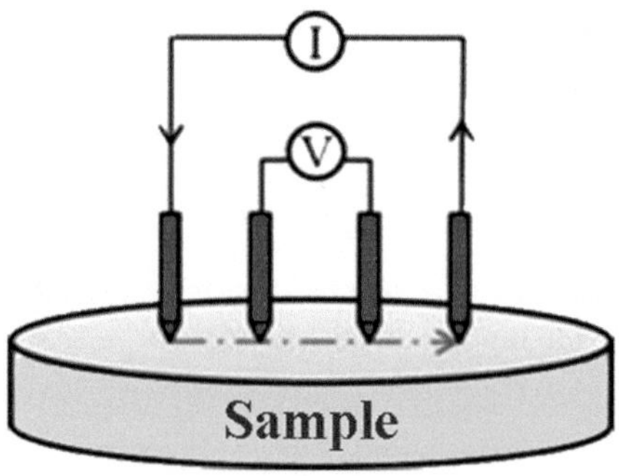

FIGURE 2.6 Schematic sketch of a four-point probe measurement.

TABLE 2.5
Measured Sheet Resistance of Different Layers Using Four-Point Probe Method

Material	Thickness (nm)	Sheet Resistance (Ω/□)	Ref.
ITO	300	7.22	[6]
AZO	711	8.25	–
BZO	900	7.23	[31]
ITO/AZO	ITO(300)/AZO(600)	4.46	[6]

Accordingly, the moving charge carriers in the material experience Lorentz force (F). Under this circumstance, the charge carriers in the material receive a pushing force in the direction of the Lorentz force (y-direction). Accordingly, the separated charge carriers form a transverse electric field that decides the continuation of the accumulation process of the charge carriers in the specific orientation (y-direction). As it is the counterbalancing force of the Lorentz force—the counterbalancing force results in the formation of a voltage, which is termed Hall voltage (V_H). The measured Hall voltage's polarity decides the semiconductor material's nature, that is, positive polarity for p-type and negative polarity for n-type semiconductor material. V_H is also used to evaluate the Hall coefficient (R_H) per the following formula [40]:

$$R_H = \frac{V_H t}{BI} \tag{2.7}$$

where t is the thickness of the test sample. On the other hand, the Hall carrier concentration (n) and the Hall carrier mobility (μ) can be evaluated using the following formulae [40]:

$$n = \frac{1}{qR_H} \tag{2.8}$$

$$\mu = \frac{|R_H|}{\rho} \tag{2.9}$$

where ρ is the resistivity of the test sample.

Figure 2.7 depicts the schematic sketch of a sample structure for Hall effect measurement. Table 2.6 shows the Hall measurement data of different layers for applications in SPV.

2.2.4 Scanning Electron Microscopy

Scanning electron microscopy (SEM), an extensively used and sophisticated technique, reveals numerous pieces of information, such as surface morphology, the size and shape of the particles, the existence of different phases, and surface defect states [41].

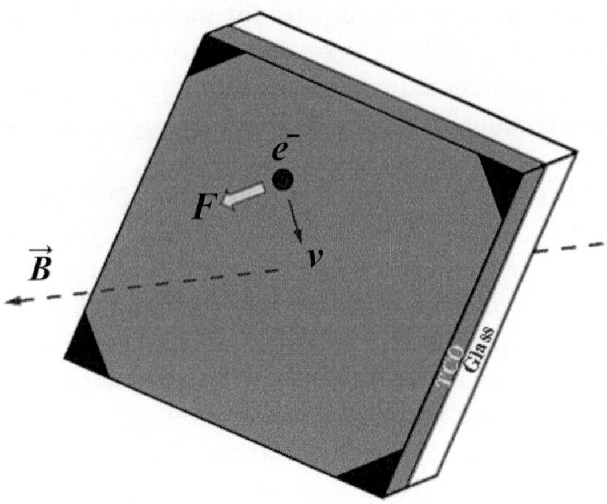

Figure 2.7 Schematic of a sample for Hall Effect measurement.

TABLE 2.6
Hall Measurement Data of Different Layers for Applications in SPV

Layer	Thickness (nm)	n (cm^{-3})	μ (cm^2/V-s)	ρ (Ω-cm)	Ref.
ITO	300	1.09×10^{22}	5.88	9.70×10^{-05}	[6]
AZO	750	1.32×10^{21}	9.79	4.83×10^{-04}	[30]
BZO	900	5.03×10^{20}	19.05	6.51×10^{-04}	[31]
ITO/AZO	ITO(300)/AZO(600)	2.67×10^{21}	9.52	2.46×10^{-04}	[6]

2.2.4.1 Working mechanism

(i) **Generation and focusing of high-energy electron beam:** An electron beam is generated using either a conventional or field-emission gun. The latter, popularly known as field-emission scanning electron microscopy (FESEM), can produce finer and more coherent beams than the conventional electron gun, that is, tungsten filament or lanthanum hexaboride. A high-energy electric field (voltage level: 1 to 30 keV) is used to accelerate the injected electrons from the electron gun to the test sample. The high-energy electrons are then guided by the electromagnetic lenses attached to the microscope unit to focus on a fine point before reaching the sample surface. The whole operation is done in a vacuum so that the direction of the electron beam can be kept in a straight line to interact with the specimen effectively. Alongside, the sample size and the thickness should be such that the same can be put in the SEM unit, and the electron beam can interact with the sample and generate the required signals. The top surface (exposed to the electron beam) is coated with conductive material (gold, carbon, platinum, etc.) for non-conducting samples to prevent the accumulation of electrostatic electrons.

(ii) **Interaction with the test sample and formation of the signals:** The high energy-focused electron beam is scanned across the specimen surface in a raster pattern. The interaction generates secondary electrons (low-energy electrons), backscattered electrons (high-energy electrons), X-rays and other signals

(iii) **Detection of the signals and image formation:** Different types of detectors are attached to the system so that the diverse signals generated due to the emission of various types of electrons can be distinguished. For example, a secondary electron detector detects the emitted secondary electrons and thus facilitates the provision of topographical imaging. Simultaneously, backscattered electron detectors assist in revealing the specimen's compositional contrast by capturing the backscattered electrons. In some SEM units, X-ray detectors are also attracted to capture the emitted X-ray to perform the elemental analysis of the sample. The last step, that is the formation of the image, is carried out in two steps: the transformation of the detected signals into electrical signals and the processing of the electrical signals to form the images. The brightness of the images depends on the intensity of the captured signals by the detectors. The effective magnification and the resolution of the SEM images may vary from machine to machine, as well as operating conditions. Advanced SEM measurement instruments are designed in such a way that they can provide other essential information such as particle size and surface roughness. Figure 2.8 demonstrates the photograph of the FESEM system installed at the DST-IIEST solar PV hub. On the other hand, Figure 2.9 represents the FESEM image of AZO films developed at different Ar:O_2 gas ratios (40:0, 40:0.5, 40:1, and 40:2 for AZO-1, AZO-2, AZO-3, and AZO-4 films, respectively) through the sputtering technique for possible applications in thin-film Si solar cells. The

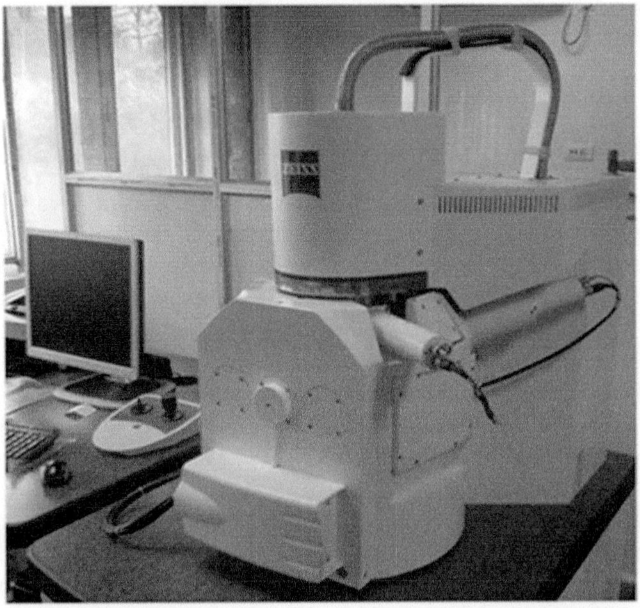

FIGURE 2.8 Photograph of FESEM system installed at DST-IIEST solar PV hub.

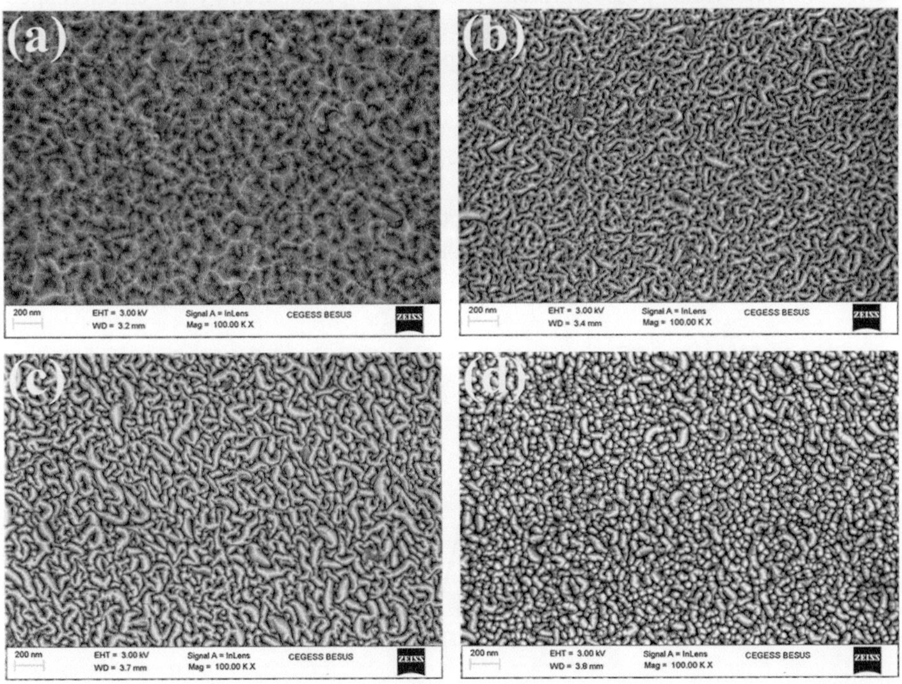

FIGURE 2.9 Topographical attributes of (a) AZO-1, (b) AZO-2, (c) AZO-3, and (d) AZO-4 films [30].

variation in surface attributes of sputtered AZO films due to the variation in the gas mixture ratio can be well understood by observing the FESEM images, as shown in Figure 2.9.

2.2.5 X-Ray Photoelectron Spectroscopy

The elemental composition of a material may be realized using X-ray photoelectron spectroscopy (XPS), one of the most popular characterization techniques [42]. As the name suggests, X-rays are utilized to remove the photoelectrons from the sample's atoms. However, unlike XRD, relatively low-energy X-rays (such as Al Kα and Mg Kα) are used to interact with the sample using the XPS technique. Therefore, the penetration/interaction of the incident X-rays is generally not possible beyond 50–100 Å of the specimen surface. That is why it is considered as the surface characterization technique. Figure 2.10 depicts the schematic of X-ray photoelectron spectroscopy.

2.2.5.1 Working mechanism

The incident X-rays react with the sample surface to remove the core electrons (the electrons of the inner cells) from the sample surface. The removal process is only possible when the energy of the incident X-rays is higher than the binding energy (BE) of the electrons of the atoms of the test sample. The removed electrons, also called photoelectrons, possess kinetic energy (KE), measured by the electron spectrometer attached to the system. After measuring the KE of the photoelectrons, one can calculate the BE of the electrons as per the following energy conversation formula [42]:

$$BE = h\upsilon - KE - \varphi_{spec} \tag{2.10}$$

where φ_{spec} is the spectrometer work function. The binding energies of the removed electrons strongly correlate with the location and intensity of the obtained peaks in the

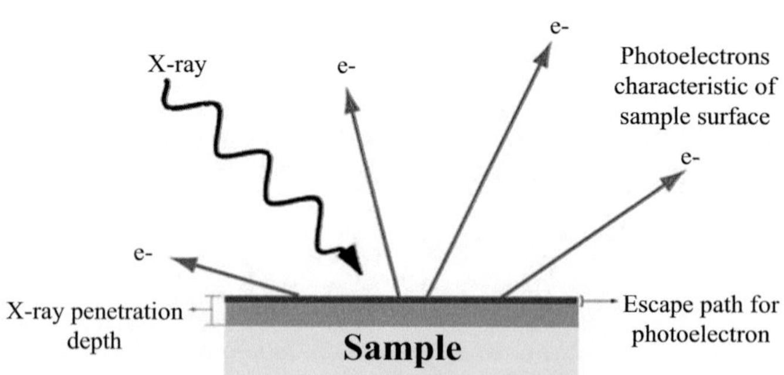

FIGURE 2.10 Schematic of X-ray photoelectron spectroscopy.

XPS spectrum. Specifically, the peak position indicates the presence of a specific element in the sample, whereas the contribution percentage of the element can be determined by observing the peak intensity. The atomic percentage of any element present in the sample surface may be realized by calculating the area under the peak (corresponding to that specific element) of the spectrum. Figure 2.11 shows

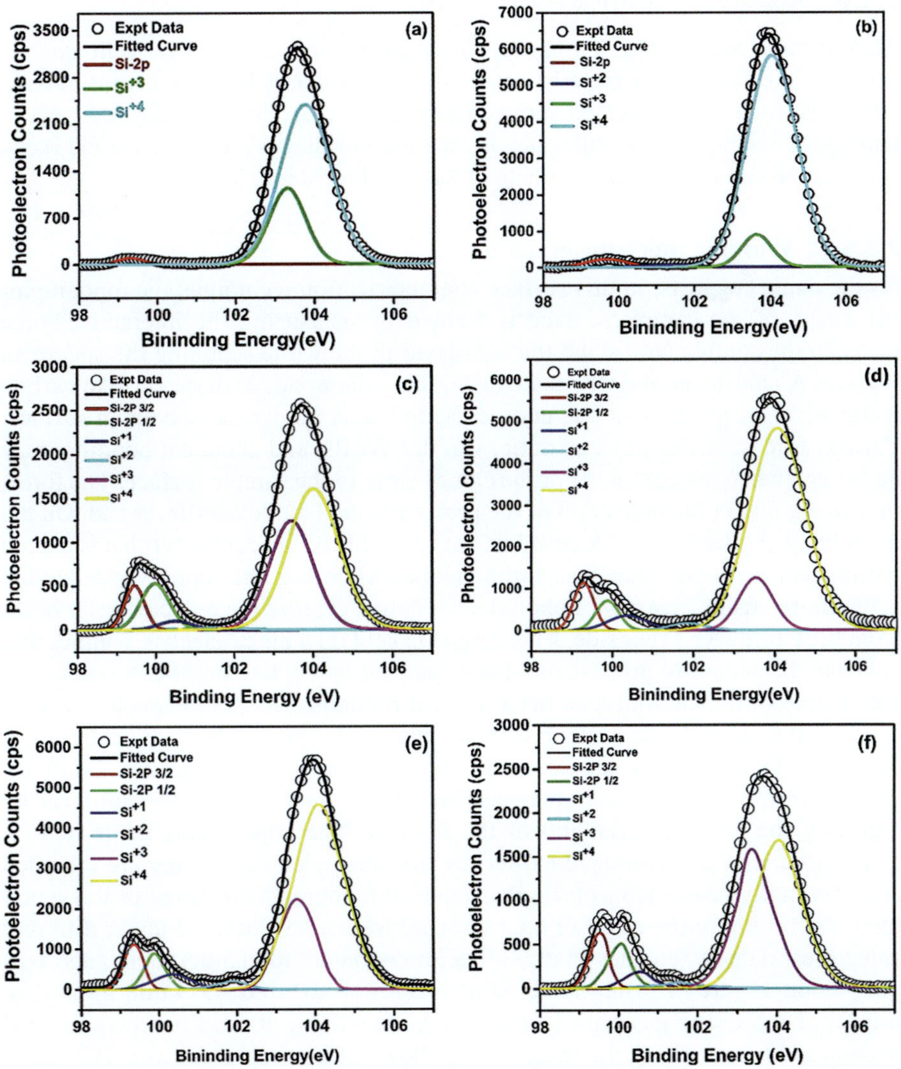

FIGURE 2.11 XPS spectra of Si–2p band (98–107 eV) showing the deconvolution of various samples. Thermal oxide on (a) p-type and (b) n-type wafers; chemical oxide on (c) p-type and (d) n-type wafers; and plasma oxide on (e) p-type and (f) n-type wafers [4].

the XPS spectra of the ultra-thin SiO_x layer grown in various routes for the application of tunneling and passivating layers in TOPCon solar cells. The oxidation state of Si determines the passivation properties of the SiO_x layer. Accordingly, an XPS study was conducted to unveil in-depth information regarding the oxidation state of the Si within the SiO_x tunneling layer grown in different routes.

2.2.6 Atomic Force Microscopy

Atomic force microscopy (AFM) is widely considered an effective technique for exposing the surface characteristics of any thin material [43]. Numerous surface attributes, such as the average surface roughness, r.m.s., surface roughness, maximum peak height, peak-to-valley height, maximum valley depth, kurtosis, and skewness, can be determined with the help of the AFM technique.

2.2.6.1 Working mechanism

As the name suggests, in this surface characterization technique, the topographical image of the sample surface is formed by measuring the interactive force between the sample atoms and the cantilever probe while scanning the specimen surface. A cantilever, usually made of Si, SiN_x, diamond, or diamond-like carbon featuring a sharp probe at one end, is used to scan across the specimen surface. Various forces, such as electrostatic, van der Waals, and chemical bonding, may be formed when the cantilever probe comes close to the sample surface. The force, in turn, results in the deflection of the probe attached to the cantilever end. On the other hand, a laser beam is focused on the back side of the cantilever that is coated with reflecting material such as a thin layer of Al or Au. The cantilever deflection is monitored using an array of photodetectors that capture the reflected laser beam from the cantilever's rear side. The sample is kept on a piezo-electric scanner that controls the scanning process in a raster pattern by controlling the movement of the sample or the cantilever in the x and y directions. The photodetector array is connected to a controller, which is connected to the piezo-electric scanner, thus forming a feedback loop (as shown in Figure 2.12). The loop ensures the maintenance of the constant force between the sample surface and the cantilever by adjusting the z-position (height) of the scanner. The adjustments made in x, y, and z directions (as mentioned earlier) by the piezo-electric scanner are recorded and eventually processed to form the required topographical image of the specimen. According to the mode of operation, AFM can be classified under different categories, namely, contact (static) mode, non-contact (dynamic) mode, and tapping mode. Figure 2.13 shows the AFM images of textured AZO thin films. The impact of the O_2:Ar gas ratio on the surface condition of the films may be well understood by observing the images. The choice of a suitable film as a TCO layer for applications in different cell architectures may become easier by analyzing the surface conditions.

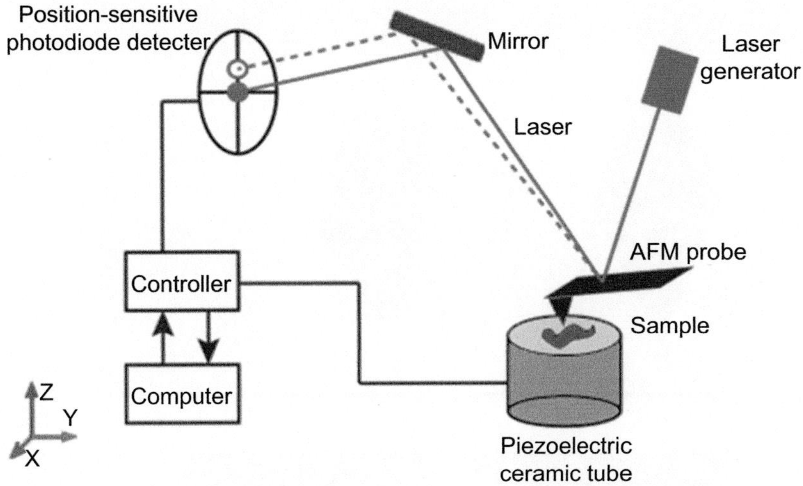

FIGURE 2.12 Schematic diagram of AFM.

2.2.7 Ultraviolet–Visible (UV–Vis) Spectroscopy

It is a well-recognized tool that is widely utilized to measure some parameters such as absorbance, transmittance, and reflectance values of a sample interacting with electromagnetic radiation of ultraviolet and visible regions [44–46]. These parameters are plotted against the wavelength of incident radiations to estimate some important properties of the sample such as the optical bandgap of a thin-film material and reflectivity of a textured surface.

2.2.7.1 Working mechanism

When a monochromatic light within the UV–vis region is applied to a sample, it interacts with the molecules. As a result, the electrons within the sample molecules move to higher energy electronic states via the absorption of photons [47, 48]. Subsequently, the transmitted light becomes less intense than that of irradiation, which is detected by the detector. This is repeated with different monochromatic light beams to get the intensity alteration in each case. If the intensities of the incident and transmitted radiation are I_0 and I, percentage transmittance (%T) is represented by $100I/I_0$, whereas absorbance (A) is obtained from $\log(I_0/I)$ [47, 49, 50]. The instrument only records the I_0 and I values for different wavelengths, and the rest of the part is taken care of by the software, which essentially calculates the aforementioned parameters (viz., absorbance, transmittance, etc.) for all the wavelengths used. Finally, the calculated data of absorbance or transmittance is plotted against the wavelength (λ) or energy ($h\nu$, where h is Planck's constant and ν is the frequency of the incident radiation) of irradiation. In the case of thin films, UV–vis

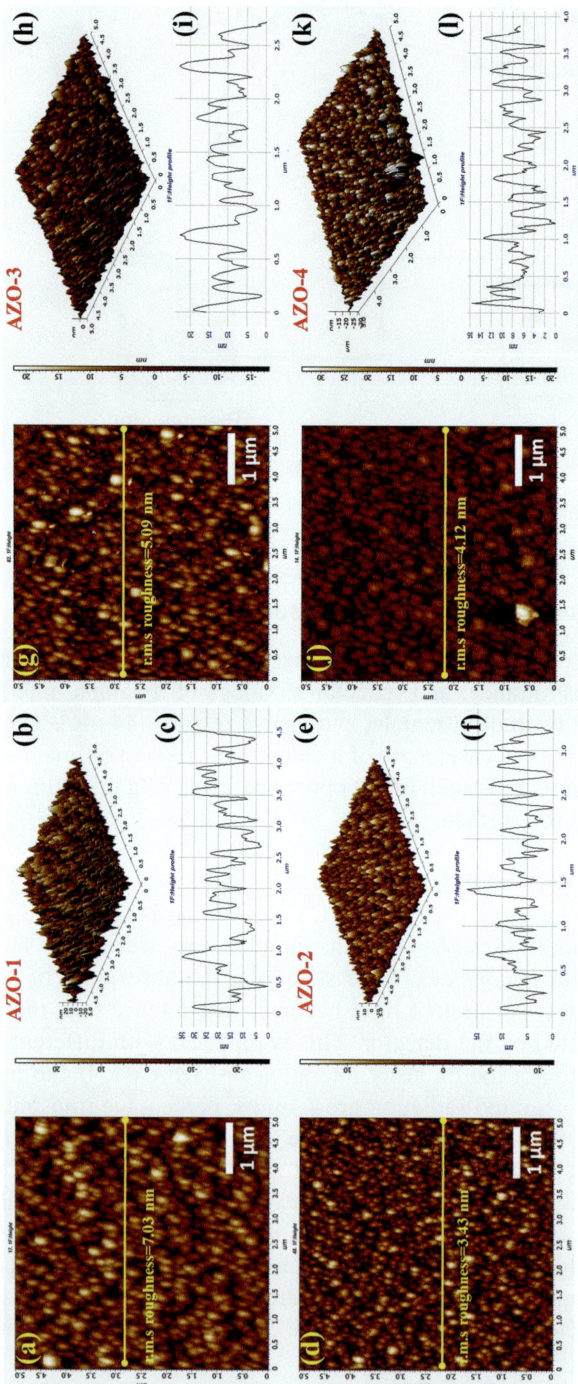

FIGURE 2.13 AFM images of AZO-1 [(a) 2D, (b) 3D image, and (c) line profile], AZO-2 [(d) 2D, (e) 3D image, and (f) line profile], AZO-3 [(g) 2D, (h) 3D image, and (i) line profile], AZO-4 [(j) 2D, (k) 3D image, and (l) line profile] [30].

spectroscopy is consistent with Lambert's law, $A = \alpha d$ (where d is the thickness of the thin-film sample and α is termed as the absorption coefficient). Data obtained from the UV–vis spectrum can be used to evaluate the optical bandgap of a material. First, a graph of $(\alpha h\nu)^n$ versus $h\nu$ (n is a constant, which denotes the nature of transition), known as the Tauc plot, is plotted. By performing some extrapolations in the plot, optical bandgap can be estimated [31, 44, 51]. The schematic of UV–vis spectrophotometer has been illustrated in Figure 2.14, whereas Figure 2.15 presents the calculated band gap of BZO films (sputtered at different Ar:B_2H_6 gas concentration ratio) using Tauc plot.

An extra arrangement may also be attached to the instrument to measure the reflectance of a sample for a range of wavelengths in the UV–vis region. Here also, the percentage reflectance is plotted against the wavelength, and a detailed analysis is carried out [45]. For example, a rough surface and a smooth surface comprising the same material will exhibit different reflectance values. A smooth surface will reflect a huge amount of radiation and allow the rest of it to transmit via sample. So, the effect of texturization in solar cells can be measured quantitatively by UV–vis spectroscopy [52].

2.2.8 Ellipsometry

Ellipsometry, a non-destructive, highly sensitive, versatile, and contactless optical technique, reveals properties like thickness, dielectric function, extinction coefficient, refractive index, and roughness of a thin film or a material [53, 54]. In-depth information regarding the optical properties of the layer is essential for choosing the appropriate material and its position in the cell architecture, along with optimization of the thickness of that material, and so on, which may be realized by conducting ellipsometry of the test sample. Figure 2.16 shows the schematic of ellipsometry, whereas the variation of the refractive index of the AZO films developed through the sputtering technique at different Ar:O_2 gas mixture ratios has been depicted in Figure 2.17.

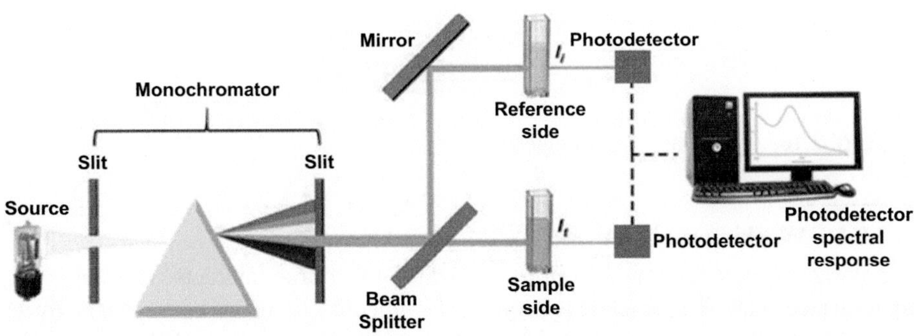

FIGURE 2.14 Schematic of UV–Vis spectrophotometer instrument.

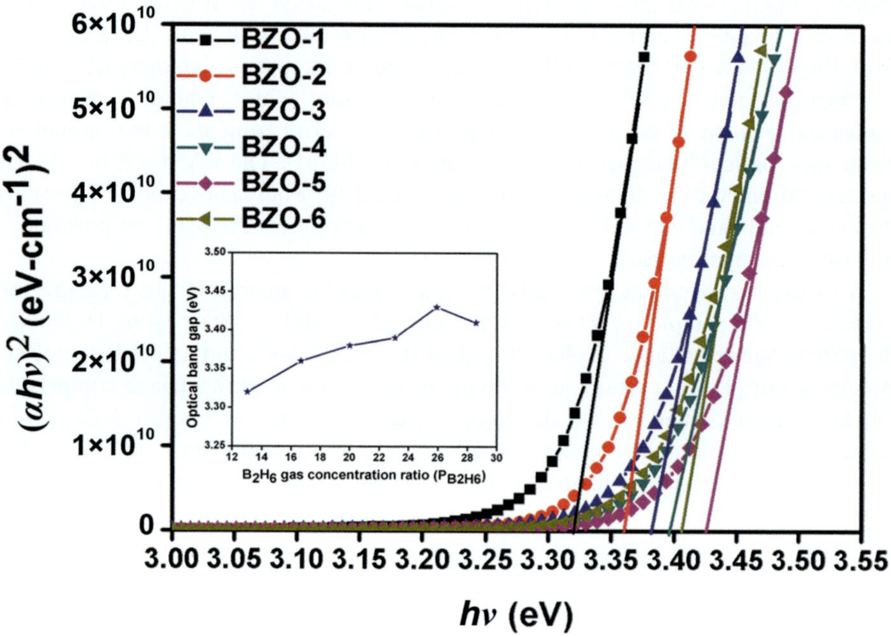

FIGURE 2.15 Calculated band gap of BZO thin films using Tauc plot [31].

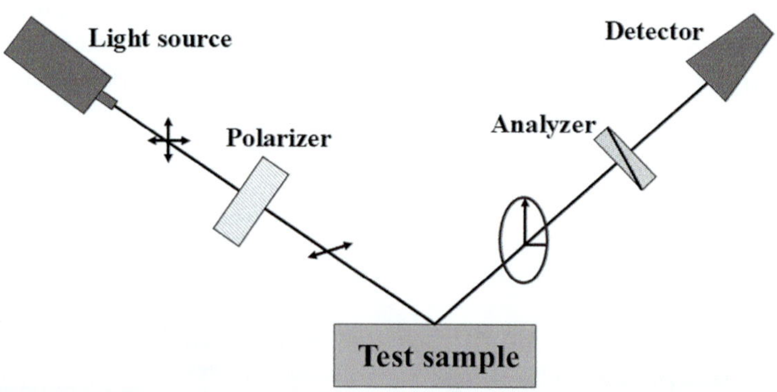

FIGURE 2.16 Schematic of ellipsometry.

2.2.8.1 Working mechanism

A polarized beam generated from a polarized light source interacts with the specimen surface at an oblique angle to produce the reflected polarized beam. An analyzer is used to realize both the alteration of the amplitude and phase difference of the reflected polarized beam in comparison with the incident polarized beam. The following equation is used to describe the alteration in the reflected polarized light [53]:

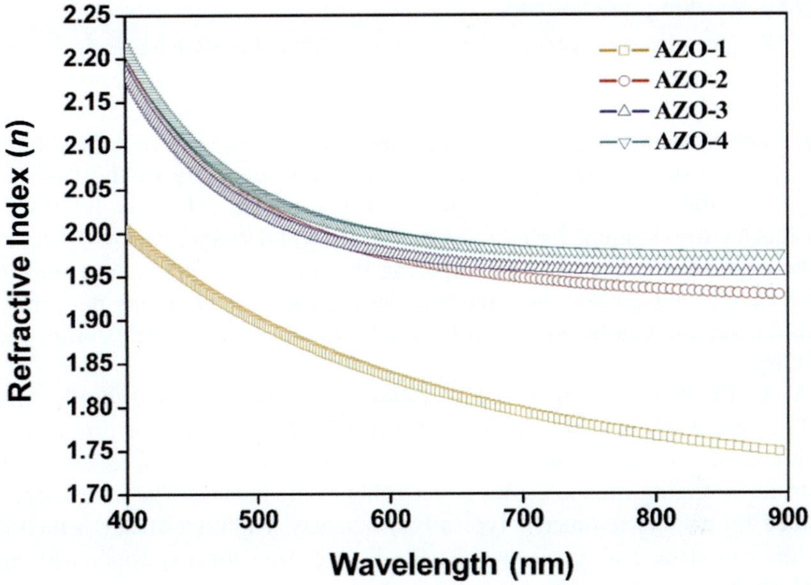

FIGURE 2.17 Impact of $Ar:O_2$ gas mixture ratio on the refractive index of sputtered AZO thin films [30].

$$\rho = \frac{r_p}{r_s} = \tan(\psi)e^{i\Delta} \tag{2.11}$$

where ρ represents the complex reflection coefficient ratio, r_p and r_s are the reflection coefficient for p- and s-polarized light (p-polarized light: the oscillation of the electric field is parallel to the plane of incidence and s-polarized light: the oscillation of the electric field is perpendicular to the plane of incidence), respectively; ψ is the amplitude ratio of the reflected p- and s-polarized light, whereas Δ is the phase difference between the reflected p- and s-polarized light. Eventually, the required information of the specimen is perceived by fitting the measured data, that is, ψ and Δ in the theoretical models of the reflected polarized light. Depending on the operational conditions, ellipsometry can be classified as spectroscopic ellipsometry, variable angle ellipsometry, and imaging ellipsometry.

2.2.9 Fourier-Transform Infrared (FTIR) Spectroscopy

The basic principle of Fourier-transform infrared (FTIR) spectroscopy is that the molecules of the test sample absorb specific frequencies of infrared light, which raises them to different vibrational levels [47]. This technique finds its wide applications in the field of material science, such as degradation and stability studies on SPVs [55] and beta-phase fraction calculation in polyvinylidene fluoride (PVDF) piezo-electric material [56].

2.2.9.1 Working mechanism

In FTIR spectroscopy, infrared radiations with a broad frequency range are directed through a sample. As the energy difference between vibrational levels aligns with the infrared radiation energy, the sample absorbs certain infrared frequencies based on the functional groups and bonds present in it, causing these entities to move to higher vibrational states. The intensity of the transmitted radiation is then measured by a detector in the system, which is definitely less than that of the incident radiation. So, the absorbed frequencies by the sample can be examined indirectly by comparing the intensity of the source and transmitted radiation. Consequently, the functional groups and bonding patterns present in the sample can be identified by analyzing the obtained absorption spectra of FTIR.

The FTIR absorption spectra also follow Lambert's law for thin-film materials. The mathematical backbone of FTIR involves the Fourier-transform technique, which converts the time-domain interferogram (raw data) into a frequency-domain spectrum, revealing the molecular fingerprints of the sample. The interferogram is obtained by an interferometer, typically a Michelson interferometer, which deals with the raw data, and, consequently, the Fourier transform is applied to get the actual spectra.

The Fourier transform $F(v)$ of a time-domain signal $f(t)$ is given by:

$$F(v) = \int_{-\infty}^{\infty} f(t) e^{-2\pi i v t} dt \tag{2.12}$$

where v is the frequency of the radiation and t represents the time.

2.2.10 Raman Spectroscopy

Raman spectroscopy provides information about the vibrational modes of molecules. The sample material is illuminated with a monochromatic light. Some photons undergo interactions with the molecules in the sample and get scattered. A small fraction of the scattered light possesses deviated energy than the incident light energy, known as the Raman effect. The scattered light is analyzed to determine the shift in energy that corresponds to the vibrational frequency of functional groups or bonds [57, 58].

2.2.10.1 Working mechanism

The basic principle of Raman spectroscopy lies in the non-elastic scattering phenomenon. A monochromatic radiation is focused on a sample to be investigated. The incident radiation interacts with the molecules, and the electrons are promoted to a virtual state. When the electrons come back to the ground state, scattering takes place in all directions. Electrons may come back to their original energy state or may reach a different vibrational level within the ground state, which leads to elastic and non-elastic scattering, respectively. The elastic scattering is termed as Rayleigh scattering, whereas the non-elastic scattering is called Raman scattering. Interestingly, in the case of Raman scattering, the photon may gain or lose energy, corresponding to the excitation or de-excitation of vibrational modes. Based on the type of excitation of vibrational mode, Raman scattering can be categorized into Stokes (the photon loses energy, resulting in a shift to a smaller frequency or energy) and anti-Stokes scattering (the photon gains energy, resulting in a shift to a larger frequency or energy), as demonstrated in Figure 2.18.

Raman spectra can be obtained by plotting the intensity of scattered light as a function of frequency shift. The bands in the spectra correspond to the vibrational frequencies of the bonds or functional groups in the sample. Each bond or functional group exhibits its unique signature band, which can be used to identify and characterize the substance. The energy shift in Raman scattering, also known as the Raman shift, is represented as the wavenumber shift ($\Delta\bar{v}$):

$$\Delta\bar{v} = \bar{v}_{\mathrm{i}} - \bar{v}_{\mathrm{s}} \tag{2.13}$$

where $\bar{v}_{\mathrm{i}}$ and $\bar{v}_{\mathrm{s}}$ are the wavenumber corresponding to the incident and scattered radiation, respectively. For stokes scattering, $\Delta\bar{v} > 0$ and for anti-Stokes scattering $\Delta\bar{v} < 0$ [57, 58].

FIGURE 2.18 Schematic sketch of Rayleigh and Raman scattering.

2.3 DEVICE CHARACTERIZATION

2.3.1 I–V Measurement

The PV performance of any solar cell architecture is measured using the I–V measurement technique [59]. Specifically, all the output PV parameters, namely, V_{oc} (open-circuit voltage), I_{sc} (short-circuit current), FF, and η (efficiency), are determined with the help of the I–V measurement technique. The dark (without illumination) I–V measurement is also conducted to explore the diode characteristics of the solar cell. The magnitude of the ideality factor, leakage current, series, and shunt resistance values can be evaluated by the dark I–V measurement [60].

2.3.1.1 Working mechanism

At first, the device to be measured is exposed to the light source (solar simulator) for the generation of the photocurrent. The test is generally conducted by maintaining the standard test conditions, which feature three crucial criteria:

(i) The input light irradiation on the solar cell must be under 100 mW/cm^2.
(ii) The working temperature has to be kept at 25°C.
(iii) The test has to be conducted under the air mass (AM) 1.5 spectrum.

The I–V curve of the solar cell is obtained by connecting the device with a variable electronic load or source measure unit that alters the voltage from 0 V (short-circuit condition) to V_{oc} across the device [59]. The increment of the voltage level leads to providing different current values, which are recorded and plotted with the corresponding voltage magnitude so that the I–V characteristics under illumination may be realized. The I_{sc} of the solar cell is obtained by shorting the two terminals of the cell with an ammeter. On the other hand, FF and η may be calculated per the following mathematical equations:

$$FF = \frac{V_m I_m}{V_{oc} I_{sc}} \tag{2.14}$$

$$\eta = \frac{V_{oc} I_{sc} FF}{P_{in}} \tag{2.15}$$

where V_m and I_m are the corresponding voltage and current magnitude at the maximum power point (P_m) as depicted in Figure 2.19; P_{in} is the input power provided by the incident photon energy.

2.3.2 Measurement of Quantum Efficiency

The quantum measurement technique is used to evaluate the effectiveness of any solar cell as it measures the ratio of the number of collected electron–hole pairs by the cell to the number of incident photons at a specific wavelength. This type of quantum efficiency (QE) is known as external quantum efficiency (EQE) [61]. In

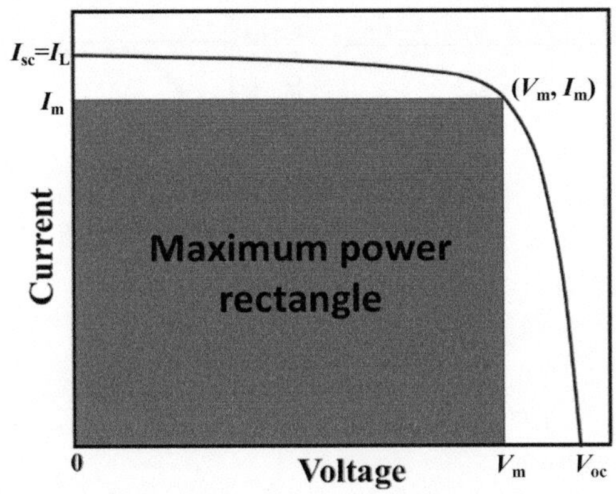

FIGURE 2.19 I–V curve of a solar cell.

contrast, the ratio of collected electron–hole pairs to the number of absorbed photons in a specific wavelength is known as internal quantum efficiency (IQE). Hence, IQE is always higher than its counterpart, that is, EQE. However, in the context of solar cells, measurement of EQE is a more realistic approach as it includes all the optical losses, that is, losses due to reflection, transmission, the absorption of the incident light by the passive layers of the device, and so on. QE, also called the spectral response, is represented in percentages.

2.3.2.1 Working mechanism

In this technique, the illuminated light generated from the light source (LED, xenon lamp) is passed through a monochromator and optical filters to test the solar cell at the desired wavelength. Before illumination, the light source is calibrated with a reference photodiode or a solar cell to determine the incident photon flux at each wavelength. A lock-in amplifier embedded in the system detects tiny fluctuations in the photocurrent by modulating the light source and synchronizing the detection. The process is repeated for each wavelength to evaluate the EQE of the device at the required wavelength range. Mathematically, EQE can be calculated as follows [61]:

$$EQE(\lambda) = \frac{I_{ph}(\lambda) \times h \times c}{e \times P_{inc}(\lambda) \times \lambda} \tag{2.16}$$

where $I_{\mathrm{ph}}(\lambda)$ is the magnitude of the photocurrent at wavelength λ, c is the velocity of the light, e is the elementary charge, and $P_{\mathrm{inc}}(\lambda)$ is the incident optical power at wavelength λ. From the EQE curve, the IQE curve of a device may also be obtained by analyzing the reflection and transmission losses of the same. Figure 2.20 illustrates the I–V characteristics and EQE measurement of 3″ × 3″ p-TOPCon solar cells fabricated at the DST-IIEST solar PV hub.

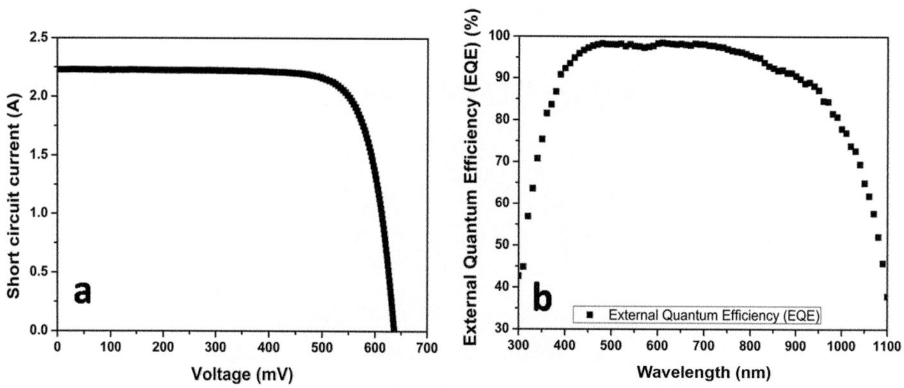

FIGURE 2.20 (a) I–V characteristics and (b) EQE measurement of 3″ × 3″ p-TOPCon solar cells fabricated at DST-IIEST solar PV hub [62].

2.4 CONCLUSIONS

In this chapter, an in-depth discussion regarding the various deposition and characterization tools utilized for conducting research activities in SPV has been made. First, a detailed discussion of the current status of the various deposition tools, namely, PECVD, ALD, sputtering, and thermal evaporation, has been carried out. Moreover, the chapter provides a brief overview and the working principle of numerous high-end characterization tools employed to realize the structural, morphological, compositional, optical, and electrical attributes of the different layers used in various advanced solar cell architectures. Additionally, various mathematical equations used to perceive the various features of the layer have also been mentioned in the corresponding characterization techniques. In the end, the working mechanism of the characterization tools, namely, I–V and EQE, has been pinpointed. The former is utilized to examine the efficiency potential of a solar cell. In contrast, the latter enlightens about the PV performance of any solar cell across the spectral range of interest.

REFERENCES

[1] Rola KP. Anisotropic etching of silicon in KOH + Triton X-100 for 45° micromirror applications. Microsyst Technol. 2017; 23:1463–1473. https://doi.org/10.1007/s00542-016-3103-0.

[2] Moynihan M, O'Connor C, Barr B, Tiffany S, Braun W, Allardyce G, Rentsch J, Birmann K. In-line and vertical texturing of mono-crystalline solar cells. In: 2010 35th IEEE Photovoltaic Specialists Conference; 2010 Jun 20–25; Honolulu, HI, USA. IEEE. pp. 1028–1033. https://doi.org/10.1109/PVSC.2010.5614629.

[3] Iqbal S, Hussain A, Wu W, Su D, Yang Y, Guo X, Zhang T. Rapid anisotropic chemical etching for quick formation of novel octagonal pyramids on silicon surface for photovoltaics. Surf Interface. 2022; 33:102205. https://doi.org/10.1016/j.surfin.2022.102205.

[4] Mandal NC, Biswas S, Acharya S, Panda T, Sadhukhan S, Sharma JR, Nandi A, Bose S, Kole A, Das G, Maity S, Chaudhuri P, Saha H. Study of the properties of SiOx layers

prepared by different techniques for rear side passivation in TOPCon solar cells. Mater Sci Semicond Process. 2020; 119:105163. https://doi.org/10.1016/j.mssp.2020.105163.

[5] Choy KL. Chemical vapour deposition of coatings. Prog Mater Sci. 2003; 48(2):57–170. https://doi.org/10.1016/S0079-6425(01)00009-3.

[6] Ghosh DK, Bose S, Das G, Mukhopadhyay S, Sengupta A. Realization of performance enhancement of thin film silicon solar cells by applying ITO/AZO bilayer TCO films as front electrode. J Mater Sci Mater Electron. 2023; 34:2189. https://doi.org/10.1007/s10854-023-11570-9.

[7] Klein S, Repmann T, Brammer T. Klein S, Repmann T, Brammer T. Microcrystalline silicon films and solar cells deposited by PECVD and HWCVD. Sol Energy. 2004; 77(6):893–908. https://doi.org/10.1016/j.solener.2004.08.029.

[8] Santana G, Morales-Acevedo A. Optimization of PECVD SiN:H films for silicon solar cells. Sol Energy Mater Sol Cells. 2000; 60(2):135–142. https://doi.org/10.1016/S0927-0248(99)00078-1.

[9] Gao T, Yang Q, Guo X, Huang Y, Zhang Z, Wang Z, Liao M, Shou C, Zeng Y, Yan B, Hou G, Zhang X, Zhao Y, Ye J. An industrially viable TOPCon structure with both ultra-thin SiOx and n+-poly-Si processed by PECVD for p-type c-Si solar cells. Sol Energy Mater Sol Cells. 2019; 200:109926. https://doi.org/10.1016/j.solmat.2019.109926.

[10] Yoon SF, Tan KH, Zhang Q, Rusli M, Ahn J, Valeri L. Effect of microwave power on the electron energy in an electron cyclotron resonance plasma. Vacuum. 2001; 61(1):29–35. https://doi.org/10.1016/S0042-207X(00)00429-2.

[11] Sharma R. Silicon nitride as antireflection coating to enhance the conversion efficiency of silicon solar cells. Turk J Phys. 2018; 42(4). https://doi.org/10.3906/fiz-1801-28.

[12] Cao Y, Zhou J, Ren Y, Xu W, Liu W, Cai X, Zhao B. Study on effect of process and structure parameters on SiNxHy growth by in-line PECVD. Sol Energy. 2020; 198:469–478. https://doi.org/10.1016/j.solener.2020.01.054.

[13] Dingemans G, Engelhart P, Seguin R, Mandoc MM, van de Sanden MCM, Kessels WMM. Comparison between aluminum oxide surface passivation films deposited with thermal ALD, plasma ALD and PECVD. In: 35th IEEE Photovoltaic Specialists Conference; 2010 Jun 20–25; Honolulu, HI, USA. IEEE. pp. 3118–3121. https://doi.org/10.1109/PVSC.2010.5614508.

[14] Mandal NC, Acharya S, Biswas S, Panda T, Sadhukhan S, Sharma JR, Bose S, Das G, Kole A, Nandi A, Maity S, Chaudhuri P, Saha H, Guha S. Evolution of PERC from Al-BSF: Optimization based on root cause analysis. Appl Phys A. 2020; 126:569. https://doi.org/10.1007/s00339-020-03747-4.

[15] Ghosh DK, Bose S, Das G, Acharyya S, Nandi A, Mukhopadhyay S, Sengupta A. Fundamentals, present status and future perspective of TOPCon solar cells: A comprehensive review. Surf Interface. 2022; 30:101917. https://doi.org/10.1016/j.surfin.2022.101917.

[16] Wang J, Ru X, Ruan T, Hu Y, Zhang Y, Yan H. Performance of heterojunction solar cells with different intrinsic a-Si:H thin layers deposited by RF- and VHF-PECVD. J Mater Sci Mater Electron. 2021; 32:25327–25331. https://doi.org/10.1007/s10854-021-06991-3.

[17] Faÿ S, Steinhauser J, Nicolay S, Ballif C. Polycrystalline ZnO: B grown by LPCVD as TCO for thin film silicon solar cells. Thin Solid Films. 2010; 518(11):2961–2966. https://doi.org/10.1016/j.tsf.2009.09.189.

[18] Kaur G, Dutta T, Xin Z, Ling ZP, Naval MJ, Saifullah M, Stangl R, Danner A. Ultra-thin LPCVD SiNx/n+poly-Si passivated contacts—A possibility? In: 2019 IEEE 46th Photovoltaic Specialists Conference (PVSC); 2019 Jun 16–21; Chicago, IL, USA. IEEE. pp. 2679–2683. https://doi.org/10.1109/PVSC40753.2019.8980624.

[19] Azzaro C, Couderc JP. Thermal modeling of tubular horizontal hot-wall low pressure chemical vapor deposition reactors. Chem Eng J Biochem Eng J. 1995; 57(1):39–52. https://doi.org/10.1016/0923-0467(94)02859-9.

[20] Huang H, Lv J, Bao Y, Xuan R, Sun S, Sneck S, Li S, Modanese C, Savin H, Wang A, Zhao J. 20.8% industrial PERC solar cell: ALD Al2O3 rear surface passivation, efficiency loss mechanisms analysis and roadmap to 24%. Sol Energy Mater Sol Cells. 2017; 161:14–30. https://doi.org/10.1016/j.solmat.2016.11.018.

[21] Hossain MA, Khoo KT, Cui X, Poduval GK, Zhang T, Li X, Li WM, Hoex B. Atomic layer deposition enabling higher efficiency solar cells: A review. Nano Mater Sci. 2020; 2(3):204–226. https://doi.org/10.1016/j.nanoms.2019.10.001.

[22] van Ommen JR, Goulas A, Puurunen RL. Atomic layer deposition. In: Kirk-Othmer Encyclopedia of Chemical Technology. John Wiley & Sons; 2021. https://doi.org/10.1002/0471238961.koe00059.

[23] Maurya R, Dsouza N, Singh AK, Bandaru N, Kanakala R, Uddin MS, Madaka R, Roy SC, Rath JK. Tuning the Fermi-level of MoOx by plasma-enhanced ALD for carrier selective contact solar cell application. Mater Today Proc. 2024; 104:53–57. https://doi.org/10.1016/j.matpr.2024.01.040.

[24] Mattox DM. Chapter 7—Physical sputtering and sputter deposition (sputtering). In: Mattox DM, editor. Handbook of Physical Vapor Deposition (PVD) Processing. 2nd ed.; 2010. p. 237–286. https://doi.org/10.1016/C2009-0-18800-1.

[25] Rozati SM, Ziabari SAM. A review of various single layer, bilayer, and multilayer TCO materials and their applications. Mater Chem Phys. 2022; 292:126789. https://doi.org/10.1016/j.matchemphys.2022.126789.

[26] Grove WR. Philosophical transactions of the Royal Society of London. Philos Trans R Soc Lond. 1852; 142:87–101.

[27] Bräuer G, Szyszka B, Vergöhl M, Bandorf R. Magnetron sputtering—Milestones of 30 years. Vacuum. 2010; 84(12):1354–1359. https://doi.org/10.1016/j.vacuum.2009.12.014.

[28] Chavan GT, Kim Y, Khokhar MQ, Hussain SQ, Cho EC, Yi J, Ahmad Z, Rosaiah P, Jeon CW. A brief review of Transparent Conducting Oxides (TCO): The influence of different deposition techniques on the efficiency of solar cells. Nanomaterials. 2023; 13:1226. https://doi.org/10.3390/nano13071226.

[29] Dikshit AK, Mandal NC, Bose S, Mukherjee N, Chakrabarti P. Optimization of back ITO layer as the sandwiched reflector for exploiting longer wavelength lights in thin and flexible (30 μm) single junction c-Si solar cells. Solar Energy. 2019; 193:293–302. https://doi.org/10.1016/j.solener.2019.09.017.

[30] Bose S, Arokiyadoss RF, Bhargav PB, Ahmad G, Mandal S, Barua AK, Mukhopadhyay S. Modification of surface morphology of sputtered AZO films with the variation of the oxygen. Mater Sci Semicond Process. 2018; 79:135–143. https://doi.org/10.1016/j.mssp.2018.01.027.

[31] Bose S, Mandal S, Barua AK, Mukhopadhyay S. Properties of boron doped ZnO films prepared by reactive sputtering method: Application to amorphous silicon thin film solar cells. J Mater Sci Technol. 2020; 55:136–143. https://doi.org/10.1016/j.jmst.2019.12.004.

[32] Tepner S, Lorenz A. Printing technologies for silicon solar cell metallization: A comprehensive review. Prog Photovoltaics. 2023; 31(6):557–590. https://doi.org/10.1002/pip.3674.

[33] Urrejola E, Peter K, Plagwitz H, Schubert G. Silicon diffusion in aluminum for rear passivated solar cells. Appl Phys Lett. 2011; 98(15):153508. https://doi.org/10.1063/1.3579541.

[34] De Rose A, Erath D, Nikitina V, Schube J, Güldali D, Minat Ä, Rößler T, Richter A, Kirner S, Kraft A, Lorenz A. Low-temperature metallization & interconnection for silicon heterojunction and perovskite silicon tandem solar cells. Sol Energy Mater Sol Cells. 2023; 261:112515. https://doi.org/10.1016/j.solmat.2023.112515.

[35] X-Ray Diffraction Basics | Chemical Instrumentation Facility; n.d. www.cif.iastate.edu/services/acide/xrd-tutorial/xrd (accessed October 23, 2024).

[36] Tchenka A, Agdad A, Samba Vall MC, Hnawi SK, Narjis A, Nkhaili L, et al. Effect of RF sputtering power and deposition time on optical and electrical properties of indium tin oxide thin film. Adv Mater Sci Eng. 2021; 2021:5556305. https://doi.org/10.1155/2021/5556305.

[37] SURAGUS Everything about Four Point Probe Measurement; n.d. www.suragus.com/en/technology/four-point-probe/ (accessed October 23, 2024).

[38] Smits FM. Measurement of sheet resistivities with the four-point probe. Bell Syst Tech J. 1958; 37:711–718. https://doi.org/10.1002/J.1538-7305.1958.TB03883.X.

[39] Schroder DK, Schueler BW, Shaffner T, Strossman GS. Electrical, physical, and chemical characterization. Handbook of Semiconductor Manufacturing Technology. 2017:28–21. https://doi.org/10.1201/9781420017663-28.

[40] 2.1.3 The Hall Effect; n.d. www.tf.uni-kiel.de/matwis/amat/mw2_ge/kap_2/backbone/r2_1_3.html (accessed October 23, 2024).

[41] Zhou W, Apkarian R, Wang ZL, Joy D. Fundamentals of Scanning Electron Microscopy (SEM). Scanning Microscopy for Nanotechnology: Techniques and Applications. 2006:1–40. https://doi.org/10.1007/978-0-387-39620-0_1.

[42] Stevie FA, Donley CL. Introduction to x-ray photoelectron spectroscopy. J Vac Sci Technol A: Vac Surf Films 2020; 38. https://doi.org/10.1116/6.0000412/1024200.

[43] Binnig G, Quate CF, Gerber C. Atomic force microscope. Phys Rev Lett 1986; 56:930–933. https://doi.org/10.1103/PHYSREVLETT.56.930.

[44] Mkawi EM, Ibrahim K, Ali MKM, Farrukh MA, Mohamed AS. The effect of dopant concentration on properties of transparent conducting Al-doped ZnO thin films for efficient Cu2ZnSnS4 thin-film solar cells prepared by electrodeposition method. Appl Nanosci (Switz). 2015; 5:993–1001. https://doi.org/0.1007/S13204-015-0400-3/FIGURES/8.

[45] Zito G, Rusciano G, Vecchione A, Pesce G, Di Girolamo R, Malafronte A, et al. Nanometal skin of plasmonic heterostructures for highly efficient near-field scattering probes. Sci Rep. 2016; 6:1–16. https://doi.org/10.1038/srep31113.

[46] de Oliveira JB, Michels FS, Silva de Pádua Melo E, Nazário CED, Caires ARL, Gonçalves DA, et al. Data on mineral composition, fatty acids, oxidative stability, UV-VIS spectra and fluorescence emission of the Dersani® and Sunflower® oils used as a cicatrizing agent. Data Brief 2019; 26:104427. https://doi.org/10.1016/J.DIB.2019.104427.

[47] Khan SA, Khan SB, Khan LU, Farooq A, Akhtar K, Asiri AM. Fourier transform infrared spectroscopy: Fundamentals and application in functional groups and nanomaterials characterization. Handbook of Mater Charact. 2018:317–344. https://doi.org/10.1007/978-3-319-92955-2_9.

[48] Lazar NE, Mazkad D, Moussadik A, El Habib Hitar M, Benzaouak A, Touach N, et al. High-performance ferroelectric photocatalysts for rapid dye degradation: ZrO2-doped LiTa0.5Nb0.5O3 under solar UV light. J Solgel Sci Technol. 2024; 110:233–245. https://doi.org/10.1007/S10971-024-06330-Y/FIGURES/7.

[49] Beer Lambert Law | Transmittance & Absorbance | Edinburgh Instruments; n.d. www.edinst.com/blog/the-beer-lambert-law/ (accessed October 23, 2024).

[50] Mayerhöfer TG, Pahlow S, Popp J. The bouguer-beer-lambert law: Shining light on the obscure. Chem Phys Chem. 2020; 21:2029–2046. https://doi.org/10.1002/CPHC.202000464.

[51] Zhong H, Pan F, Yue S, Qin C, Hadjiev V, Tian F, et al. Idealizing tauc plot for accurate bandgap determination of semiconductor with ultraviolet-visible spectroscopy: A case study for cubic boron arsenide. J Phys Chem Lett. 2023; 14:6702–6708. https://doi.org/10.1021/ACS.JPCLETT.3C01416/SUPPL_FILE/JZ3C01416_SI_001.PDF.

[52] Zhong H, Pan F, Yue S, Qin C, Hadjiev V, Tian F, et al. Idealizing tauc plot for accurate bandgap determination of semiconductor with ultraviolet-visible spectroscopy: A case study for cubic boron arsenide. J Phys Chem Lett. 2023; 14:6702–6708. https://doi.org/10.1021/ACS.JPCLETT.3C01416/SUPPL_FILE/JZ3C01416_SI_001.PDF.

[53] Spectroscopic Ellipsometry: Basic Concepts—HORIBA; n.d. www.horiba.com/ind/scientific/technologies/spectroscopic-ellipsometry/spectroscopic-ellipsometry/ (accessed October 23, 2024).

[54] Fujiwara H. Spectroscopic ellipsometry: Principles and applications. Spectroscopic Ellipsometry: Principles and Applications. 2007:1–369. https://doi.org/10.1002/9780470060193.

[55] Neugebauer H, Brabec CJ, Hummelen JC, Janssen RAJ, Sariciftci NS. Stability studies and degradation analysis of plastic solar cell materials by FTIR spectroscopy. Synth Met. 1999; 102:1002. https://doi.org/10.1016/S0379-6779(98)01184-9.

[56] Cai X, Lei T, Sun D, Lin L. A critical analysis of the α, β and γ phases in poly(vinylidene fluoride) using FTIR. RSC Adv. 2017; 7:15382–15389. https://doi.org/10.1039/C7RA01267E.

[57] Orlando A, Franceschini F, Muscas C, Pidkova S, Bartoli M, Rovere M, et al. A comprehensive review on raman spectroscopy applications. Chemosensors. 2021; 9:262. https://doi.org/10.3390/CHEMOSENSORS9090262.

[58] Long DA. The Raman effect: A unified treatment of the theory of Raman scattering by molecules. John Wiley & Sons, Ltd; 2002: p. 597. https://doi.org/10.1002/0470845767.

[59] IV Measurement Overview | Sciencetech Inc.; n.d. https://sciencetech-inc.com/page/i-v-measurement (accessed October 30, 2024).

[60] Green MA. Solar Cells: Operating Principles, Technology, and System Applications. Martin A. Green; 1982: xiv, p. 274.

[61] Ananda W. External quantum efficiency measurement of solar cell. In: QiR 2017–2017 15th International Conference on Quality in Research (QiR): International Symposium on Electrical and Computer Engineering 2017; 2017 December. pp. 450–456. https://doi.org/10.1109/QIR.2017.8168528.

[62] Mandal NC. Efficiency improvement of silicon solar cell through advancing processes and technologies [PhD thesis]. Shibpur: Indian Institute of Engineering Science and Technology (IIEST); 2020.

3 Global Penetration and Recent Developments in Semiconductor Devices for Solar Harvesting

A Review

Asif Nazar and Naqui Anwer

3.1 INTRODUCTION

The journey of solar photovoltaic (PV) power generation has seen many phases, each phase instrumental in opening the doors and contours for the next phases. This journey is both tumultuous and encouraging, simultaneously. It all began with the discovery of the PV effect by Becquerel in 1839. Another significant development was the development of the first practical solar cell, a crystalline silicon cell, with an efficiency of 6% by Bell Laboratories in the year 1954 [1]. PV cells' use on the space satellite in the year 1958 was another milestone. The use of PV cells in space satellites gave rise to healthy competition between silicon cell technology and thin-film CdS cells. The reason for competition between these two technologies was to overcome the vulnerability due to radiation damage in silicon cell technology. Apart from these developments and the use of space satellites, there was a relatively calm and lull in this period [2]. The first oil crisis of 1973 brought impetus to various non-conventional energy sources including solar PV technology. This resulted in the development of the first amorphous silicon cell by Wronski and Carlson. However, the efficiency of this amorphous silicon cell by Wronski and Carlson in 1976 was between 1% and 2% [3]. From here, the focus shifted towards higher efficiency and cost reduction, unfolding the use of different materials and combinations of materials. Mono-crystalline silicon, polycrystalline silicon, and amorphous silicon solar cells got impetus during the 1980s and 1990s.

The cost of wafer-based silicon PV cells is highly skewed towards material costs. These material costs include the cost of a silicon wafer, glass cover sheet, and other necessary encapsulations. In the late 1990s, the material costs for the first-generation solar PV were estimated to be 70% of the total manufacturing costs of the cell [4]. The cost associated with crystalline solar cells is an important concern. To address this concern, thin-film solar cell that uses less material has got

DOI: 10.1201/9781003515784-3

attention. The advantages attributed to thin-film solar cell technology compared to crystalline solar cells include a reduction in materials used, reduced cost, higher absorption coefficient, and flexibility in different applications [5]. It is not the case that it is all rosy about thin-film technology. There are disadvantages too, like faster degradation and the material used; some are scarce and in some cases toxic. GaAs, copper indium gallium selenide (CIGS), CdTe, and a:Si are some of the thin-film solar technologies in vogue.

To reduce net carbon emissions, a lot of efforts are made in the development of novel solar PV cells. The objective is to reduce cost as well as environmental impact. Organic solar cells are the product of those very objectives. Key concerns addressed are material availability, non-toxicity of the material, low-temperature processing, and capability of being tailored for different purposes like building-integrated PVs [6, 7]

Amidst advancement and development in solar PV technology, there has been a significant increase in the installed capacity of solar PV. In 2000, the worldwide installed capacity of solar PV was less than 1 GW, around 0.1% of all renewable generation [8]. In the year 2023, solar PV's installed capacity surged to around 1,412 GW and has a whopping share of 36.5% amongst all renewable generation [9]. There are associated challenges with a substantial increase of solar PV in the energy mix of a country. Grid balancing is one of the challenges associated with increased penetration of intermittent energy resources like solar PV. For grid balancing, in view of the increased penetration of solar PV, there are various mechanisms available. Electrical energy storage is one of the mechanisms of grid balancing [10]. Battery energy storage systems and hydrogen storage are some of the examples of electrical energy storage that are gaining momentum [11].

The rest of the chapter unfolds in the following manner. Section 3.2 deals with the classification of solar PV cells. Section 3.3 delineates PV technology. Section 3.4 discusses the efficiency of PV cells and modules. Section 3.5 gives an overview of the global demand for solar PV. Section 3.6 is about solar PV production. Section 3.7 covers energy storage technologies like battery energy storage systems and hydrogen storage. Section 3.8 pertains to future challenges and the decomposition of solar PV waste. And, finally, Section 3.9 provides the conclusion.

3.2 CLASSIFICATION OF SOLAR PV CELLS

The development of technologies of various solar PV cells can be observed parallelly as well as sequentially. In fact, there are overlaps. Mono-crystalline solar PV cells were developed in 1954, hydrogenated amorphous silicon technology came up in 1976, and the work on organic solar PV began in the 1950s [12]. So the classification of solar PV cells either has been carried out in generations [13, 14] or is based on materials used [15] or the junction applied [16]. There may also be other ways of classification. So, for the sake of completeness, all these approaches will be presented here. The idea behind it is to show the whole gamut of the evolution of technology.

3.2.1 Classification Based on Generation

The first-generation solar PV cells are synonymous with wafer-based technology. Crystalline solar cells, mono-crystalline and polycrystalline, are the first-generation solar PV cells. The first-generation solar cells use silicon crystals that are of very high purity. Apart from that, the process involved in the production is of advanced technology. Thin-film technology leads the second generation of solar PV cells. The following types of solar PV cells are available in the second generation:

- Amorphous silicon (a-Si)
- Cadmium telluride (CdTe)
- CIGS
- Gallium arsenide (GaAs)

The third-generation solar PV cells represent a myriad of organic materials and other technologies. It is listed here:

- Organic PV cells
- Dye-sensitized solar cells
- Perovskite solar cells
- Quantum dot solar cells

3.2.2 Classification Based on Materials

When it comes to the classification of solar PV cells based on the materials used, classification unfolds three broad categories of materials:

1. Silicon-based
2. Semiconductor compound-based
3. Novel materials

The third-generation PV cells and novel materials-based PV cells share the same PV cell technology. Organic, perovskite, and dye-sensitized solar cells represent the third generation as well as novel material-based classification.

The classification based on silicon-based materials includes crystalline solar cells and amorphous silicon. Chalcogenides (CdTe, CIGS, CZTS, etc.) and III–V groups (GaAS, InP, GaP) fall into the semiconductor compound-based classification.

3.2.3 Classification Based on Junction Applied

This category of classification of solar PV cells is based on the junction applied. Silicon, gallium arsenide, cadmium telluride, CIGS, CZTS, and so on are put into p–n junction-based solar cells. P3HT:PCBM, organic semiconductor solar cells, fall under donor–acceptor junction-based solar cells. Dye-sensitized solar cells can be counted under hybrid junction-based solar cells.

These different approaches to classification, generation-based classification, material-based classification, and junction-based classification, enumerate various solar PV cell technologies. Materials, processes, efficiencies, and costs are the main motivators in the research and development of these solar PV cells.
This classification has been summarized in Table 3.1.

Mainly, two factors led to the introduction of second-generation PV cells. One was material, and the other was energy consumption in the production of first-generation crystalline silicon PV cells. In other words, the reasons were economic and environmental. The effort for second-generation PV cells began well in the 1970s. The second-generation PV cell technology is that of thin-film solar PV technology. In solar thin film, a photoactive layer as thin as 1–10 μm can be used. The cost of thin-film technology-based solar cells is less compared to crystalline cells, and the production process is shorter, albeit efficiency is less compared to crystalline solar cells [17]. The hallmark of third-generation PV cells is their simplicity in production processes, introduction of novel materials like organic and semi-organic materials, cost reduction in material and processes, and a try for higher efficiencies. Though most third-generation solar PVs are in the R&D phase, these are good candidates for the future [18].

3.3 PV TECHNOLOGY

In this section, the underlying technology of different generations of solar PV cells has been explained to understand the evolution of technology and the need for the evolution of technology.

TABLE 3.1
Classification of Solar PV Cells

Generation-Based Classification	Material-Based Classification	Junction-Based Classification
First generation: • Mono-crystalline PV cells • Polycrystalline PV cells	Silicon-based: • Crystalline solar cells • a-Si cells	p–n junction-based: • Si • GaAs • CdTe • CIGS
Second generation: • a-Si • CdTe • GaAS • CIGS	Semiconductor compound-based: • CdTe • CIGS • CZTS	Donor/acceptor-based: • Organic solar cells • P3HT/PCBM
Third generation: • Organic • Dye-sensitized solar cells • Perovskite	Novel material-based: • Organic • Perovskite • Dye-sensitized solar cells	Hybrid junction-based: • Dye-sensitized solar cells

3.3.1 Crystalline Silicon Cell

Crystalline silicon cell manufacturing processes involve the following broad stages of production:

- Raw material
- Material preparation
- Crystal growth methods
- Slicing

Quartz sand (SiO_2) is the raw material for silicon production. Solar-grade silicon (purity 99.999%) is needed. First of all, metallurgical-grade silicon (purity 98%) is obtained by processing quartz sand and coal in an electric arc oven. This metallurgical-grade silicon is processed in several reactors to get solar-grade silicon.

Now it is the turn of crystal growth. Depending on the method of crystal growth, one can have either mono-crystalline ingots or multi-crystalline ingots. Czochralski process/float-zone/Bridgman–Stockbargers techniques result in mono-crystalline ingots, and Bridgman columnar growth results in multi-crystalline ingots.

These ingots are sliced into wafers to be doped with p–n impurities and soldered with conducting surfaces [19, 20].

3.3.2 Thin-Film Solar PV

Thin-film solar PV module involves the following steps:

- Glass substrate coated with transparent conductive oxide (TCO) layer
- Deposition of absorber layer
- Deposition of back metal contact

A thin-film solar cell is fabricated on a substrate coated with TCO. The substrate may be made up of glass, stainless steel, or polyimide. TCO has a specific function. TCO provides continuous and transparent metal contact. The continuous and transparent metal contact allows the light to enter into the cell. The continuous metal contact on the front surface is needed for efficient charge collection. TCO fulfils this function by providing continuous and transparent metal contact. Further, TCO may be used at the back side of the cell for enhancing optical properties [21, 22].

The next step is that of deposition of the absorber layer. Film deposition techniques can be broadly divided into two categories: Physical vapour deposition and chemical vapour deposition. Evaporation and sputtering fall under physical vapour deposition, whereas low-pressure chemical vapour deposition and plasma-enhanced chemical vapour deposition come under chemical vapour deposition. Deposition of Ag, Al, SiO_2, and amorphous Si can be done using evaporation [21].

The following types of solar PV cells are available in the second generation:

- Amorphous silicon (a-Si)
- Cadmium telluride (CdTe)

- CIGS
- Gallium arsenide (GaAs)

3.3.2.1 Cadmium Telluride Solar Cell

The following is the structure of CdTe thin-film solar cell [23]:

- Ordinary glass substrate or soda lime glass substrate
- TCO used in SnO_2:FTO (fluorine-doped tin oxide), In_2O_3:Sn (indium-doped tin oxide—ITO), or Cd_2SnO_4 (cadmium tin oxide)
- n-type CdS as a buffer layer (deposition done using chemical bath deposition [CBD], sputtering, or vacuum evaporation)
- p-type CdTe absorber layer (deposition can be done using physical vapour deposition [PVD], etc.)

Al, Au, Ni, Pt, or Mo can be deployed as metal contact (back contact).

The schematics of the CdTe solar cell are shown in Figure 3.1.

3.3.2.2 CIGS Solar Cell

The following is the structure of the CIGS thin-film solar cell [24]:

- Metal contact of molybdenum (Mo) sputtered on a glass substrate
- Sputtering to deposit individual copper, indium, and gallium and then selenize at a temperature above 550°C.
- Incorporation of n-CdS using CBD technique
- Deposition of intrinsic zinc oxide and n-type Al-doped zinc oxide
- Deposition of Al/Ni front metal contact using sputtering technique

The schematics of the CIGS solar cell are shown in Figure 3.2.

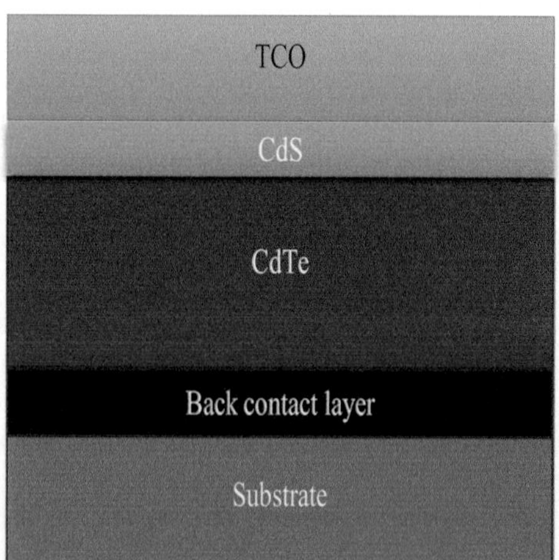

FIGURE 3.1 Schematics of CdTe solar cell.

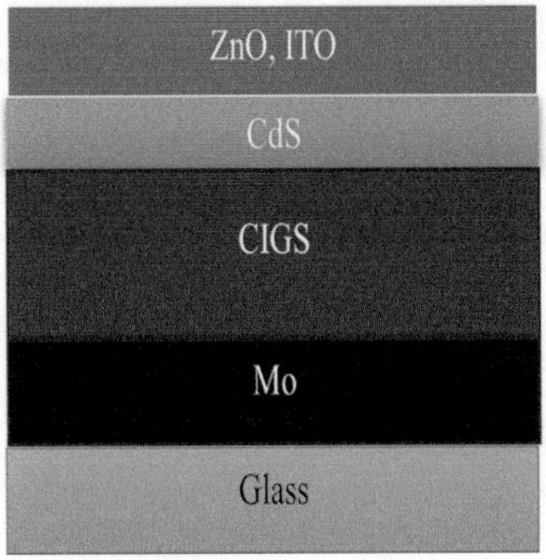

FIGURE 3.2 Schematics of CIGS solar cell.

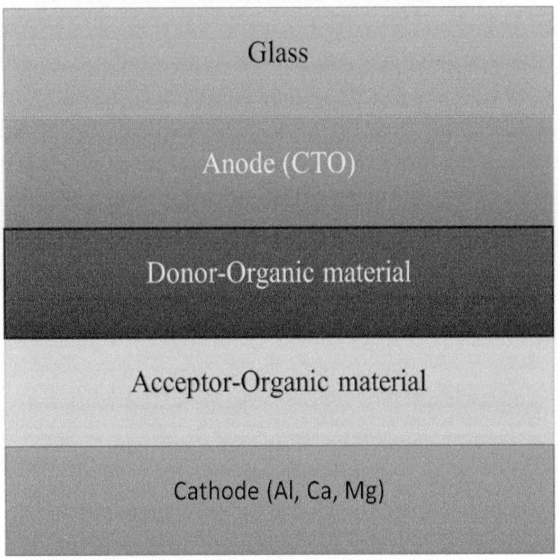

FIGURE 3.3 Schematics of simple bilayer heterojunction.

3.3.3 Third-Generation Solar PV Technology

3.3.3.1 Organic Solar Cell

An organic solar cell is made up of polymers or organic materials. Organic solar cells use conductive organic polymer to absorb light energy and generate electricity. Donor/acceptor materials are mixed, and they serve the purpose of a photoactive layer. There are electrode (cathode and anode) interfaces for carrier collection. Additional carrier transport layers, electron transport layer, and hole transport layer are inserted between the photoactive layer and electrodes to reduce ohmic contact resistances at the electrode interface [25–28]. The schematics of an organic solar cell are shown in Figure 3.3.

3.3.3.2 Dye-Sensitized Solar Cells

There are mainly four components in dye-sensitized solar cells. These are working electrodes, sensitizers (dye), electrolytes, and counterelectrodes. The incident photon excites dye molecules. The excited dye releases electrons in the conduction band of the TiO_2 nanoparticle, in this way creating a potential difference. This electron is free to move through an external circuit and reach the counterelectrode. At the interface of the counterelectrode and electrolyte, the electron takes part in the redox reaction and is supplied back to the dye molecules [29, 30]. The schematics of dye-sensitized solar cells are shown in Figure 3.4.

3.3.3.3 Perovskite Solar Cells

Perovskites are materials that have an ABX_3 crystal structure. Here A stands for the organic/inorganic cation group (e.g. caesium: Cs^+, methylammonium: MA^+), B is for the metal cation group (e.g. Sn^{2+}, Pb^{2+}), and X is halide group (F, Cl, Br) [31].

There are three main layers in perovskite solar cells. These are the perovskite absorber layer, hole transport layer, and electron transport layer. The perovskite absorber layer is sandwiched between these two transport layers. These materials are located between two electrodes called back contact and front contact. The perovskite absorber layer absorbs solar radiations that result in electron–hole pair generation. The hole transport layer and electron transport layer allow a selective number of charge carriers to flow to their electrodes. Conventionally gold is utilized as the metal back contact [32].

The schematics of perovskite solar cells are shown in Figure 3.5.

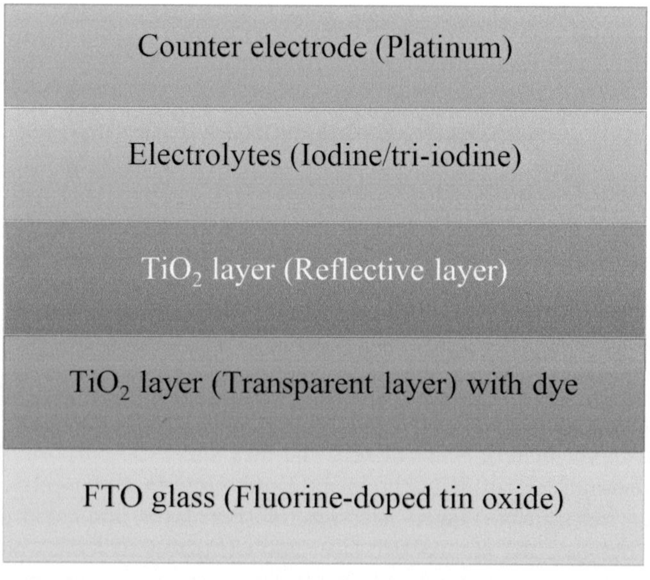

FIGURE 3.4 Schematics of a dye-sensitized solar cell.

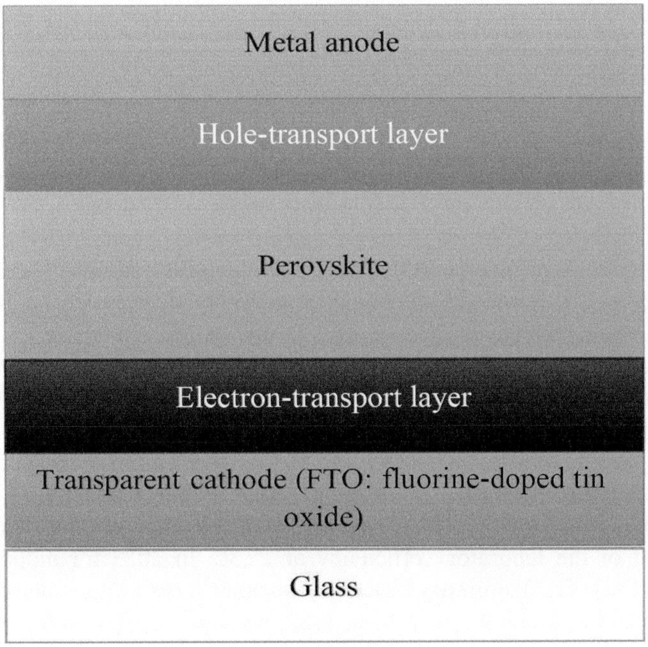

FIGURE 3.5 Schematics of a perovskite solar cell.

3.4 EFFICIENCY OF PV CELLS AND MODULES

The efficiency of solar PV cells can be defined as:

$$Efficiency\, of\ PV\ cell = \frac{Amount\ of\ electrical\ energy\ coming\ out\ of\ the\ cell}{Energy\ from\ light\ shining\ on\ it}$$

$$Efficiency\, of\ PV\ cell = \frac{Electrical\ power\ output}{Solar\ power\ input} = \frac{P_{out}}{P_{in}}$$

$$Efficiency\, of\ PV\ cell = \frac{V_{oc} * I_{sc} * FF}{G * A}$$

Where V_{oc} = Ope circuit voltage

I_{sc} = short circuit current

FF = Fill Factor

G = solar irradiation

A = area of PV cell

A solar PV cell is typically measured under standard test conditions (STCs). STC is defined as:

$G = 1{,}000\ W/m^2$, solar cell temperature = 25°C, and air mass (AM) = 1.5

As per this definition, the limit to the theoretical efficiency of solar PV can be around 90%. However, the theoretical limit to crystalline silicon PV cells is around 33%, known as the Shockley and Queisser limit. This limit is attributable to the use of absorber technology and incident light wavelength. The absorber used, based on bandgap energy, ascertains the possible wavelength of light that it can absorb and convert into electricity. Further, many free electrons thus created may not make it to the metal contact. Thus, the theoretical limit for crystalline silicon PV cells is around 33%. In the year 2023, the cell efficiency for mono-crystalline and multi-crystalline silicon wafer-based technology reached 27.3% and 24.4%, respectively. The theoretical maximum attainable efficiency to achieve efficiency in laboratory conditions for silicon crystalline PV cells is narrowing rather rapidly, a sign of the maturity phase of technology [33–35].

However, this laboratory efficiency has not been achieved overnight. Concerted efforts of decades reflect this achieved efficiency. The contribution of material quality, process techniques, and improved solar cell architectures are behind the improvement of the laboratory efficiency of 27.3% in 2023 for mono-crystalline silicon cells [36]. The laboratory efficiency of mono-crystalline silicon cells in the year 1954 was 6%, increased to 17% in 1983, became 22.7% in the year 1998, and further attained 24.7% in the year 2005 [37–39]. So overall it is a remarkable journey in attaining efficiency of 27.3% in the year 2023 for mono-crystalline silicon cells, which began in 1954 with 6%. For mono-crystalline silicon, the wafer thickness has come down to 150 μm. Techniques employed for higher efficiency include passivation layers, rear contacts, and advanced surface texturing. Deposition of passivation layers uses techniques of either PVD or chemical vapour deposition (CVD). For homojunction solar cells, plasma-enhanced CVD and low-pressure CVD result in better efficiency. Improved solar cell architecture uses an Al-back surface field, and presently passivated emitter rear cell technology is in vogue [36].

Solar PV cells under thin-film technology like CIGS and CdTe have attained an efficiency of 23.4% and 21%, respectively, by the year 2023. Further, perovskite, from emerging technology or third-generation technology, attained a lab cell efficiency of 25% [40]. The high-concentration multi-junction solar cells have attained a lab efficiency of 47.6% till 2023. An interactive best research cell efficiency chart for different solar PV technologies is available on the NREL website [41].

Organic solar cells with an efficiency of 18% have been achieved [42].

The efficiency of commercial modules for the year 2023, falling under various technologies, is summarized here:

- Mono-crystalline wafer-based silicon modules (22%)
- CdTe module (20%)

In the past decade, there has been a considerable increase in the efficiency of commercial solar PV modules. Mono-crystalline solar PV module's efficiency has risen from 16% to 22%. Likewise, the efficiency of the commercial CdTe modules saw a more than 10% jump in efficiency [33].

3.5 GLOBAL PENETRATION OF SOLAR PV

In 2023, the total installed electricity capacity worldwide reached 9,080 GW, with fossil fuels accounting for 4,673 GW (51%), renewables for 3,833 GW (42%), and nuclear power for 382 GW (4%). In comparison, the total installed capacity in the year 2000 was 3,480 GW, with fossil fuels contributing 2,288 GW (65%), renewables 726 GW (20%), and nuclear power 352 GW (10%). Over this period, the share of fossil fuels has declined from 65% in 2000 to 51% in 2023, while the share of renewables has more than doubled, increasing from 20% to 42%. Meanwhile, the share of nuclear power has decreased from 10% to 4%. Figure 3.6 illustrates the global installed electricity capacity (GW) by energy source from 2000 to 2023 [43].

When renewables are categorized into hydroelectric and non-hydroelectric sources, the share of hydroelectricity in total installed capacity decreased from 19% in 2000 to 13% in 2023. In contrast, the share of non-hydroelectric renewables saw a significant rise, increasing from 1.67% in 2000 to 28.83% in 2023. This shift was primarily driven by the large-scale installation of solar and wind energy. The share of wind power grew from 0.5% in 2000 to 11.2% in 2023, while solar power expanded from just 0.03% in 2000 to 15% in 2023. Figure 3.7 illustrates the percentage share of installed electricity capacity worldwide by energy source from 2000 to 2023 [43].

In 2023, global electricity production reached 29,479 TWh, nearly doubling from 15,277 TWh in the year 2000. Electricity generation from coal increased from 5,809 TWh in 2000 to 10,467 TWh in 2023, although its share in total production declined from 38% to 35.5% over the same period.

The share of nuclear power in electricity production decreased from 16.6% in 2000 to 9% in 2023, while oil-based electricity generation saw a sharp decline from 8.7% to 2.7%. Similarly, the share of hydroelectricity dropped from 17.2% to 14.3%.

Conversely, the share of natural gas in electricity generation increased from 18% in 2000 to 22.5% in 2023. Renewable sources such as wind and solar experienced

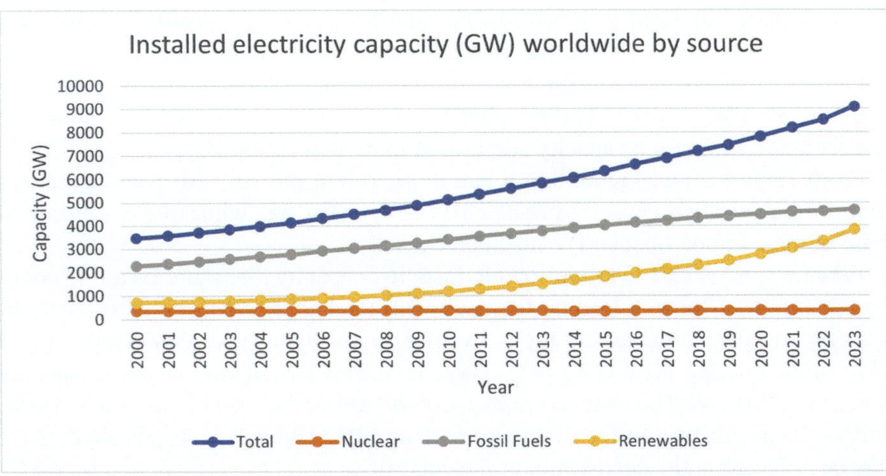

FIGURE 3.6 Installed electricity capacity (GW) by source.

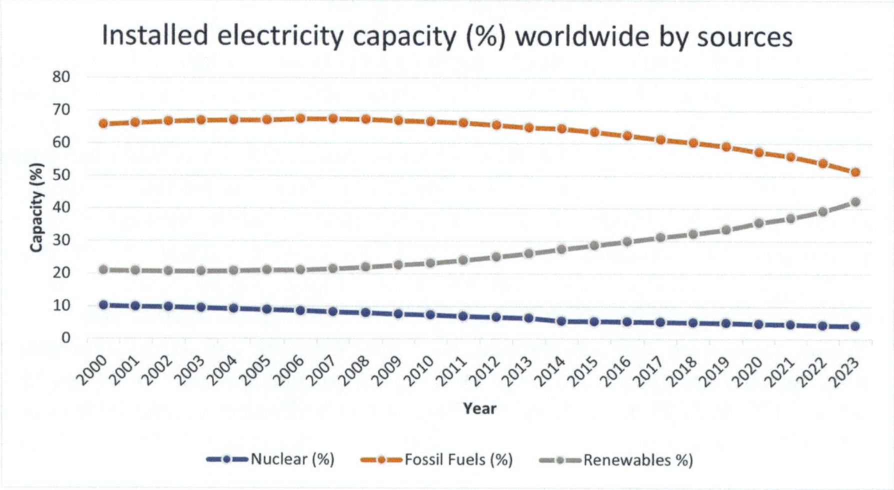

FIGURE 3.7 Installed electricity capacity (%) worldwide by sources.

significant growth, with wind power rising from 0.2% to 7.8% and solar power expanding from a negligible 0.01% to 5.5%.

This shift in the share of different electricity generation sources over the years is summarized in Table 3.2 [44].

3.5.1 Overview of Solar PV Capacity Addition

In 2000, the worldwide installed capacity of solar PV was around 1 GW, around 0.16% of all renewable generation [8]. In the year 2023, solar PV's installed capacity surged to around 1,624 GW and has a whopping share of 36.5% amongst all renewable generation [9]. Growth of worldwide solar PV installed capacity from the year 2000 to 2023 can be gauged through the following data in Table 3.3 [8, 9].

Figure 3.8 shows global cumulative installed capacity (GW) and yearly capacity addition (GW) for solar PV. There is significant capacity addition every year, and the past six years saw a total capacity addition of more than 1,000 GW.

From Table 3.3 and Figure 3.8, it can be seen that the year 2010 saw a marked jump in the installed capacity of solar PV. Every year has witnessed a substantial share of the addition of solar PV installed capacity.

When a look is cast over worldwide installed capacity, the share of a few countries dominates. China is leading the pack with total solar PV installed capacity of 662 GW in 2023, followed by the US (169.5 GW), India (95.3 GW), Japan (91.4 GW), and Germany (81.6 GW) [45]. These countries have a total installed capacity of nearly 1,100 GW. The installed capacity of the rest of the world is around 524 GW. This is the asymmetry in the distribution of installed capacity, and this asymmetry may be reflected in the energy mix of the countries.

TABLE 3.2
Percent Share of Electricity Generation Sources

Source	Percent Share 2000	Percent Share 2023
Coal	38.03	35.51
Natural gas	17.97	22.46
Hydro	17.21	14.3
Nuclear	16.63	9.11
Wind	0.20	7.82
Solar	<0.01	5.53
Bioenergy	1	2.3
Oil	8.66	2.67
Other renewables	0.29	0.30

TABLE 3.3
Worldwide Cumulative Installed Capacity of Solar PV

Year	Installed Capacity (GW)	Yearly Capacity Addition (GW)
2000	1.288	
2001	1.615	0.327
2002	2.069	0.454
2003	2.635	0.566
2004	3.723	1.088
2005	5.112	1.389
2006	6.660	1.548
2007	9.183	2.523
2008	15.844	6.661
2009	23.185	7.341
2010	40.336	17.151
2011	70.469	30.133
2012	100.504	30.035
2013	138.856	38.352
2014	178.391	39.535
2015	229.300	50.909
2016	306.500	77.2
2017	404.500	98
2018	509.300	104.8
2019	633.700	124.4
2020	772.200	138.5
2021	940.000	167.8
2022	1,177.000	237
2023	1,624	447

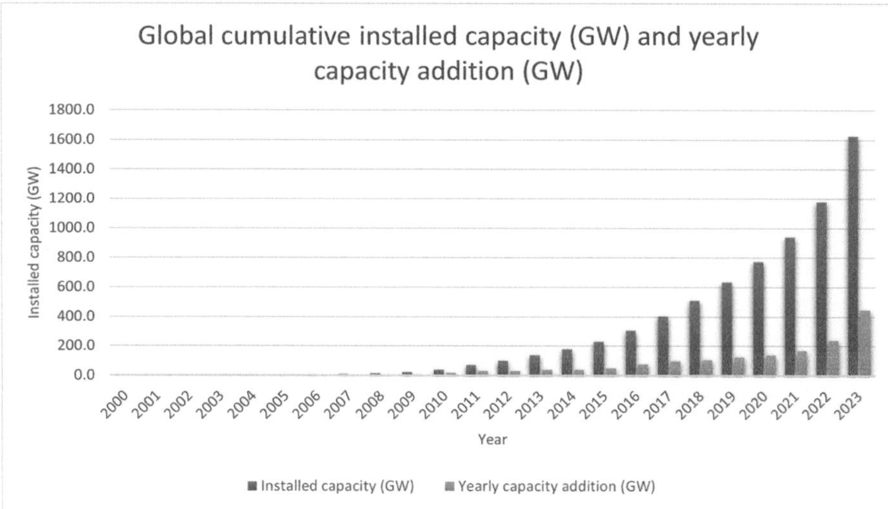

FIGURE 3.8 Year-wise global PV cumulative installed capacity (GW) and yearly capacity addition (GW).

The growth and deployment of solar PV can be viewed in three phases through some representative countries. The first phase is pre-2000, the second phase is between 2000 and 2013, and the third phase is from 2014 onwards. Germany has dominated in cumulative solar PV installation between 2005 and 2014, followed by Japan. From 2014 onwards, China's growth has been phenomenal, and its dominance in the cumulative installed capacity of solar PV is remarkable. Table 3.4 shows the installed capacity of solar PV from 2014 to 2023 for China, the US, Germany, Japan, India, and the world. The same has been plotted in a graph (Figure 3.9).

The pre-2000 phase largely focuses on technological development and cost reduction of solar cells. The US, Japan, and Germany have been dominant players during this phase. The technological advancement made in this period provided the necessary impetus for the post-2000 phase. The governments of Germany and Japan provided steady and increasing support through various incentives and subsidies, bringing momentum to the capacity addition of solar PV systems. This was the period when Japan and Germany were proven leaders in the installed capacity of solar PV. The US approach has been a bit different from Germany and Japan. The US relied on federal tax credits, state renewable portfolio standards (RPSs), and innovative private-sector financing [47, 48].

3.5.2 Growth Story of China in Solar PV

The genesis of new and renewable energy in China lies in the formation of "Ten countermeasures for improving environment and development" against the backdrop of the UN Conference on Environment and Development in 1992 and its affirmation

TABLE 3.4
Year-Wise Cumulative Installed Capacity of Countries [46]

	Cumulative Installed Capacity in GW					
	China	US	India	Japan	Germany	World
2014	28.1	18.3	4	23.3	38.2	177
2015	43.5	25.6	5.7	34.4	39.7	227.1
2016	78.1	40.3	9.98	42.8	41.2	303
2017	131	51	18.3	49	42	402.5
2018	176.1	62.2	27.5	56	45.4	500
2019	204.7	75.9	35.25	63	49.2	627
2020	253.4	93.2	39.70	71.4	53.9	760.4
2021	308.5	123	50	78.2	59.2	942
2022	414.5	141.6	63.40	84.9	67.2	1,185
2023	662	169.5	73	91.4	81.6	1,600

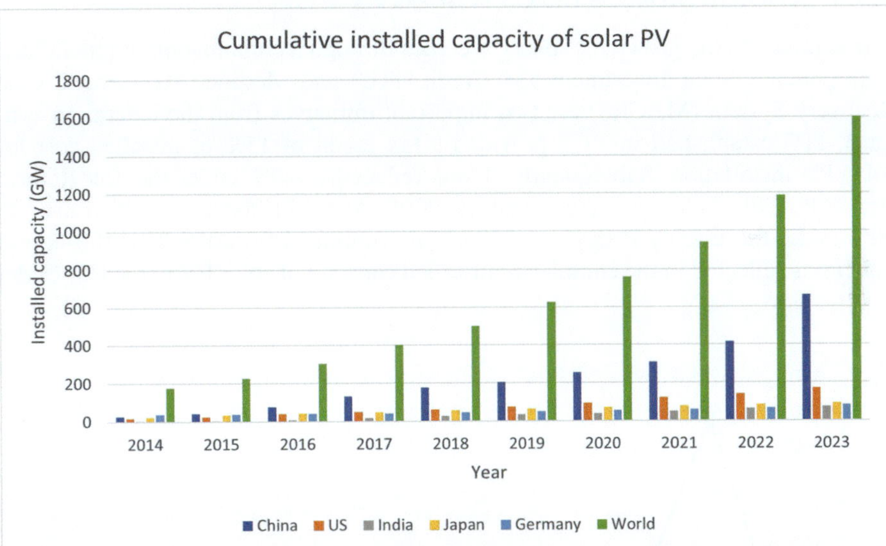

FIGURE 3.9 Year-wise cumulative installed capacity (GW) of representative countries.

in the "China 21 century agenda" in 1994. This new and renewable energy includes solar, wind, biomass, tidal, and ocean energy [49]. In the year 1985, the PV installed capacity in China was 180 kW, and this was for various applications like communication systems, agriculture use, and so on [50]. The solar PV installed capacity in China in the year 1998 was 3 MW, and this solar PV was not connected to the national grid [51]. The year 2000 witnessed the installation of a production line of crystalline silicon solar cells in China, and the total installed capacity of solar

PV systems neared around 22 MW by the year 2002 [52]. Renewable Energy Law 2005 and its amendment of 2009 under the 11th five-year plan paved the way for the growth of solar PV in China [53]. The growth of solar PV installation in China can be seen in two phases. The 11th five-year plan has been a milestone in nurturing and providing a trajectory for the growth of solar PV. Thus, the installation of solar PV can be understood by analysing the growth story in the pre-11th five-year plan and the post-11th five-year plan. The installed capacity of solar PV during the 11th five-year plan saw momentum; it increased from 80 MW (2006) to 800 MW (2010). It was significant compared to an installed capacity of 19 MW (2000) and 70 MW (2005) in the pre-11th five-year plan. The post-11th five-year plan saw unprecedented growth of 1,200%, that is, from 3.3 GW (2011) to 43.18 GW (2015). This remarkable growth can be attributed to the global financial crisis of 2008 that led to the growth of domestic solar PV capacity addition with various policy initiatives [53–55]. This period coincides with the lessening of the share of cumulative installed capacity of Germany, Japan, and the US. The annual growth rate from the year 2012 to 2023 for China, Germany, Japan, and the US has been plotted in Figure 3.10 [56].

3.5.3 Growth Story of the US in Solar PV

The stimulus to the growth of solar PV comes through a combination of federal and state policies. Solar Investment Tax Credit (ITC) and Modified Accelerated Cost Recovery System (MACRS) are two important initiatives from the federal government. ITC established in 1978 provided a tax credit of 15% of installed cost for solar PV installation. Subsequently, it was reduced to 10% under the Tax Reform Act 1986. This state was maintained till 2005. Again in 2005, ITC was raised to 30% under the Energy Policy Act 2005. This continued till 2016. This flip-flop in policies is reflected in the cumulative installed capacity of solar PV in the US. Under

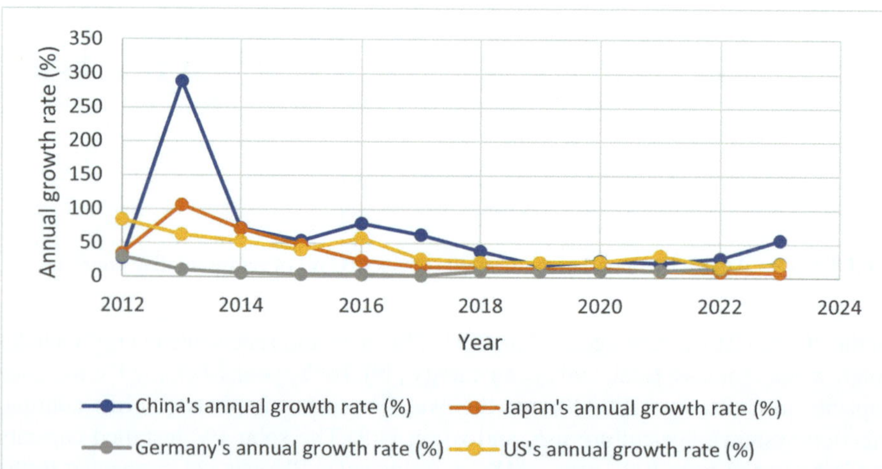

FIGURE 3.10 Annual growth rate (%) for solar PV in China, Japan, Germany and the US.

the Tax Reform Act 1986, MACRS was introduced. MACRS is a depreciation rule for commercial entities that allows business entities to qualify for rapid five-year tax depreciation [57–59].

Two important instruments used by states are RPS and net metering. Under RPS, load-serving entities must provide a part of their electricity supply through solar PV. Net metering allows customer-sited PV generators to offset electricity provided by load-serving entities with kWh supplied by their PV system [57–59]. The cumulative installed capacity and annual growth rate of the US from 2000 to 2010 have been shown in Table 3.5 and the corresponding graph in Figure 3.11.

TABLE 3.5
Cumulative Installed Capacity of Solar PV in the US from 2000 to 2010 (compiled from various sources)

Year	Installed Capacity	Annual Growth Rate (%)
2000	18 MW	
2001	29 MW	61
2002	52 MW	79
2003	97 MW	87
2004	155 MW	60
2005	234 MW	51
2006	339 MW	45
2007	508 MW	50
2008	819 MW	61
2009	1.6 GW	95
2010	3 GW	88

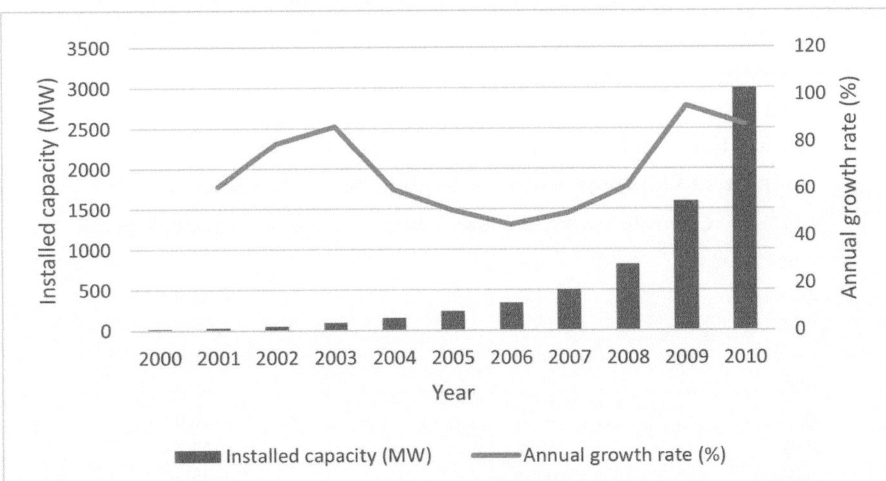

FIGURE 3.11 Cumulative installed capacity and the annual growth rate (%) for solar PV in the US.

3.5.4 Growth Story of Germany in Solar PV

The research and development programmes in the domain of solar PV have been rooted in the case of Germany since 1974. In the 1990s, German federal and state governments incentivized homeowners for rooftop solar PV installation under "1000-roof" projects. Despite being a demonstration project, rooftop solar PV installation proved to be the cornerstone for future growth and a model for adaptation by various countries [60, 61]. The programme was able to install 5.3 MW of solar PV till the year 1993. This was the result of the Electricity Feed-in-Law of 1999 [62]. The status of grid-connected solar PV installed capacity and the annual growth rate from 1993 to 2002 have been tabulated in Table 3.6 [63].

Renewable Energy Sources Act (RESA) 2000 and a further amendment to it in 2004 replaced the Electricity Feed-in-Law of 1991 [62]. The purpose of RESA was to double the generation of renewable energy sources for electricity (RES-E) by 2010. Though RESA retained the feed-in tariff as the central feature, the long-term contract was the highlight of the RESA. In the year 2003, Germany was the second largest in terms of installed solar PV capacity (350 MW). Japan was the world leader with an 860 MW solar PV installed capacity in 2003 [62].

Germany's cumulative installed capacity from 2003 to 2013, the leader in cumulative installed capacity of solar PV, has been tabulated in Table 3.7 [64].

3.5.5 Growth Story of Japan in Solar PV

Three important technology projects, namely, new energy, energy conservation, and environmental technology, were initiated in the years 1974, 1978, and 1989, respectively, under "Sunshine Project," "Moonlight Project," and "Research and Development Project on Environmental Technology," respectively. The Sunshine Project was mainly for new energy like solar energy, geothermal energy, and the

TABLE 3.6
Cumulative Installed Capacity and Annual Growth Rate of Solar PV in Germany from 1993–2002 [64] and Other Sources

Year	Cumulative Installed Capacity (MW)	Annual Growth Rate (%)
1993	8.5	
1994	11.6	36.4
1995	16.1	38.8
1996	24.2	50.3
1997	35.2	45.4
1998	44.7	26.9
1999	58.0	29.7
2000	100.0	72.4
2001	178.0	78.0
2002	260.6	46.4

TABLE 3.7
Cumulative Installed Capacity and Annual Growth Rate of Solar PV in Germany from 2003 to 2013 [64] and Other Sources

Year	Cumulative Installed Capacity (GW)	Annual Growth Rate (%)
2003	0.435	
2004	1.1	153
2005	2.1	91
2006	2.9	38
2007	4.2	45
2008	6.1	45
2009	10.6	74
2010	17.9	69
2011	25.4	42
2012	33.0	30
2013	36.3	10

utilization of hydrogen as an energy carrier. The "New Sunshine Programme" of 1993 unifies the three earlier projects, namely, Sunshine Project, Moonlight Project, and R&D Project on Environmental Technology. The New Sunshine programme comprises three technological systems:

1. Innovative technology development
2. International large-scale collaborative research
3. Collaborative R&D on appropriate technology.

The aim of the New Sunshine Programme was to promote a comprehensive long-term R&D programme. The increase in demand was envisaged through a reduction in cost using technological development. The government subsidies for the promotion of solar PV came into effect through the PV field test in 1992 and the residential PV system monitoring programme in 1994. This spurred the growth of solar PV to 300 MW in the year 2000 [65, 66]. This installed capacity was the highest compared to the US (18 MW) and China (19 MW). Thus, in early 2000, Japan emerged as a global leader in terms of cumulative installed capacity as well as in the production of solar cells [67]. The emphasis during this period has been on the R&D of solar cells.

The termination of the solar PV subsidy programme in 2006 saw a subsequent decrease in the demand for solar PV installed capacity in 2007 and 2008. In 2009, many policy interventions were introduced to increase the solar PV installed capacity and regain a leadership position in the installation of solar PV. The policy interventions include bringing back subsidies for household installation and implementation of solar feed-in tariff [67]. With the dramatic increase in the share of cumulative installed capacity since 2015, Japan's share decreased in subsequent years. The story of the development of solar PV in Japan crisscrosses through the US and Germany.

A glimpse of it can be had with the cumulative installed capacity of Japan, Germany, and the US from 2000 to 2015 in Table 3.8.

3.5.6 Growth Story of India in Solar PV

3.5.6.1 Enabling Policies in Solar PV Growth

The development of solar PV in India can broadly be divided into three phases. Under the 6th Five-Year Plan (1980–1985), the Department of Non-Conventional Energy Sources was formed to fund research, development, and demonstration of renewable energy technologies like solar and wind. The National Solar Photovoltaic Energy Demonstration Programme helped Central Electronics Limited (CEL), a Government of India Enterprise, manufacture solar cells and modules, albeit in kW capacities. This was the first phase of the development of solar PV in India.

The second phase began with several initiatives taken by the central and state governments. Some of the key initiatives that contributed to the growth of solar PV in India include:

- Indian Electricity Act, 2003
- National Electricity Policy, 2005
- National Action Plan on Climate Change, 2008

TABLE 3.8
Cumulative Installed Capacity of Solar PV in Japan, the US, and Germany from 2000 to 2015

	Cumulative Installed Capacity		
	US	Japan	Germany
2000	18 MW	300 MW	100 MW
2001	29 MW	453 MW	178 MW
2002	52 MW	637 MW	260 MW
2003	97 MW	860 MW	435 MW
2004	155 MW	1.13 GW	1.1 GW
2005	234 MW	1.4 GW	2.1 GW
2006	339 MW	1.7 GW	2.9 GW
2007	508 MW	1.9 GW	4.2 GW
2008	819 MW	2.1 GW	6.1 GW
2009	1.6 GW	2.6 GW	10.6 GW
2010	3 GW	3.6 GW	17.9 GW
2011	5.2 GW	4.9 GW	25.4 GW
2012	8.1 GW	6.6 GW	33 GW
2013	12 GW	13.6 GW	36.3 GW
2014	16.4 GW	23.3 GW	38.2 GW
2015	22.5 GW	34.2 GW	39.8 GW

- Semiconductor Policy, 2007
- Karnataka Semiconductor Policy, 2010

Under the National Action Plan on Climate Change (2008), eight national missions were launched, one of which was the National Solar Mission. The National Solar Mission was launched in 2010 to create policies that would facilitate the large-scale deployment of solar PV across the country. Initially, the target for solar PV power was 20 GW by 2022. This mission paved the way for the scaling of solar PV. By 2014, the installed capacity of solar PV had reached nearly 4 GW, a substantial increase compared to just 65 MW in 2010.

The third phase began in 2015 with the revised target under the National Solar Mission, which provided a significant boost to solar PV deployment. The revised target was 100 GW by 2022, with 40 GW from rooftop projects and 60 GW from utility-scale projects. The years 2021 and 2022 saw a series of policy interventions, including:

- Production-Linked Incentive (PLI) Scheme
- Import Duties
- Approved List of Models and Manufacturers (ALMM)

The PLI Scheme, launched in 2021, aimed to boost the domestic manufacturing of solar PV components. To ensure the use of high-quality solar PV components in projects, the ALMM was introduced in 2021 [68–72].

The ALMM played a crucial role in adding 20.8 GW of solar module manufacturing capacity and 3.2 GW of solar cell manufacturing capacity in 2023. As a result, by December 2023, India's total manufacturing capacity stood at 64.5 GW for solar modules and 5.8 GW for solar cells. Mono-crystalline modules accounted for 67.5% of the total manufactured modules [73].

To encourage local production, basic customs duties of 40% on imported solar modules and 20% on imported solar cells were implemented on April 1, 2022 [74].

3.5.6.2 Overview of Solar PV in India

The total installed electricity capacity in India in 2023 was 428 GW. The share of nuclear, fossil fuels, and renewables was 7.5 GW (1.7%), 240 GW (56%), and 180.8 GW (42.2%), respectively. India's installed electricity capacity stood at 105 GW in March 2002. In 2002, the share of nuclear, fossil fuels, and renewables was 2.7 GW (2.6%), 74.4 GW (70.9%), and 27.9 GW (26.5%), respectively.

In terms of percentage share, the contribution of fossil fuels declined from 70.9% in 2002 to 56% in 2023, while the share of renewables increased from 26.5% in 2002 to 42.2% in 2023 [75].

When renewables are classified into hydroelectric and non-hydroelectric sources, the percentage share of hydroelectricity in 2002 was 25%, which decreased to 11% in 2023. In contrast, the percentage share of non-hydroelectric renewables increased significantly from 1.5% in 2002 to 31.3% in 2023. This sharp rise in the share of non-hydroelectric renewables was primarily driven by the installation of solar and wind power. In 2023, the percentage share of wind power in installed capacity was 10.5%, while solar power accounted for 17.1% [75].

India has witnessed tremendous growth in the installed capacity of solar PV. In 2014, the installed capacity of solar PV in India was 3.99 GW, which increased to 73.32 GW by 2023, reflecting a growth rate of 1,725%. The vision to promote solar energy is enshrined in the Jawaharlal Nehru National Solar Mission, launched in 2010. This mission gained significant momentum in 2015, and India now ranks third globally in solar energy capacity [76–78]. The annual growth rates of India, Germany, and Japan from 2014 to 2023 are illustrated in Figure 3.12. As evident from the graph, India's annual growth rate outperforms that of Germany, the US, and Japan.

3.6 SOLAR PV CELL PRODUCTION

Through concerted R&D, Japan, Germany, and the US were leaders in solar PV manufacturing until 2008. In 2008, China became the largest producer of solar cells. In the year 2023, the contribution of China in solar PV module production was 86%. The share of the European Union and the US/Canada was 2% each [79, 80].

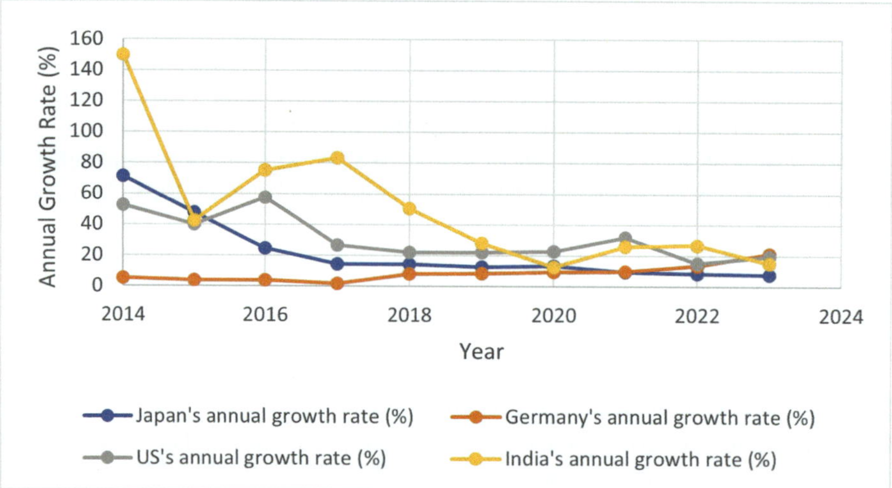

FIGURE 3.12 Annual growth rate of India, the US, Japan, and Germany.

TABLE 3.9
Market Share of PV Cells [79, 80]

	Mono-Crystalline Silicon	Polycrystalline Silicon	Crystalline Silicon	Thin Film
2015 [79]	24%	69%	93%	7% (3% a-Si, 2.5% CdTe, and <2% CIGS)
2023 [80]			97%	3%

The contribution of crystalline silicon has been substantial in the installed capacity of solar PV. This can be found in the market share of crystalline silicon PV cells and thin-film solar PV cells. The share of these two types of PV cells has been tabulated in Table 3.9.

The year 2023 saw 485 GW of mono-crystalline silicon PV production. Thin-film solar PV production for 2023 was 13 GW [80].

3.7 SOLAR PV AND STORAGE

With the substantial penetration of intermittent sources of renewable energy like solar PV in the energy mix of a country, grid balancing emerges as an important challenge. When it comes to grid balancing strategy, there are four main options. These can be the use of:

1. Flexible generation
2. Transmission interconnection
3. Demand side management
4. Electrical energy storage

3.7.1 Battery Energy Storage Systems

Amongst electrical energy storage, one of the alternatives is battery energy storage. In recent years, there has been a substantial addition of battery energy storage capacity in different countries. Pumped hydro storage as a mainstay option has geographical limitations. Battery energy storage by virtue of its time response and flexibility in providing various ancillary services necessary for grid balancing is emerging as one of the key players [81].

China had an installed capacity of 1 GW of battery energy storage systems in 2018. The US and Germany had battery energy storage systems installed capacity of 0.87 and 0.37 GW, respectively, for the year 2018 [82]. In the year 2023, the total installed capacity of battery energy storage systems in China stood at 31.4 GW [83]. For the US battery energy storage systems installed capacity for the year 2023 was 16 GW [84]. For the year 2023, the installed capacity of battery energy storage systems for Germany was 7.6 GW [85]. Compared to the year 2018, the installed capacity in these three countries has increased multiple times, showing its importance in grid balancing and other usage. Thus, linking the increase of installed capacity of solar PV with battery energy storage systems.

3.7.2 Solar PV and Hydrogen Storage for Climate Change Mitigation

The path of solar PV energy and hydrogen crisscrosses at broadly four junctions. These are:

1. Energy-to-energy
2. Energy-to-gas

3. Energy-to-fuel
4. Energy-to-feedstock

In the energy-to-energy route, energy is converted into hydrogen, and hydrogen is used in fuel cells to produce energy. In the energy-to-gas route, energy is converted to hydrogen through electrolysis, and this hydrogen can be stored directly or moved for other purposes. In the energy-to-fuel route, energy is converted to hydrogen through electrolysis, and that hydrogen can be used in fuel cell electric vehicles. In the energy-to-feedstock route, energy is used to convert hydrogen for feedstock to produce chemical compounds [86].

Green hydrogen is a clean source of fuel for the entire gamut of applications. Thus it is an important candidate in reducing carbon emissions. Hydrogen can be produced using direct solar PV and electrolyser systems. The electrolysers that can be used are:

- Proton exchange membrane
- Alkaline water electrolyser
- High-temperature steam

Electrolytes and electrodes vary for different electrolysers. Green hydrogen production using solar PVs is an important alternative. Conventional crystalline silicon PV, as well as advanced or novel solar PV technology, can be used for the production of green hydrogen with the potential to reduce carbon emissions.

Hydrogen is being generated using a lot of fossil fuels for industrial applications, and that is resulting in a significant contribution to greenhouse gases. Various factors are responsible for the cost of green hydrogen production. A few that can be mentioned are the type of solar PV cell used, the type of electrolyser used, the availability of solar radiation, installation costs, and so on [87].

Two sectors, transportation and industry, are witnessing the dependency on fossil fuels. One of the routes to attain green hydrogen is the solar PV route. The transportation sector is responsible for 28% of greenhouse gases. So the transition from fossil fuel-based transportation systems to electric vehicles is taking shape. Battery electric vehicles (BEV) will be a suitable candidate if the BEV is charged with non-fossil fuel-based electricity. Otherwise, it has its own consequences on the environment. Fuel cell-based electric vehicles using hydrogen can be a candidate for clean sources of transportation [88–90].

3.8 FUTURE CHALLENGES AND WASTE DECOMPOSITION TECHNIQUE (PV CELL)

At the end of 2016, the estimated volume of solar PV waste was between 0.043 and 0.25 million tonnes worldwide. The solar PV waste is forecasted to be about 1.7–87 million tonnes by 2030 and about 60–78 million tonnes by 2050 [91]. As of now crystalline silicon PV panels are dominant in total solar PV installed capacity. So solar PV waste generated will have a dominant contribution from crystalline silicon PV panels for many decades as the expected life of solar panel is typically 25 years. Recycling and reusing end-of-life solar panels are likely to reduce actual waste substantially [92].

A typical crystalline silicon solar panel contains the following components [92]:

- Glass (70%)
- Aluminium (18%)
- Adhesive sealant (5%)
- Silicon (3.5%)
- Plastic (1.5%)
- Other materials (2%)

The following three waste decompositions can be employed for the recycling process of crystalline silicon solar PV panels [93, 94]:

- Physical processes
- Thermal process
- Chemical process

The physical process involves the dismantling of PV panel components. From dismantled panels, the aluminium frame, junction box, and cables are removed. Further panels are crushed for the next process. By thermal process, 90% of the glass is recovered. Using different solvents and solutions, different elements of solar panels can be recovered under chemical processes [95].

However, there are challenges to the recycling process. These challenges are in terms of material recovery, environmental impact, and economic viability. To overcome these challenges, policies, regulations, and guidelines for handling PV panels at the end of their lifetime are required. European Union Waste from Electrical and Electronic Equipment (EU WEEE) directives are an important step in solar PV recycling. The UK is the first country to adopt EU WEEE followed by Germany. Many solar PV manufacturing companies and developed countries are aware of the need to recycle and reuse solar PV waste, hence devising devise policies and regulations proactively. Developing nations have yet to take the initiative in this regard. Earlier, India did not classify solar PV waste under e-waste or hazardous waste categories, leaving it unregulated for recycling and reuse before the E-Waste (Management) Rules, 2022 came into effect. Further, the onus of disposal of solar PV waste is on developers, and they are supposed to dispose of it per e-waste management and handling rules. Central Pollution Control Board (CPCB) is in the process of making guidelines for safe handling and disposal of solar modules, panels, and cells as per the E-Waste (Management) Rules, 2022 for different stakeholders. China is working in the direction of bringing policies in the domain of PV recycling, but specific regulations are not in place [96–99].

3.9 CONCLUSIONS

Solar PV has now shown significant penetration in various countries' energy mix. Significant milestones in the development of solar PV technology are one of the reasons for its acceptance as an alternative to fossil fuels. This way it is driving climate change policies in mitigating greenhouse gases. Though crystalline silicon PV modules are dominating the PV market with a larger chunk of share beyond 90%, other technologies are constantly showing encouraging signs that will be helpful in finding bottlenecks to the existing technology. Grid balancing is seen as one of the challenges with the substantial addition of intermittent resources like solar PV and wind energy. Against this backdrop, the role of electrical energy storage is envisaged. Battery energy storage systems have evinced interest in the recent past. Hydrogen storage is all set to impact various sectors including transportation and industry. Solar PV has to play a significant role in the production of green hydrogen. Hydrogen used in different processes and in transportation in the future will have to be produced from non-fossil fuels to be environmentally friendly. Thus, the role of solar PV can be seen in terms of power generation and in the facilitation of green power generation through other routes. However, the future challenges in terms of disposal of the end-of-life PV cells can't be ignored.

REFERENCES

[1] Crossley PA, Noel GT, Wolf M. Review and evaluation of past solar cell development efforts. 1967 Jun 1.
[2] Mytton RJ. Comparative review of silicon and thin film solar cells for space applications. Physics in Technology. 1973;4(2):92.
[3] Wronski CR, Carlson DE. Amorphous silicon solar cells. Clean Electricity from Photovoltaics. 2001 Jun 4:199.
[4] Green MA. Third generation photovoltaics: Solar cells for 2020 and beyond. Physica E: Low-Dimensional Systems and Nanostructures. 2002 Apr 1;14(1–2):65–70.
[5] Adeyinka AM, Mbelu OV, Adediji YB, Yahya DI. A review of current trends in thin film solar cell technologies. International Journal of Energy and Power Engineering. 2023;17(1):1.
[6] Riede M, Spoltore D, Leo K. Organic solar cells—The path to commercial success. Advanced Energy Materials. 2021 Jan;11(1):2002653.
[7] Duan L, Hoex B, Uddin A. Progress in semitransparent organic solar cells. Solar RRL. 2021 May;5(5):2100041.
[8] Cumulative installed solar PV capacity of worldwide for 2000 to 2023, www.statista.com/statistics/280220/global-cumulative-installed-solar-pv-capacity/
[9] IEA, photovoltaic power system programme. Annual Report 2023, https://iea-pvps.org/annual-reports/annual-report-2023/
[10] Moseley PT, Garche J, editors. Electrochemical energy storage for renewable sources and grid balancing. Newnes; 2014 Oct 27.
[11] Barthélémy H, Weber M, Barbier F. Hydrogen storage: Recent improvements and industrial perspectives. International Journal of Hydrogen Energy. 2017 Mar 16;42(11):7254–7262.
[12] Bi P, Zhang S, Wang J, Ren J, Hou J. Progress in organic solar cells: Materials, physics and device engineering. Chinese Journal of Chemistry. 2021 Sep;39(9):2607–2625.
[13] ElKhamisy K, Abdelhamid H, El-Rabaie ES, Abdel-Salam N. A comprehensive survey of silicon thin-film solar cell: Challenges and novel trends. Plasmonics. 2024 Feb;19(1):1–20.
[14] Marques Lameirinhas RA, Torres JP, de Melo Cunha JP. A photovoltaic technology review: History, fundamentals and applications. Energies. 2022 Mar 1;15(5):1823.
[15] Oni AM, Mohsin AS, Rahman MM, Bhuian MB. A comprehensive evaluation of solar cell technologies, associated loss mechanisms, and efficiency enhancement strategies for photovoltaic cells. Energy Reports. 2024 Jun 1;11:3345–3366.

[16] Liu F, Wang W, Wang L, Yang G. Working principles of solar and other energy conversion cells. Nanomaterials and Energy. 2013 Jan;2(1):3–10.
[17] Ayvaz BŞ, Bayrak A. Electricity generation methods from solar energy. Mühendis ve Makina. 2024 Jul 1;(716):521–557.
[18] Malinowski M, Leon JI, Abu-Rub H. Solar photovoltaic and thermal energy systems: Current technology and future trends. Proceedings of the IEEE. 2017 Apr 28;105(11):2132–2146.
[19] Linden L, Bottaro D, Moskowitz J, Ocasio W. The solar photovoltaics industry: The status and evolution of the technology and the institutions (MIT Energy Laboratory Report No. MIT-EL-77-021). Massachusetts Institute of Technology, Energy Laboratory. Prepared for the U.S. Department of Energy; 1977.
[20] King RR, Sinton RA, Swanson RM. One-sun, single-crystalline silicon solar cell research. Sandia National Lab.(SNL-NM), Albuquerque, NM (United States); Stanford Univ., CA (United States). Solid-State Electronics Lab.; 1991 Jun 1.
[21] Solanki CS. Solar photovoltaics: Fundamentals, technologies and applications. Phi Learning Pvt. Ltd.; 2015 May 9.
[22] Ullal HS. Overview and challenges of thin film solar electric technologies. National Renewable Energy Lab. (NREL), Golden, CO (United States); 2008 Dec 1.
[23] Ullal HS. Overview and challenges of thin film solar electric technologies. National Renewable Energy Lab. (NREL), Golden, CO (United States); 2008 Dec 1.
[24] Lee TD, Ebong A. Thin film solar technologies: A review. In 2015 12th International Conference on High-capacity Optical Networks and Enabling/Emerging Technologies (HONET). 2015 Dec 21 (pp. 1–10). IEEE.
[25] Chawla R, Singhal P, Garg AK. Photovoltaic review of all generations: Environmental impact and its market potential. Transactions on Electrical and Electronic Materials. 2020 Oct;21(5):456–476.
[26] Akhtaruzzaman M, Selvanathan V, editors. Comprehensive guide on organic and inorganic solar cells: Fundamental concepts to fabrication methods. Academic Press; 2021 Nov 18.
[27] Cui Y, Xu Y, Yao H, Bi P, Hong L, Zhang J, Zu Y, Zhang T, Qin J, Ren J, Chen Z. Single-junction organic photovoltaic cell with 19% efficiency. Advanced Materials. 2021 Oct;33(41):2102420.
[28] Tress W, Tress W. Organic solar cells. Springer International Publishing; 2014.
[29] Sharma K, Sharma V, Sharma SS. Dye-sensitized solar cells: Fundamentals and current status. Nanoscale Research Letters. 2018 Dec;13:1–46.
[30] Sugathan V, John E, Sudhakar K. Recent improvements in dye sensitized solar cells: A review. Renewable and Sustainable Energy Reviews. 2015 Dec 1;52:54–64.
[31] Park NG. Perovskite solar cells: An emerging photovoltaic technology. Materials Today. 2015 Mar 1;18(2):65–72.
[32] Osman BA, Abdolkader TM, Ahmed IS. A review of perovskite solar cells. International Journal of Materials Technology and Innovation. 2021 Nov 1;1(2):48–66.
[33] Norton MG. A green new deal. In A modern history of materials: From stability to sustainability (pp. 187–206). Cham: Springer International Publishing; 2023 Feb 12.
[34] Lazaroiu AC, Gmal Osman M, Strejoiu CV, Lazaroiu G. A comprehensive overview of photovoltaic technologies and their efficiency for climate neutrality. Sustainability. 2023 Nov 24;15(23):16297.
[35] Photovoltaics Report, Fraunhofer Institute for Solar Energy Systems, ISE with the support of PSE Projects GmbH. 2024 Jul 29, www.ise.fraunhofer.de
[36] Di Sabatino M, Hendawi R, Garcia AS. Silicon solar cells: Trends, manufacturing challenges, and AI perspectives. Crystals. 2024 Feb;14(2):167.
[37] Blakers A. High efficiency crystalline silicon solar cells. Festkörperprobleme 30: Plenary Lectures of the Divisions Semiconductor Physics Thin Films Dynamics and Statistical Physics Magnetism Metal Physics Surface Physics Low Temperature Physics

Molecular Physics of the German Physical Society (DPG), Regensburg, 1990 Mar 26 to 30; 1990:403–423.
[38] Knobloch J, Noel A, Schaffer E, Schubert U, Kamerewerd FJ, Klussmann S, Wettling W. High-efficiency solar cells from FZ, CZ and MC silicon material. In Conference Record of the Twenty Third IEEE Photovoltaic Specialists Conference-1993 (Cat. No. 93CH3283-9). 1993 May 10 (pp. 271–276). IEEE.
[39] Harmon C. Experience curves of photovoltaic technology, IR-00-014, IIASA, Laxenburg, Austria; 2000.
[40] Photovoltaics Report, Fraunhofer Institute for Solar Energy Systems, ISE with the support of PSE Projects GmbH. July 29, 2024, www.ise.fraunhofer.de
[41] Energy NT. Best Research-Cell Efficiency Chart. 2021. www. nrel.gov/pv/cellefficiency.html.2019.
[42] Cui Y, Yao H, Zhang J, Xian K, Zhang T, Hong L, Wang Y, Xu Y, Ma K, An C, He C. Single-junction organic photovoltaic cells with approaching 18% efficiency. Advanced Materials. 2020 May;32(19):1908205.
[43] US Energy Information Administration. Electricity data browser. www.eia.gov/international/data/world/electricity/electricity-capacity?pd=2&p=0000000000000000000007vo7&u=0&f=A&v=mapbubble&a=-&i=none&vo=value&&t=C&g=00000000000000000000000000000000000000000000000001&l=249-ruvvvvvfvtvnvv1vrvvvvfvvvvvvfvvvou20evvvvvvvvvvnvvvs0008&s=315532800000&e=1672531200000}
[44] Ritchie H, Rosado P. Electricity mix. Our World in Data. 2024 Jan. https://ourworldindata.org/electricity-mix
[45] Cumulative solar photovoltaic capacity globally as of 2023, by select country, www.statista.com/statistics/264629/existing-solar-pv-capacity-worldwide/
[46] IEA. PVPS snapshot reports, https://iea-pvps.org/snapshot-reports/
[47] Sahu BK. A study on global solar PV energy developments and policies with special focus on the top ten solar PV power producing countries. Renewable and Sustainable Energy Reviews. 2015 Mar 1;43:621–634.
[48] Shum KL, Watanabe C. Photovoltaic deployment strategy in Japan and the USA—An institutional appraisal. Energy Policy. 2007 Feb 1;35(2):1186–1195.
[49] Ping ZJ. China's new and renewable energy situation. Wind Energy. 2001;3226:253GW.
[50] Si-cheng W. Present situation of photovoltaic applications in China. In Seventh EC Photovoltaic Solar Energy Conference: Proceedings of the International Conference, Held at Sevilla, Spain. 1986–1987 Oct 27–31 (pp. 56–60). Springer Netherlands.
[51] Junfeng L, Wan YH, Ohi JM. Renewable energy development in China: Resource assessment, technology status, and greenhouse gas mitigation potential. Applied Energy. 1997 Mar 1;56(3–4):381–394.
[52] Yang H, Wang H, Yu H, Xi J, Cui R, Chen G. Status of photovoltaic industry in China. Energy Policy. 2003 Jun 1;31(8):703–707.
[53] Zhao ZY, Zhang SY, Hubbard B, Yao X. The emergence of the solar photovoltaic power industry in China. Renewable and Sustainable Energy Reviews. 2013 May 1;21:229–236.
[54] Zhang S, He Y. Analysis on the development and policy of solar PV power in China. Renewable and Sustainable Energy Reviews. 2013 May 1;21:393–401.
[55] Rongqiang C, Jianqiang W, Qinghao Y, Shiquan Y, Yang S, Jiabing D, Le Y, Fanying M, Lin X, Wenzhong S. Development of solar photovoltaic in China. In 2007 Asia Optical Fiber Communication and Optoelectronics Conference. 2007 Oct 17 (pp. 599–601). IEEE.
[56] Cumulative installed capacity of solar power in China from 2012 to 2023, www.statista.com/statistics/279504/cumulative-installed-capacity-of-solar-power-in-china/
[57] Byrne J, Kurdgelashvili L. The role of policy in PV industry growth: Past, present and future. Handbook of Photovoltaic Science and Engineering. 2011 Mar 29:39–81.

[58] Tabassum S, Rahman T, Islam AU, Rahman S, Dipta DR, Roy S, Mohammad N, Nawar N, Hossain E. Solar energy in the United States: Development, challenges and future prospects. Energies. 2021 Dec 4;14(23):8142.
[59] Oliver M, Jackson T. The market for solar photovoltaics. Energy Policy. 1999 Jul 1;27(7):371–385.
[60] Branstetter LJ, Vidal RC, Bruch VL, Zurn R. Outlook for renewable energy technologies: Assessment of international programs and policies. Sandia National Lab.(SNL-NM), Albuquerque, NM (United States); 1995 Feb 1.
[61] Runci PJ. Energy R and D in Germany. Pacific Northwest National Lab.(PNNL), Richland, WA (United States); 1999 Nov 1.
[62] Lauber V, Mez L. Three decades of renewable electricity policies in Germany. Energy & Environment. 2004 Jul;15(4):599–623.
[63] Stubenrauch F, Jülich P. National survey report of PV power applications in Germany.
[64] Altenhöfer-Pflaum G, Horbelt R. National Survey Report of PV Power Applications in Germany. IEA-PVPS, Tech. Rep. 2015.
[65] Tatsuta M. New sunshine project and new trend of PV R&D program in Japan. Renewable Energy. 1996 May 1;8(1–4):40–43.
[66] Yamaguchi M. Present status and prospects of photovoltaic technologies in Japan. Renewable and Sustainable Energy Reviews. 2001 Jun 1;5(2):113–135.
[67] Li A, Xu Y, Shiroyama H. Solar lobby and energy transition in Japan. Energy Policy. 2019 Nov 1;134:110950.
[68] Kapoor K, Pandey KK, Jain AK, Nandan A. Evolution of solar energy in India: A review. Renewable and Sustainable Energy Reviews. 2014 Dec 1;40:475–487.
[69] Rohankar N, Jain AK, Nangia OP, Dwivedi P. A study of existing solar power policy framework in India for viability of the solar projects perspective. Renewable and Sustainable Energy Reviews. 2016 Apr 1;56:510–518.
[70] Sharma NK, Tiwari PK, Sood YR. Solar energy in India: Strategies, policies, perspectives and future potential. Renewable and Sustainable Energy Reviews. 2012 Jan 1;16(1):933–941.
[71] Gulia J, Garg FV. Photovoltaic manufacturing outlook in India. JMK/IEEFA Report. 2022 Feb, https://jmkresearch.com/renewable-sector-published-reports/photovoltaic-manufacturing-outlook-in-india
[72] Wandhe DP. An Overview on Production Linked Incentive (PLI) Scheme by the Government of India. Available at SSRN 4693578. 2024.
[73] Executive Summary. State of solar PV manufacturing in India 2024. Mercom India. 2024.
[74] Ministry of New and Renewable Energy (MNRE). Imposition of Basic Customs duty (BCD) on solar PV cells & modules/panels. 2021, https://mnre.gov.in/img/documents/uploads/file_f-1615355045648.PDF
[75] Central Electricity Authority. Installed capacity report, https://cea.nic.in/installed-capacity-report/?lang=en
[76] Renewable Energy Statistics, 2023–24. Ministry of new and renewable energy, https://cdnbbsr.s3waas.gov.in/s3716e1b8c6cd17b771da77391355749f3/uploads/2024/10/20241029512325464.pdf
[77] Sharma A, Srivastava K, Kar SK. Jawaharlal Nehru national solar mission in India. Energy Sustainability Through Green Energy. 2015:47–67.
[78] Islam MB, Huda MB, Rather NA, Eslamian S. Potential of solar energy in India. In: *Handbook of Uncertainty in Eurasian Forecasting (HEF)*. Saeid Eslamian, Mir Bintul Huda and Faezeh Eslamian (editors), Nova Science publisher Inc. ISBN: 978-1-68507-966-6978 DOI: https://doi.org/10.52305/IWAW5705
[79] Shubbak MH. Advances in solar photovoltaics: Technology review and patent trends. Renewable and Sustainable Energy Reviews. 2019 Nov 1;115:109383.
[80] Photovoltaics Report, Fraunhofer Institute for Solar Energy Systems, ISE with the support of PSE Projects GmbH. 2024 Jul 29, www.ise.fraunhofer.de

[81] Rancilio G. Battery energy storage systems for ancillary services provision, Thesis, Department of Energy, Politecnico Di Milano, p. 1154, 2018 Mar.
[82] Nazar A, Anwer N. Electrical energy storage influencing shift in grid balancing approach. In Latest Trends in Renewable Energy Technologies: Select Proceedings of NCRESE 2020. 2021 (pp. 77–83). Springer, Singapore.
[83] Bloomberg, China almost quadrupled its new storage capacity. 2025 Jan 25, www.bloomberg.com/news/articles/2024-01-25/china-almost-quadrupled-its-new-energy-storage-capacity-in-2023-lrsvdt3p
[84] US Energy Information Administration. US battery storage capacity expected to nearly double in 2024. 2024 Jan 9, www.eia.gov/todayinenergy/detail.php?id=61202
[85] Fraunhofer ISE, www.ise.fraunhofer.de/en/press-media/press-releases/2024/public-electricity-generation-2023-renewable-energies-cover-the-majority-of-german-electricity-consumption-for-the-first-time.html
[86] Hassan Q, Abdulateef AM, Hafedh SA, Al-samari A, Abdulateef J, Sameen AZ, Salman HM, Al-Jiboory AK, Wieteska S, Jaszczur M. Renewable energy-to-green hydrogen: A review of main resources routes, processes and evaluation. International Journal of Hydrogen Energy. 2023 May 29;48(46):17383–17408.
[87] Kalaiselvan N, Mathimani T. Solar-driven green hydrogen generation for revolutionizing the future of zero-carbon energy. Fuel. 2024 Nov 1;375:132538.
[88] Dash SK, Chakraborty S, Roccotelli M, Sahu UK. Hydrogen fuel for future mobility: Challenges and future aspects. Sustainability. 2022 Jul 6;14(14):8285.
[89] Ishaq H, Dincer I, Crawford C. A review on hydrogen production and utilization: Challenges and opportunities. International Journal of Hydrogen Energy. 2022 Jul 22;47(62):26238–262364.
[90] Abdalla AM, Hossain S, Nisfindy OB, Azad AT, Dawood M, Azad AK. Hydrogen production, storage, transportation and key challenges with applications: A review. Energy Conversion and Management. 2018 Jun 1;165:602–627.
[91] Weckend S, Wade A, Heath GA. End of life management: Solar photovoltaic panels. National Renewable Energy Laboratory (NREL), Golden, CO (United States); 2016 Jun 1; No. NREL/TP-6A20-73852.
[92] Ferreira Hiranobe M, Silva Souza D, da Silva Filgueira M, Araújo Kawall S, Deltrejo Ribeiro G, Toshiyuki Hiranobe C. Recycling photovoltaic modules: A sustainable approach to renewable energy. GeSec: Revista de Gestao e Secretariado. 2024 Dec 15;15(12).
[93] Babaei A, Nasr Esfahani A. A review of photovoltaic waste management from a sustainable perspective. Electricity. 2024 Oct 14;5(4):734–750.
[94] Preet S, Smith ST. A comprehensive review on the recycling technology of silicon based photovoltaic solar panels: Challenges and future outlook. Journal of Cleaner Production. 2024 Mar 8:141661.
[95] Chen PH, Chen WS, Lee CH, Wu JY. Comprehensive review of crystalline silicon solar panel recycling: From historical context to advanced techniques. Sustainability. 2024 Jan;16(1):60.
[96] Farrell CC, Osman AI, Doherty R, Saad M, Zhang X, Murphy A, Harrison J, Vennard AS, Kumaravel V, Al-Muhtaseb A, Rooney DW. Technical challenges and opportunities in realising a circular economy for waste photovoltaic modules. Renewable and Sustainable Energy Reviews. 2020 Aug 1;128:109911.
[97] Sharma A, Mahajan P, Garg R. End-of-life solar photovoltaic panel waste management in India: Forecasting and environmental impact assessment. International Journal of Environmental Science and Technology. 2024 Jan;21(2):1961–1980.
[98] Chowdhury MS, Rahman KS, Chowdhury T, Nuthammachot N, Techato K, Akhtaruzzaman M, Tiong SK, Sopian K, Amin N. An overview of solar photovoltaic panels' end-of-life material recycling. Energy Strategy Reviews. 2020 Jan 1;27:100431.
[99] Ndalloka ZN, Nair HV, Alpert S, Schmid C. Solar photovoltaic recycling strategies. Solar Energy. 2024 Mar 1;270:112379.

4 Using Artificial Intelligence and Computer Vision for Solar Cell Wafer Inspection

K. Deepa, Vijay Kumar Gill, and Bhanu Pratap Chaudhary

4.1 SOLAR CELL WAFER INSPECTION

Solar wafers made from thin slices of crystalline silicon play an important role in energy conversion in photovoltaic (PV) cells [1]. The wafers are crucial components in the solar industry, and to ensure the quality and performance of the cell it is necessary to examine the solar wafers. Any defect in the wafer may affect the overall efficiency and performance of the PV cells. Cracks, scarps, and pits are the most common issues that may affect the electrical properties of the solar panels [2]. The inspection process includes different stages starting from basic assessments to complex image processing techniques. These examinations enhance the life span and efficiency of the solar panels. Apart from this, manufacturers can confirm that every wafer encounters strict quality standards, leading to higher energy output, and extended lifespan, leading to high energy conversion, reduced production costs, reduced warranty claims, increased customer satisfaction, and so on. Traditionally the inspection methods relied on manual and visual checks, which are highly time-consuming and susceptible to human errors and becoming more apparent with time. It is necessary to incorporate advanced inspection techniques and tools like artificial intelligence (AI) and computer vision (CV) to assess the wafer quality in current manufacturing processes as they provide innovative solutions in the inspection. Integrating AI can handle large amounts of data quickly and accurately. It can also detect and define defects in the right way. Integrating imaging techniques helps detect minor problems that can affect the quality of wafers and solar panels Traditionally, wafer inspection has depended on manual methods and simple visual checks, which are time-consuming and prone to human mistakes. The limitations of the traditional standard inspection techniques are becoming more and more apparent with time. To assess

DOI: 10.1201/9781003515784-4

wafer quality effectively and consistently, advanced inspection tools are required in view of current manufacturing processes. AI and CV are leading in technology. They provide innovative solutions for solar cell wafer inspection. AI can quickly and accurately look at large amounts of data. This helps find and categorize defects right away. Better imaging techniques, along with CV, help spot small problems that can harm the quality of wafers and how well solar panels work.

4.1.1 Overview of the Inspection Process

To ensure the quality and efficiency of the PV cells, the inspection process is mandatory.

The following are the steps involved in the solar wafer verification process for quality control.

4.1.1.1 Initial Assessment of Solar Wafers

Upon receiving solar wafers from the factory, a comprehensive initial evaluation is conducted to verify their quality. The first step is to examine shipping documentation to confirm that the shipment matches the order specifications. Skilled individuals perform a detailed visual inspection in well-lit conditions using magnifying tools to detect defects like cracks, chips, scratches, and surface contamination. Inspectors follow clean room protocols during this inspection, wearing gloves and masks to avoid contamination. Special tools are used to carefully handle wafers, and all the details are recorded in inspection logs along with photographic proof. Wafers are categorized based on the severity of defects. Those identified as severely defective will face rejection while others undergo further testing [3]. This process helps ensure that only high-quality wafers are utilized in subsequent production phases. However, the process is time-consuming, which may result in delayed production as thorough visual inspections and documentation require a labor-intensive process. Moreover, the process can be subjective as it relies on the expertise of staff, which may result in inconsistencies during the process. Visual inspections mainly identify surface-level defects, and subsurface issues may go unnoticed. Inspecting wafers manually raises the likelihood of physical damage, such as scratches or contamination. Furthermore, the requirement for trained inspectors and specialized tools may increase the operational cost.

4.1.1.2 Pretreatment

During the pretreatment process, the wafers are cleaned with industrial soaps. The process is closely connected to the inspection process in the production of solar cells. Cleaning eliminates the presence of residues and also ensures the following processes like doping, and texturing yields better results, which may impact the overall performance of the solar cell [4]. However, cleaning may affect the wafer inspection as it changes the condition of the wafer surface. For accurate inspection, reliable performance, and proper processing, it is necessary that the wafer must be clean. In a clean wafer, defects like scratches and micro-cracks can be easily identified. Additionally, films or any other semiconductor layers can be easily attached without defects if the wafer is clean. If the wafer is not completely clean, proper adhesion is not feasible as dust and contaminants may lead to defects. A clean wafer is necessary to prevent defects and ensure that the electronic components function accurately. Further thorough cleaning aids to reduce wastes and cost, improving the production efficiency.

The pretreatment cleaning step is essential to manufacture high-quality wafers as it helps detect impurities or defects and meet the required performance standards.

4.1.1.3 Geometric Inspection

Geometric inspection of wafers is another important process in semiconductor production, which verifies the accurate geometric and dimensional necessities that will be used in the fabrication of integrated circuits. This test checks for thickness, evenness, and consistency of wafers. It also detects issues like chipping, bending, surface irregularities, or major cracks that might impact the solar cell performance. A 2D geometric inspection test can be used to detect chips or major cracks. In general, wafers that do not meet geometric criteria are rejected. In this test features like diameter, thickness, roughness, flatness, and bow warp are considered. These factors are evaluated by tests like optical inspection and automated vision systems or through coordinate measuring machines [5]. Adhering to industry standards, such as Semiconductor Equipment and Materials International standards, ensures the appropriateness and consistency in the production process. However, external factors like temperature and humidity and assessing minute details make the process complex. This test results in reducing integrated circuits (IC) defects, improving production, and ensuring the suitability of the wafer for the next steps, leading to effective semiconductor production.

4.1.1.4 Advanced Imaging Techniques

Photoluminescence imaging is one of the techniques used for examining solar wafers to detect minor defects like micro-cracks, grain boundaries, and dislocations, which are difficult to detect through manual visual inspection techniques. Here a high-energy light from a laser source is used to excite the wafer, which in turn excites electrons in the material. The excited electrons will emit light energy and return to the initial state. Information about the internal structure of the material/wafer can be obtained from the light energy. The photoluminescence signal can be mapped with the crystal properties, and defects in crystal lattice, impurities, and doping levels can be studied [6]. The process is effective in multi-crystalline wafer structures, which may impact the efficiency of solar cells.

4.1.1.5 Electrical Property Testing

This method can quickly assess the quality of wafers and ensure their electrical performance criteria, which enhances the overall yields and reliability of the PV modules produced. Incorporating non-contact methods also enables quick and precise detection without damaging the wafer. Eddy current techniques and capacitance techniques are frequently used non-contact methods in electrical property testing.

- **Eddy current technique:** The bulk resistivity of the wafer is one of the key factors that determines the electrical conductivity capability. Eddy currents are induced through varying magnetic fields, and the electronic response of the wafer is measured. This is a quick and precise method to assess the wafer resistivity, which directly impacts the solar cell efficiency after production. Eddy current testing is preferred for its speed and non-destructive characteristics, making it ideal for inline inspection during mass manufacturing.

- **Capacitance-based testing:** It is used to determine the thickness of the wafer and its total thickness variation. The technique operates by applying an electric field and measuring the capacitance between two electrodes placed near the wafer. Differences in capacitance can provide information about the wafer's thickness and uniformity. It is essential to maintain a consistent wafer thickness to ensure that all regions of the wafer are treated evenly during later processing steps, such as doping and coating. Thickness variations can lead to uneven disparities in electrical properties and reduce the efficiency of solar cells. Table 4.1 summarizes the various solar wafer inspection techniques.

4.1.2 AI and CV in Solar Cell Wafer Inspection

The integration of AI and CV in solar cell wafer inspection has transformed quality control and defect detection in the PV sector. These technologies allow for inspections

TABLE 4.1
Comparison of Solar Wafer Inspection Techniques

Inspection Technique	Purpose/Tools Used	Advantages	Disadvantages
Initial Assessment	Visual inspection for cracks, chips, scratches, and surface contamination	• Simple and inexpensive • Identifies visible flaws • Provides instant feedback	• Requires significant time and labor • Prone to inconsistency • Limited to surface defects • May damage wafers during handling
Pretreatment	Eliminates residues that might impact performance, ensuring subsequent processes	• Ensures accurate inspection by improving surface quality • Reduces faulty wafers at later stages • Improves efficiency	• Time-consuming process • Requires specialized chemicals and equipment
Geometric Inspection	Verifies that wafers meet predetermined dimensional standards	• Provides precise dimensional data • Detects cracks, warps, and other defects • Ensures uniformity for further processing	• Sensitive to temperature variations • Requires complex setup and calibration • Limited in detecting minute irregularities
Photoluminescence Imaging	Uses light to stimulate electrons, exposing micro-cracks, dislocations, and variations in doping	• Non-invasive method • Identifies internal flaws and contaminants • Beneficial for multi-crystalline wafers	• Requires specialized LED/laser equipment • Limited by image resolution • Ineffective for surface contamination

TABLE 4.1 (*Continued*)
Comparison of Solar Wafer Inspection Techniques

Inspection Technique	Purpose/Tools Used	Advantages	Disadvantages
Electrical Property Testing	Evaluates electrical resistance and thickness using eddy current and capacitance methods	• Quick, accurate, and non-contact • Suitable for inline inspection on a production line • Ensures consistent electrical properties	• Limited to assessing electrical properties • Requires sensor calibration • Sensitive to external environmental factors

that are more efficient, accurate, and reliable in comparison to traditional methods. In the past, solar cell inspections were done manually/visually, a process that is both time-consuming and prone to human mistakes. The results of manual inspections are commonly inconsistent, making it challenging to ensure that solar panels are in perfect condition before installation. A major concern for solar farm operators is the high cost linked with maintenance and replacements.

4.1.2.1 Typical Defects Found in Solar Panels

- **Micro-cracks:** Very small fractures that can form during the production process or handling, usually not visible without a microscope, but detrimental to functionality.
- **Cracking glass:** Damage to the glass protection layer that can expose cells to environmental damage.
- **Cross cracks:** Patterns resembling trees that indicate structural weaknesses in the solar cells.
- **Hotspots:** Regions that become excessively hot due to defects, which may result in possible malfunctions.

The integration of AI and CV techniques in solar wafer inspection not only streamlines the identification of defects but also improves overall manufacturing productivity. Through this process, manufacturers can achieve higher throughput, cost reduction, and improved quality control. As these technologies keep developing, they provide greater efficiencies and innovations in the renewable energy sector.

4.2 ROLE OF AI IN WAFER INSPECTION

Industrial AI combines specialized expertise with AI technology, integrating various functions and features into different AI-powered applications in industrial sectors. Incorporating AI technologies in the industry accelerates and enhances processes by enabling autonomous and semi-autonomous tasks to manage operations. These systems aim to create self-optimizing manufacturing facilities, ensuring continuous efficiency and improvement [7]. AI plays a dual function in the solar wafer sector: AI acts as a key enabler for digitizing manufacturing processes, providing the technology required to optimize operations and control process parameters in the solar wafer manufacturing process. AI technologies are implemented to increase the efficiency of industrial processes by handling complex scenarios within pre-existing systems. AI in industrial sectors offers an opportunity for organizations to optimize resources and enhance revenue, leading to significant impact. AI is fundamentally changing the way wafer inspection is conducted, revolutionizing the semiconductor and solar sectors. Utilizing the recently implanted AI programs enables the solar cell manufacturer to analyze extensive datasets efficiently and instantaneously to obtain online results of the inspection including details that an inspector could miss. While evaluating the defects, the aspects involving the variation or inconsistency of the defects can be easily identified when AI is utilized. An AI system will consistently offer a single solution and will produce the same results while the process is being executed. The results can be generated quickly with greater efficiency by incorporating AI techniques. Data analysis includes using statistical and computational methods to comprehend and interpret data. Effective data analysis is essential in the solar energy field to maximize performance, ensure reliability, and prolong the life of solar panels. With rapid advancements in technology, specifically AI, the ability to analyze complex datasets has greatly improved. Data analysis plays a crucial role in detecting abnormalities in the solar wafers that could lead to failures in the system. There are two main types of data analysis methods for identifying issues in solar cells: image-based techniques and electrical testing techniques. Every technique uses various methods to detect errors, emphasizing different facets of the solar panels. The method has already been adopted by leading industries globally. Table 4.2 provides an overview of different industries utilizing AI for associated purposes [8, 9, 10, 11].

4.2.1 AI Advancements in Solar Wafer Inspection

Nowadays, AI-powered techniques are more and more essential in the solar wafer inspection process. This transformation involves various aspects ranging from the production of solar panels to upgrading energy management systems. AI includes diverse technologies aimed at addressing real-world challenges in different sectors. AI, as a field of science and technology, develops intelligent machines, algorithms, and programs that can complete tasks that typically demand human-like intelligence, such as learning, recognizing patterns, and making decisions [12]. ML, a branch of AI, focuses on empowering machines to perform tasks that usually require human intelligence. By learning from mistakes, machines improve their skills and performance as they gain more experience over time through a gradual process [13]. Deep learning (DL), a subset of machine learning (ML), utilizes multi-layers of algorithms, such as input, hidden, and output layers, to represent the varying degrees of abstraction. It is utilized in voice synthesis, image processing, recognizing handwriting, detecting objects, predicting analytics, and making decisions, allowing for sophisticated data analysis and pattern recognition [14]. Automated optical inspection systems, when paired with ML algorithms, allow for rapid detection and sorting of defects in wafers. The presence of shelf AI tools, along with personalized models designed for specific production lines, is a crucial aspect of today's PV manufacturing needs [15].

TABLE 4.2
AI Use in Solar Energy Companies

Company	Country	AI Application
Applied Materials	USA	Uses the Vericell Solar Wafer Inspection System for accurate defect detection, analysis, and process optimization, achieving a high throughput of up to 16,000 wafers per hour.
Solar Analytics	Australia	Implements AI for monitoring solar panel systems. Machine learning (ML) algorithms analyze real-time data for fault detection and proactive maintenance, improving efficiency, reducing downtime, and lowering maintenance costs.
Equinox Drones	USA	Employs AI-powered drones equipped with thermal and RGB sensors for rapid solar panel inspections, covering large areas and detecting hot spots or temperature irregularities.
CGI	Canada	Utilizes drones with thermal and RGB cameras to inspect PV plants. Deep learning algorithms enable fault detection with up to 90% accuracy.
Scopito	Denmark	Offers cloud-based software for visual data management in solar inspections, enhancing efficiency through AI-based fault detection.
Glint Solar	Norway	Uses AI for site prospecting. ML algorithms analyze satellite data to identify optimal locations for solar installations.
Smart Helio	Czech Republic	Develops AI solutions for performance monitoring of solar panel installations.
EasyFlow	USA	Controls AI-powered drone inspections to detect defects in solar farms, significantly accelerating the inspection process. Inspections are 50 times faster than manual methods.

4.2.1.1 Comparison of AI, ML, and DL

The comparative analysis in Table 4.3 underscores the importance of selecting the appropriate technique for specific problem domains.

4.2.2 ML for Solar Wafer Inspection

The popularity of ML techniques for solar wafer inspection has grown as they can analyze complicated data, spot patterns, and improve defect detection accuracy. The incorporation of ML methods in examining solar wafers has become more critical as semiconductor production demands high accuracy and efficiency. Different ML techniques are being used to detect faults, classify the quality of wafers, and improve manufacturing procedures. Methods like rule-based systems (RBSs), evolutionary computation (EC), fuzzy logic, and knowledge-based systems provide robust alternatives to ML for solar wafer inspection.

4.2.3 Rule-Based System for Solar Wafer Inspection

An RBS uses a set of predefined rules (logical “if-then” statements) to make decisions or inferences [16]. In solar wafer inspection, these systems automate defect

TABLE 4.3
Comparison of AI, ML, and DL

Characteristics	AI	ML (a Subset of AI)	DL (a Deeper ML Branch)
Description	Mimics human intelligence for decision-making	Enables machines to learn from data	Uses neural networks
Possibility	Uses ML, NLP, robotics, and computer vision	Improves algorithms through experience	Uses artificial neural networks (ANN)
Learning approach	Employs rules and reinforcement learning	Uses regression and clustering for data analysis	Employs deep neural networks to learn directly from raw data
Complexity	Ranges from simple rules to complex models	Requires feature extraction for optimization	Uses multiple layers for automatic pattern extraction
Data requirement	Works with small datasets and predefined rules	Needs large labeled datasets for training	Requires vast datasets and high computational power

detection and quality control through the utilization of expert knowledge embedded in rules. This method is especially beneficial in cases where the defects and evaluation processes are clearly outlined. Figure 4.1 depicts the basic components of an RBS system [17].

- **Knowledge base:** Contains all information about solar wafer inspection like guidelines for evaluating criteria during inspections (e.g., allowable levels of faults), information about wafer features and data obtained from different inspection techniques (e.g., visual inspection and laser scanning).
- **Inference engine:** Utilizes rules to make deductions from input information. It is the main element that enforces the rules on the input data. It derives conclusions from the given facts and rules. For example, the system could determine that a wafer is faulty if it contains a crack that is longer than a certain length.
- **Memory:** Holds data temporarily while analyzing it. For example, it may contain the results of various measurements taken during the inspection [18].

RBS can be used for examining solar wafers to detect and categorize different types of defects, like surface imperfections (scratches and cracks, along with dirt and impurities). RBS is also capable of examining thickness variations and identifying problems related to uneven doping levels and contamination. RBS can also be implemented to detect issues related to electricity like short circuits and open circuits, along with problems with light reflection and anti-reflective coating. It can also evaluate thermal strain and deformation, such as thermal fracturing and shape distortion, along with production flaws like uneven coating and weak attachment. Incorporating RBS allows manufacturers to automate defect detection, thus improving quality control in solar wafer production [15].

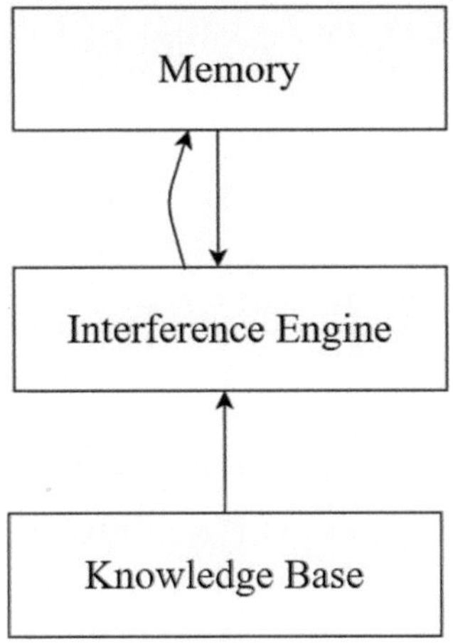

FIGURE 4.1 Components of rule-based system.

4.2.4 Few Examples

4.2.4.1 Example 1: RBS to identify Scratches as Critical Scratch

Rule: A scratch is considered critical if it is longer than 1 mm and deeper than 10 μm.

Output: Critical scratch (yes/no)

Example: Input:

- Scratch length: 1.5 mm
- Scratch depth: 15 μm

The length measures 1.5 mm, which exceeds 1 mm, and the depth measures 15 μm, which exceeds 10 μm.

Output: Since both criteria are met, the result will be "Yes, it is a critical scratch."

```
# Function to determine if a scratch is critical def
is_critical_scratch(scratch_length, scratch_depth):
          if scratch_length > 1.0 and scratch_depth > 10.0:
          # length in
    ,→ mm, depth in micrometers return "Yes, it is a
        critical scratch."
    else:
      return "No, it is not a critical scratch."

# Example input scratch_length = 1.5
# mm scratch_depth = 15.0 # m

# Evaluate the rule result = is_critical_scratch(scratch_
length, scratch_depth) print(result) # Output: Yes, it is a
critical scratch.
```

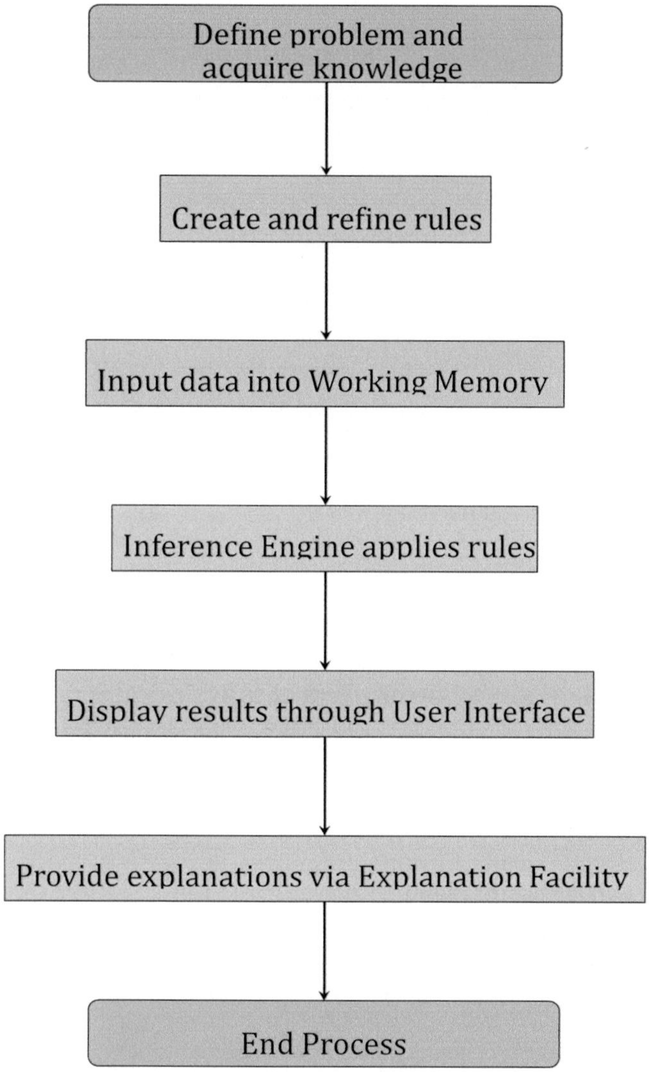

FIGURE 4.2 RBS process flow chart [17].

4.2.4.2 Example 2: Identify cracks and their type

The crack may be a hairline crack, micro-crack, or macro-crack. **Hairline cracks** are extremely thin and narrow, hardly noticeable to the naked eye. It is usually a small line or crack on the outer layer. It is, generally, less than 0.5 mm in length and usually under 0.1 mm in width. It typically does not penetrate the substance. It occurs as a result of small pressures and changes in temperature, causing expansion or contraction. However, it can indicate underlying structural vulnerabilities if not resolved. **Micro-cracks** are tiny cracks that are slightly more visible than hairline cracks yet still small in size. They might not be seen without the use of magnification. Generally, they vary from 0.5 to 1 mm in both length and width. They may develop from various causes such as manufacturing defects, fatigue, or minor impacts. When propagated, they could potentially cause more serious problems. **Macro-cracks** are larger cracks that are easily visible and can have a major impact on the structural stability of the substance. Generally, they exceed 1 mm in both length and width. They are caused as a result of high levels of stress, impact, or environmental conditions. They need immediate attention as they can result in material or structural failure.

- **Rule 1: Hairline Crack**
  If the crack length is not more than 0.5mm and the crack width is not more than 5 μm, then it is a hairline crack.
  IF crack length ≤ 0.5 mm AND crack width ≤ 5.0 μm THEN classify as "Hairline Crack" (1)
- **Rule 2: Micro-crack**
  If the crack's size is over 0.5 mm but under or equal to 2.0 mm, and the width is above 5 μm but under or equal to 20 μm, then it is a micro-crack.
  IF 0.5 < crack length ≤ 2.0 mm AND 5.0 μm < crack width ≤ 20.0 μm THEN classify as "Micro-Crack" (2)
- **Rule 3: Macro-crack**
  If the crack length exceeds 2.0 mm and the crack width is more than 20 μm, then it is a macro-crack.
  IF crack length > 2.0 mm AND crack width > 20.0 μm THEN classify as "Macro-Crack" (3)

```
# Function to determine the type of crack based on length and
width def classify_crack(crack_length, crack_width):
  if crack_length <= 0.5 and crack_width <= 5.0:
          return "\textcolor{red}{Hairline Crack}" # Return
          statement in red
  elif 0.5 < crack_length <= 2.0 and 5.0 < crack_width <= 20.0:
    return "\textcolor{red}{Micro Crack}" # Return statement
    in red
  elif crack_length > 2.0 and crack_width > 20.0:
    return "\textcolor{red}{Macro Crack}" # Return statement
    in red
  else:
          return "\textcolor{red}{Not classified}" # Return
          statement in red

# Example input crack_length = 2.5 # mm crack_width = 25.0 # m
# Evaluate the crack type crack_type =
classify_crack(crack_length, crack_width)
# Print the result print(f"The crack is classified as:
{crack_type}.") # Output classification
```

4.2.5 EC for Solar Wafer Inspection

EC is a branch of AI that replicates natural selection to address optimization issues. Within the solar wafer inspection, EC can improve defect detection, boost manufacturing efficiency, and optimize quality control procedures in PV cell manufacturing. Techniques like a genetic algorithm (GA), particle swarm optimization, and differential evolution can be incorporated.

4.2.5.1 GA for Defect Detection

GAs are being more commonly used for detecting defects in solar wafers, utilizing their optimization abilities to improve the efficiency of PV systems. GAs can be applied for defect detection in solar wafers by improving image processing methods or ML algorithms to better identify defects like cracks, inclusions, scratches, contamination, and grain boundaries [19]. The aim of the GA is to optimize image processing parameters or select features to categorize images as faulty or not faulty.

4.2.5.2 Steps Involved in Solar Wafer Inspection

- **Representation of chromosomes:** Each chromosome in the population represents a group of parameters used in image processing (such as threshold values, edge detection configurations, and filters).
- **Fitness function:** Evaluation of each chromosome's performance is done by the fitness function, which assesses how well the parameters classify defects (based on metrics like accuracy or F1 score).
- **Selection:** The top-performing chromosomes are chosen to be parents for the succeeding generation.
- **Crossover and mutation:** Involves creating new offspring by combining parental traits (crossover) and adding random changes (mutation) to discover innovative solutions [20].

  Problem: Use image processing techniques to inspect solar wafers for defects such as cracks, inclusions, and scratches. Utilize a GA to fine-tune image preprocessing, feature extraction, and classification parameters. The process consists of three key stages: image preprocessing, feature extraction, and classification.
- **Image preprocessing:** Enhance image quality before extracting features to ensure more accurate defect detection. Some of the usual methods include reducing noise by using filters such as Gaussian or median filters, contrast enhancement by adjusting brightness to emphasize defects, and conversion to binary format for simplified analysis using thresholding.
- **Feature extraction:** Involves identifying relevant attributes in images required for detecting defects. Typical characteristics include detecting edges using methods such as Sobel or Canny edge detection. Utilize texture analysis techniques, such as gray-level co-occurrence matrix (GLCM) or local binary patterns, to extract information on surface imperfections. Extract geometric properties (such as area and perimeter) to distinguish defects based on shape features.
- **Classification:** Classify the images based on the extracted features.

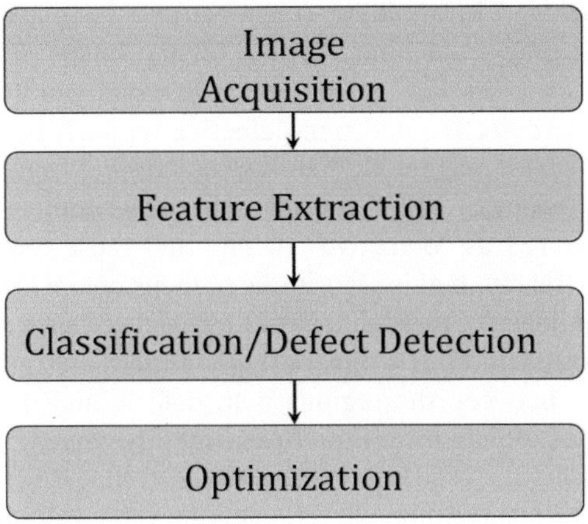

FIGURE 4.3 Image processing pipeline for defect detection.

4.2.5.3 Optimization with GA

- **Individual representation:** Each individual represents a set of image processing and classification parameters. Generate 100 random individuals with parameter values within specified ranges.
  - **Filter type:** defines the filter that is used. (0: Gaussian 1: Median 2: Bilateral)
  - **Filter size:** size of the filter kernel. (e.g., integer values like 3, 5, 7)
  - **Threshold value:** The threshold chosen for binarization (e.g., float values between 0 and 1, e.g., 0.4 and 0.5)
  - **Edge detection method:** The algorithm used for edge detection (0: Canny 1: Sobel)
  - **Texture feature method:** The technique used to extract texture features (0: Local Binary Patterns (LBP) 1: GLCM)
  - **Classifier type:** The type of classifier used (0: support vector machine (SVM) 1: Decision Tree 2: Random Forest)
- **Fitness evaluation:** The fitness evaluation step is crucial while optimizing image processing parameters for defect detection in solar wafers using a GA. It evaluates the effectiveness of each individual in identifying defects.
  - **Apply image preprocessing parameters:** Each individual of the population represents a distinct set of image preprocessing parameters like filtering and thresholding. A Gaussian filter or a median filter can be used for this. The selection of filters depends on the qualities of the images, noise types, and desired features for defect identification. Every filter has its own unique function, and the GA determines the most effective combination for a specific task by evolving different parameter sets over multiple generations. Table 4.3 presents comprehensive details about the filters [21].

  - **Thresholding:** Thresholding is a technique in image processing that converts a gray-scale image into a binary image. This is especially beneficial for tasks like defect detection when you need to differentiate between defective and non-defective regions. For example, In an 8-bit gray-scale image with pixel values ranging from 0 to 255, setting a threshold value of 128 will categorize pixels with values greater than or equal to 128 as "defective" (white) and those below 128 as "non-defective" (black). It aids algorithms to focus on areas of interest (likely containing defects), making it easier for analysis such as feature extraction or classification. It works best when there are significant intensity differences between the regions with defects and those without [22]. This enables the algorithm to effectively differentiate between the two categories.
- **Feature extraction:** Feature extraction enables the differentiation between defective and non-defective wafers. By extracting a comprehensive set of texture and shape features, the effectiveness of the defect detection algorithm will be enhanced. Each feature gives distinct data that, when combined, can significantly improve the categorization and recognition of defects in solar wafers. This composite approach allows for a better understanding of defects, resulting in improved quality control measures in manufacturing processes. The following are some common features that can be extracted [23].
  - **Texture features:** They basically describe how pixels are positioned in an image, capturing the variations in patterns, intensity, or color changes. These characteristics help in identifying the surface attributes of materials by measuring the way pixel values variation with respect to neighboring pixels. This will aid in identifying the different surface defects like scratches, inclusions, and cracks. Table 4.4 outlines the various texture feature extraction methods [24].
  - **Shape features:** The detection and classification of defects in solar wafers depend significantly on shape features. By analyzing the geometric properties of defects like cracks, scratches, or inclusion boundaries, they assist in identifying structural irregularities. Table 4.5 outlines the various techniques used for edge detection methods.
  - **Spectral and other features:** In advanced solar wafer inspection, along with spatial and texture features, utilizing spectral features and additional mathematical techniques aids in identifying hidden or subtle patterns on the wafer surface. These features play an essential role in understanding the nature, structure, and composition of defects. Table 4.6 outlines the various techniques and characteristics that can be extracted in the context of solar wafer textures to identify defects [25].
- **Selection:** In the GA, selection is an important stage where it aims to pick the best individuals (solutions) to transmit their genes (parameters) to the next generation. In the solar wafer inspection field, individuals represent a set of fine-tuned image processing and classification settings, and the

TABLE 4.4
Texture Feature Extraction Methods

Texture Feature Extraction Method	Description	Extracted Features
Gray-Level Co-occurrence Matrix (GLCM)	Examines the frequency of pixel pairs with specific gray-level values at a particular distance and angle	• Contrast • Energy • Homogeneity • Correlation
Local Binary Patterns (LBP)	Compares each pixel with its neighboring pixels to encode local patterns by assigning a binary code	• LBP histogram • Texture uniformity • Local texture variations
Wavelet Transform	Breaks down the image into different frequency fragments, capturing both larger and smaller details	• Approximation coefficients • Detail coefficients • Energy in different frequency bands
Run-Length Matrix (RLM)	Calculates the distances of adjacent pixels with matching same intensity, capturing texture continuity	• Short-run emphasis • Long-run emphasis • Gray-level non-uniformity
Histogram of Oriented Gradients (HOG)	Examines the distribution of gradient orientations in specific areas of a picture	• HOG descriptor values • Edge directions • Shape information
Gabor Filters	Uses a set of Gabor filters at different orientations and sizes to capture characteristics of texture	• Gabor filter responses • Texture energy • Orientation selectivity
Fourier Transform	Examines the image in the frequency domain to record periodic designs and textures	• Fourier coefficients • Frequency domain features

fitness value indicates how effectively these settings detect defects. The most suitable individuals are chosen for reproduction to guarantee that high-performing traits (combinations of parameters) are maintained and enhanced throughout successive generations. Table 4.7 summarizes various selection methods for solar wafer inspection [26].

- **Crossover:** Crossover combines the genetic information (parameters) of two parents of two parental individuals to produce new offspring. It is a crucial tool in GA for examining solar wafers, enabling the creation of new offspring. Table 4.8 summarizes the various methods of crossover.
- **Mutation:** Mutation introduces small random changes to an individual to maintain diversity in the population and prevent premature convergence. The following table summarizes the various types of mutation adopted.
- **Termination:** The termination step decides when the GA should stop evolving further. Having appropriate termination criteria ensures a balance

TABLE 4.5
Edge Detection Methods

Method	Description	Working	Application
Canny Edge Detector	An advanced edge detection algorithm known for its accuracy and ability to detect edges effectively	• Uses Gaussian smoothing for noise reduction • Calculates gradients to detect edges • Applies non-maximum suppression to reduce edge thickness • Uses hysteresis thresholding to refine edges	Detects fine, continuous cracks in solar wafers, ensuring accurate boundary localization
Sobel Operator	A basic edge detection method that calculates the gradient magnitude in both vertical and horizontal directions	• Calculates gradients using small convolution kernels in the x and y directions • Highlights areas with sudden intensity changes • Easily affected by noise due to the filter's basic design	Identifies scratches or inclusion boundaries where pixel intensities change sharply
Laplacian of Gaussian (LoG)	Combines Gaussian smoothing with the Laplacian operator for edge detection based on second-derivative analysis	• Uses Gaussian filter for smoothing and noise reduction • Applies Laplacian operator for detecting rapid intensity changes (zero crossings)	Suitable for detecting rounded inclusions or identifying defective cell boundaries with circular features

TABLE 4.6
Feature Extraction Methods

Feature	Purpose	Working	Application
Spectral power density (SPD)	Analyzes how power is distributed across frequency components to identify periodic patterns or oscillations	Uses fast Fourier transform (FFT) to transform the image into the frequency domain and measure intensity at different frequencies	Identifies periodic defects like repeated micro-cracks or uniform scratches
Reflectance spectroscopy	Measures reflectance spectra to detect material inconsistencies or impurities	Shines light on the wafer and measures the reflected spectra across different wavelengths (UV, visible, infrared)	Detects delamination, surface contamination, or coating issues

TABLE 4.6 (*Continued*)
Feature Extraction Methods

Feature	Purpose	Working	Application
Fractal analysis	Analyzes the self-similarity and complexity of defects by treating them as fractals	Uses fractal dimension calculations to quantify the complexity of defect shapes	Identifies and quantifies complex cracks or irregular defect patterns
Wavelet packet transform (WPT)	Decomposes the image into frequency sub-bands to analyze localized frequency information	Passes the image through wavelet filters at multiple scales, creating a multilevel decomposition	Identifies localized defects like small cracks or edge irregularities with better precision than FFT
Independent component analysis (ICA)	Separates independent components to isolate distinct defects from mixed data	Decomposes the image or spectral data into statistically independent components, each representing a different defect type	Useful for multidefect analysis, especially when defects overlap or occur together

TABLE 4.7
Selection Methods in Evolutionary Algorithms

Selection Method	Working	Application in Solar Wafer Inspection
Roulette wheel selection	• The likelihood of being chosen by an individual is directly related to their level of fitness. A metaphorical roulette wheel assigns space based on fitness. • Fitter individuals have higher selection chances.	• Apt for individuals with a wide range of fitness levels. For example, individuals with better crack detection accuracy have a higher chance of selection.
Tournament selection	• Chooses a random subset of individuals. • The most physically fit individual in the group is selected for reproduction. • Continues until a sufficient number of parents are chosen.	• Ensures the selection of high-quality parameter sets while maintaining diversity. • Useful in evolving robust solutions for identifying scratches, cracks, or contamination.
Rank-based selection	• Individuals are ranked based on their fitness values (from greatest to least). • The probability of selection depends on ranking rather than the actual fitness value. Even low-ranked individuals have some chance of selection.	• Efficient when there are minor differences in fitness levels, to prevent a few individuals from dominating the population. • Guarantees that parameters used in tasks such as texture analysis or edge detection are not discarded too soon.

TABLE 4.8
Crossover Methods

Crossover Method	Working	Application
Single-point crossover	• One crossover point is randomly selected on the parameter set. • The first part of the offspring comes from one parent, and the other part comes from another parent.	• Example: One set of parameters (e.g., filter size) comes from Parent 1, while classification type comes from Parent 2.
Two-point crossover	• Two crossover points are selected, and the middle section is exchanged between the parents.	• Allows better parameter mixing, for example, edge detection methods from one parent and support vector machine (SVM) kernel type from the other.
Uniform crossover	• Every parameter is chosen randomly from either parent with equal probability.	• Useful for creating diverse offspring for tasks that involve multiple parameters such as Gaussian filtering, thresholding, and feature extraction.
Arithmetic crossover	• Utilizes a weighted average of parental parameter values.	• Useful for continuous parameters like filter size or SVM hyperparameters, helping to fine-tune performance.

between computational time and achieving a highly effective solution for identifying and categorizing defects in solar wafers. Table 4.7 summarizes the various types of termination adopted.

4.2.5.4 Example 2: Using GA for Scratch Identification in Solar Wafer Inspection

1. **Problem Statement**—Identify scratches on a solar wafer image using GA.
   GA Parameters
   i Population size: 100
   ii Generations: 50
   iii Crossover probability: 0.8 iv mutation probability: 0.1
   v Fitness function: Scratch detection accuracy
2. **Image Preprocessing**
   i Convert image to gray scale
   ii Apply Gaussian filter $\sigma = 2$
   iii Threshold image (threshold=127)
3. **Initialization:** Generate 100 random individuals, each representing a potential scratch detection solution. The individuals in the table represent different methods to set the detection parameters. For example, individual 1, with a scratch width of 5, length of 10, and threshold of 150, could undergo testing with images of solar wafers. If it shows high accuracy, it could be retained or adjusted for more testing. Meanwhile, individual 2, having a scratch width of 3, length of 15, and threshold of 120, could offer

an understanding of the impact of parameter variations on detection performance. Through the comparison and contrast of these individuals, GA will gradually focus on the most efficient parameter set for the task.

4. **Fitness evaluation:** Evaluate the fitness of every individual in GA for scratch detection to determine the performance of each parameter set. The fitness value for each individual is calculated as [27]:

$$fitness = \left(\frac{true\ positives}{true\ positives + false\ positives} \right) * (1 - \frac{false\, negatives}{true\, negatives + false\, negatives})$$

True positives: Number of areas correctly identified as scratches.

- **False positives:** Number of non-defective areas incorrectly identified as scratches.
- **True negatives:** Number of correctly identified non-defective areas.
- **False negatives:** Number of actual scratches that were not identified.

The equation combines two key components:

Precision (First Part) and Recall (Second Part).

The first part evaluates the accuracy of identifying scratches by comparing the number of true positive scratches to the total number of identified scratches (true positives + false positives). High precision indicates that the majority of identified scratches are actual defects, which is important in cases where false positives can be expensive or misleading. This second part calculates the percentage of actual defects that were correctly identified (true positives) out of all actual defects (true positives + false negatives). A higher recall indicates the method's ability to detect

TABLE 4.9
Mutation Methods in Evolutionary Algorithms for Solar Wafer Inspection

Mutation Method	Working	Application
Bit-flip mutation	• A random parameter value is flipped (e.g., filter type switched from Gaussian to Median).	• Enables the exploration of different parameter combinations, such as experimenting with different edge detection techniques.
Gaussian mutation	• Applies a slight random value sampled from a Gaussian distribution to numerical parameters.	• Useful for fine-tuning continuous variables such as filter size or classifier hyperparameters.
Swap mutation	• Two parameter values are randomly swapped within the same individual.	• Example: Swap texture method and shape descriptor settings to explore new configurations.
Non-uniform mutation	• The mutation magnitude lessens over generations, focusing on exploration at first and fine-tuning subsequently.	• Aids in maintaining a balance between exploring and exploiting during the progression of the genetic algorithm.

TABLE 4.10
Scratch Measurements

Individual	Scratch Width	Scratch Length	Threshold
1	5	10	150
2	3	15	120
...	...	...	...

TABLE 4.11
Fitness Values of Individuals

Ind.	Scratch W.	Scratch L.	Thres.	TP	FP	TN	FN	Fitness
1	5	10	150	30	5	40	10	0.857
2	3	15	120	28	3	45	12	0.825
3	4	20	130	32	4	38	8	0.872
4	6	10	160	35	6	42	5	0.895
5	7	15	140	29	8	37	11	0.813
...	...	...	...	...	...	...	...	...
50	5	10	150	30	10	50	5	0.682

actual scratches effectively. For the fitness function provided in Table 4.11, a confusion matrix is generated for each individual. Let's consider after running the scratch detection algorithm on a set of images for individual 1 (width = 5, length = 10, threshold = 150) the result obtained is true positives = 30; false positives = 10; true negatives = 50; false negatives = 5, the fitness calculation will be

$$\text{Precision} = \frac{\text{TP}}{\text{TP} + \text{FP}} = \frac{30 + 10}{30} = \frac{40}{30} = 0.75$$

$$\text{Recall} = 1 - \frac{\text{FN}}{\text{TN} + \text{FN}} = 1 - \frac{5}{50 + 5} = 1 - \frac{5}{55} \approx 0.909$$

$$\text{Fitness} = 0.75 \times 0.909 \approx 0.682$$

This process will be repeated for individual 2 and other individuals in the population, which will result in different confusion matrix values based on how effectively they detect scratches on the solar wafer images. The fitness values computed for each individual help determine which set of parameters performs best in identifying scratches on solar wafers. The GA will favor individuals with higher fitness scores for reproduction in the next generation, driving the optimization process toward improved detection accuracy. This systematic evaluation and refinement ensure that the final selected parameters are optimal for the scratch detection task, balancing precision and recall effectively. The calculated fitness values for each individual assist in determining the optimal parameters set for detecting scratches on solar wafers. The GA will select individuals with high fitness scores to reproduce in the next generation, focusing toward enhanced detection accuracy.

6. **Selection:** After calculating the fitness values, the next process is selection. Let's consider that the selection method involves identifying the top 50 individuals with the best fitness values measured by scratch detection accuracy. Once fitness values are calculated for all individuals, they are ordered from highest to lowest according to their fitness scores. The top 50 individuals are chosen for the next generation. In the example table, individuals 1 to 50 are selected according to their fitness values. These chosen individuals will act as parents for the next generation.
7. **Crossover:** Crossover combines the characteristics of two chosen individuals to generate fresh offspring. This procedure is similar to natural reproduction, enabling the algorithm to investigate new parameter combinations for detecting scratches. Let's consider individuals 1, 3, and 4 from Table 4.12; these individuals are selected due to their higher fitness values. We will use a single-point crossover. In single-point crossover, a crossover point is randomly chosen. Assume the crossover occurs after the second parameter (Scratch Length).
   - **Parent 1:** (Scratch Width = 5, Scratch Length = 10, Threshold = 150)
   - **Parent 2:** (Scratch Width = 4, Scratch Length = 20, Threshold = 130)

The offspring after crossover are

- **Offspring 1:** (Scratch Width = 5, Scratch Length = 20, Threshold = 150)
- **Offspring 2:** (Scratch Width = 4, Scratch Length = 10, Threshold = 130)

In the scratch detection scenario, the offspring produced through crossover may have better scratch detection capabilities. For instance, if offspring 1 (with Scratch Length = 20) demonstrates a higher rate of detecting scratches in validation tests, it would suggest that the crossover operation has been effective for scratch detection.

8. **Mutation:** Mutation facilitates the exploration of a wider range of parameters. By making small changes to scratch detection parameters, GA can dynamically improve the ability of fault detection. The process helps to refine the algorithm to effectively identify scratches while minimizing false

TABLE 4.12
Offspring Parameters after Mutation

Individual	Scratch Width	Scratch Length	Threshold
Offspring 1 (after mutation)	5	21	145
Offspring 2 (after mutation)	5	10	135

positives. Let's consider the new offspring generated in the previous crossover process.

- **Offspring 1:** (Scratch Width = 5, Scratch Length = 20, Threshold = 150)
- **Offspring 2:** (Scratch Width = 4, Scratch Length = 10, Threshold = 130)

4.2.5.5 Mutation Parameters

- **Mutation Probability:** 0.1 (10% chance of mutation for each parameter).
- **Adjustment Range:**
  - Scratch Width: Can vary by ±1 pixel
  - Scratch Length: Can vary by ±1 pixel
  - Threshold: Can vary by ±5 intensity levels

4.2.5.6 Mutation Process

- **Scratch Width:**
  - Current Value: 5
  - Mutation Decision: Randomly choose not to mutate (no change).
  - **Result:** Scratch Width remains 5.
- **Scratch Length:**
  - Current Value: 20
  - Mutation Decision: Randomly mutate by +1 pixel.
  - New Scratch Length: 20 + 1 = 21 pixels.
- **Threshold:**
  - Current Value: 150
  - Mutation Decision: Randomly mutate by –5.
  - **New Threshold:** 150–5 = 145.

4.2.5.7 Result after Mutation for Individual 1

- Scratch Width: 5 pixels
- Scratch Length: 21 pixels
- Threshold: 145 intensity level

Mutation Details for Offspring 2

- **Scratch Width:**
  - Randomly mutate by +1.
  - **New Scratch Width:** 4 + 1 = 5.
- **Scratch Length:**
  - Randomly choose not to mutate (no change).
  - **Result:** 10 pixels.
- **Threshold:**
  - Randomly mutate by +5.
  - New Threshold: 130 + 5 = 135.
- **Termination:** Stop after 50 generations.

  The use of GA significantly improves the process of scratch detection on solar wafers by optimizing the parameters used to identify the defects. Through the use of feature extraction methods, GA develops possible solutions by means of selection, crossover, and mutation in an iterative process. This step-by-step approach allows the algorithm to focus on the optimal settings, ensuring that the parameters for scratch width, length, and threshold are finely tuned. The fitness evaluation, which is based on positive detection rates and minimizes false positives and negatives, ensures that the resulting models are reliable. Incorporating GA not only enhances the accuracy of scratch detection but also contributes to better quality and efficiency in solar cell manufacturing, making it a very effective solution for maintaining the reliability of solar wafers.

4.2.6 Neural Networks for Defect Detection

It is a branch of ML and DL that offers an efficient solution for solar wafer defect detection. The use of neural networks (NNs) in solar wafer inspection improves efficiency, quality, and throughput during the manufacturing process. Identifying surface defects is critical for ensuring the efficiency of solar wafers. Manual inspection or traditional image processing techniques often fail to recognize minor defects, which is where DL models, particularly NNs, come into play. NN processes large data from images, infrared information, and electroluminescence (EL) scans, providing better accuracy when compared to traditional CV techniques [28]. They have the ability to learn patterns and characteristics from data, allowing them to:

- Identify defects with high accuracy, reducing manual inspection time and increasing throughput.
- Identify minor defects that could be overlooked by human inspectors or traditional automated systems.
- Minimize incorrect defect detections to avoid unnecessary rework and waste.

4.2.7 Key Components in NN for Defect Detection

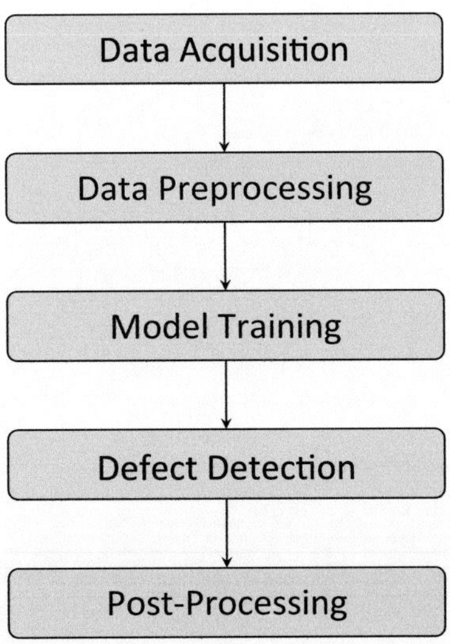

4.2.7.1 Data acquisition

Data acquisition is the first step to collecting high-quality images or scan data of solar wafers. The purpose is to capture detailed information about all surface details, including defects like cracks, stains, or micro-cracks [29]. The effectiveness of the NN heavily depends on the quality, resolution, and consistency of this data. Various data acquisition techniques are summarized in Table 4.13. Data formats and quality

```
# Function to determine if a scratch is critical def
is_critical_scratch(scratch_length, scratch_depth):
            if scratch_length > 1.0 and scratch_depth > 10.0: # length in
    ,→ mm, depth in micrometers return "Yes, it is a
        critical scratch."
    else:
        return "No, it is not a critical scratch."

# Example input scratch_length = 1.5
# mm scratch_depth = 15.0 # m

# Evaluate the rule result = is_critical_scratch(scratch_length,
scratch_depth) print(result) # Output: Yes, it is a critical scratch.
```

```
# Function to determine the type of crack based on length and width def
classify_crack(crack_length, crack_width):
    if crack_length <= 0.5 and crack_width <= 5.0:
                return "\textcolor{red}{Hairline Crack}" # Return statement in red
    elif 0.5 < crack_length <= 2.0 and 5.0 < crack_width <= 20.0:
        return "\textcolor{red}{Micro Crack}" # Return statement in red
    elif crack_length > 2.0 and crack_width > 20.0:
        return "\textcolor{red}{Macro Crack}" # Return statement in red
    else:
                return "\textcolor{red}{Not classified}" # Return statement in red

# Example input crack_length =
2.5 # mm crack_width = 25.0 #
m

# Evaluate the crack type crack_type =
classify_crack(crack_length, crack_width)

# Print the result print(f"The crack is classified as: {crack_type}.") # Output classification
```

TABLE 4.13
Overview of Data Acquisition Methods for Solar Wafer Inspection

Data Acquisition Method	Technique	Description	Purpose
Visual inspection	High-resolution cameras	Captures images of the wafer surface. Lighting configurations like bright field (even light) and dark field (angled light) are used.	Identifies surface defects like scratches, stains, chipping, and discoloration based on appearance under different lighting conditions.
Spectroscopy (photoluminescence (PL) imaging)	PL	Uses laser-induced light emission to detect material defects.	Reveals material properties and defects like micro-cracks and impurities inside the wafer.
Spectroscopy (EL imaging)	Electroluminescence (EL)	An electric current passed through the wafer illuminate's regions with electrical failures.	Identifies inactive regions or internal defects like internal micro-cracks and electrical failures.
Scan data (advanced microscopy)	Scanning electron microscope (SEM)	Produces extremely precise surface pictures on a microscopic scale.	Useful for identifying sub-micron defects such as fine cracks, tiny particles, and surface irregularities.

considerations play a vital role in solar wafer inspection. Images can be captured as 2D formats such as JPEG or TIFF, which provide detailed visual representations, while 3D point clouds are used for capturing the surface topography of wafers. High-quality data is essential for exactly spotting minor defects, as lower quality can hide critical details. Furthermore, it is vital to reduce noise that may come from environmental factors and sensor constraints to improve the performance of the model. Filtering can help to enhance the quality of data supporting NN to effectively learn from the defect patterns [30].

4.2.7.2 Data Preprocessing

Preprocessing the data aids the NN to effectively perform the quality assessment based on the consistency of the input data. To make minor defects more visible image enhancement is mandatory. This involves defining the brightness, contrast, and sharpness, which is done in the data preprocessing stage. Noise removal techniques are also applied to remove unwanted pieces that could interfere with defect detection. Normalization scales the pixel values or data points to a standard range (e.g., 0–1), ensuring that no single feature dominates the learning process and helping the model converge faster during training. Data augmentation is used to artificially increase the dataset by introducing transformations like rotation, flipping, zooming, or cropping, allowing the NN to generalize better by learning from a broader range of circumstances. This action helps reduce the chances of overfitting. The techniques are listed in Table 4.14.

TABLE 4.14
Overview of Data Preprocessing Techniques for Defect Detection

Technique	Description	Purpose
Histogram equalization	Adjusts the intensity distribution of an image to improve contrast	Enhances the visibility of minor defects
Filtering	Uses Gaussian or median filters for image smoothing and noise reduction	Reduces noise and retains essential image details
Feature extraction	Identifies and extracts relevant features such as edges, textures, or patterns from the data	Highlights key characteristics for better defect detection

4.2.7.3 Model Training

Training is an essential process when the NN learns and updates itself to identify the defects from well-labeled data. It involves four steps: dataset splitting, model selection, hyperparameter tuning, and training. To achieve reliable accuracy and adaptability, the dataset is divided into three subsets as follows: *training set* (approximately 80%), *validation set* (approximately 10%), and *testing set* (approximately 10%). The training set helps the model learn defect patterns and establish associations between input data and corresponding labels. The results from validation are used in training to analyze the performance of the NN model and to refine the parameters like learning rate and batch size to avoid overfitting [31]. The overall performance after training is analyzed from the testing set, and thus efficiency is ensured. For the defect detection system to be successful, it is essential to choose the right NN architecture. Convolution neural networks (CNNs) can be preferred for image-based defect detection as they can automatically import features like edges, textures, and patterns from the wafer images. They are also effective in detecting surface-based defects like cracks, stains, and pits. On the other hand, recurrent neural networks can be preferred for handling sequential data involving time-series data tracking like wafer degradation and electrical property analysis over time. The overall performance of the NN model depends on hyperparameter tuning in which the learning rate controls the capability of model update in each iteration. A high learning rate can speed the training process but results in overshooting, whereas batch size determines the number of training samples handled parallelly before weight updating. Batch size plays a crucial role in determining the overall efficiency of NN effectiveness. Epoch refers to the number of times that the sample in the training dataset has been fed through the process. Too many epochs may lead to overfitting. Proper tuning of the learning rate, batch size, and epoch leads to quick convergence, reduced errors, and bypass over- and underfitting. During the training phase, the NN model learns to reduce the gaps between the original and predicted outputs. To optimize the model's internal parameters, gradient descent and backpropagation techniques are used. Backpropagation with the help of the chain rule calculates the gradient of the loss function enabling the network to update its weight accordingly. Gradient descent is an iterative method which helps to compute the optimal set of weights and updates opposite to that of gradient [32]. A well-trained NN model can precisely detect defects in solar wafers such as fragile or barely visible ones.

4.2.7.4 Defect detection

Trained NN models are implemented to identify defects and classify solar wafers accordingly. It incorporates a systematic methodology to prepare input data, execute trained models, and inspect outputs based on accuracy. The process involves three steps: input data preparation, model inference, and defect localization. During the input data preparation, the new wafer is preprocessed to ensure that it is in the required format for analysis. Processes like image enhancement, standardization of pixel values, normalization, and resizing of the image are done to meet the requirements. In model inference, the preprocessed input data is run against the trained model. The model analyzes the new wafer images and generates predictions about the defects. After the inference, the defect localization focuses on identifying specific spots and types of defects on wafers [33]. Thus, the model's prediction is evaluated, the exact areas of defects are located, and the error types such as cracks, scratches, and stains are characterized.

4.2.7.5 Post-Processing

This is the last stage which improves the predictability and accuracy of the defect by refining the results obtained in the previous stages. It starts filtering the model output to remove false positives and allows only true positives. The error results are categorized based on the levels of severity, and further action is taken. Effective visualization techniques are then employed to clearly present

TABLE 4.15
Key AI Technologies in Inspection and Inspection Techniques

Company/ Group	Key Technology	Improvement (Cost, Accuracy, etc.)	Specific Advantage
Intel	Deep learning	Rapid fault diagnosis; higher production yield	End-to-end automation
VIA Technologies	Edge AI	High recall rate 100%; operational cost reduction	Live fault detection
Applied Materials	ExtractAI	Inspection costs lowered by 3×	Hybrid optical and e-beam inspection system
Micron Technology	SmartSight	Increased output with minimal waste	Predictive analytics
Google Cloud	AutoML Vision	Validation rate up to 95%; faster model training	Instant deployment on factory production lines
Siemens	Factory AI	Performance improved by 6%	Streamlined prototyping

the locations and types of defects, using tools like heat maps, scatter plots, or 3D visualizations. Techniques like thresholding set probability thresholds for defect detection, reducing false positives and enhancing reliability, while clustering groups similar defects to reveal hidden patterns and relationships. By carefully executing these components of post-processing [34], NN can accurately detect defects, eventually improving the overall quality and efficiency of the solar wafer manufacturing process.

4.2.8 Companies/Industries/Research Groups Who Opted or Working with This Field

AI and ML are undertaking a revolution in wafer inspection for semiconductor manufacturing by improving the accuracy and efficiency of defect detection. Several companies and research groups have adopted AI/ML techniques into their inspection systems, which have led to significant improvements in cost reduction, energy efficiency, and production yield.

Enlight optical inspection system from Applied Materials is combined with the SEM Vision G6/G7 e-beam review system and Extract AI technology. This collaboration enables quick identification and sorting of wafer defects. The Extract AI software uses DL to identify yield critical defects and differentiate between signal and noise. This method has led to a three times lessening in the cost of inspection and a marked rise in energy conservation through optimal inspection processes [35]. Intel has adapted AI-enabled DL and image analysis to increase yield analysis. Every wafer in each lot undergoes a 100% automated end-to-end defect identification examination. The process also cuts down on time by directly integrating with current workflows [36]. To improve defect detection capabilities, increase yield, and reduce rework, KLA's eSL10 e-beam patterned wafer defect inspector utilizes DL algorithms to identify small defect signals among pattern and process noise. This AI-driven system has increased yield and reduced rework and production costs. Furthermore, the eSL10 offers better defect detection, real-time data processing, and high-resolution imaging, leading to a more ecologically friendly and sustainable manufacturing process [37].

4.3 CONCLUSION AND FUTURE WORK

The rapid advancement of AI and CV has greatly different industrial processes, especially in the field of solar cell wafer inspection. With the increasing popularity of high-efficiency solar panels growing, it is essential to accurately detect defects in solar wafers. The integration of AI techniques, specifically GA and NN, offers a robust framework for addressing these challenges, improving the overall quality and reliability of solar energy production and establishing a new standard for quality control in the solar sector. The GA-based optimization techniques have played a key role in improving feature extraction methods. GA techniques allow for more efficient processing of imaging data by balancing detection accuracy with computational complexity. Meanwhile, the use of NN structures, specifically CNNs, has shown impressive results in defect detection surpassing traditional rule-based methods.

Additionally, hybrid approaches that combine GA and NN techniques not only enhance the robustness of defect detection but also set the stage for new AI applications in the field. As we approach toward Industry 4.0, the use of AI for inspecting solar cell wafers will enable smart manufacturing practices. Incorporating advanced AI-based technologies will increase production efficiency and drive the industry's evolution. AI-powered automated inspection systems will significantly improve defect detection rates, reducing waste and improving the overall quality of solar cells. To fully utilize these advancements, future research should focus on scaling AI-powered inspection systems for mass production. Exploring the possibilities of edge computing and the Internet of things can help improve real-time data processing and analysis, which will enhance inspection workflow. The incorporation of AI and CV into solar cell wafer inspection represents a critical advancement for the PV manufacturing industry. By integrating these technologies, the solar industry can enhance efficiency, lower operational costs, and promote innovation for a sustainable energy future. As the solar energy field continues to develop committing to AI-driven solutions will be key for the industry's growth and stability in a changing market. Using these advanced techniques will not only boost manufacturing capabilities but will also play a key role in the global shift to renewable energy. The future of AI in this field with developments in DL will be promising. It will enable more accurate defect detection while minimizing errors. The incorporation of Edge AI will allow real-time processing, which will reduce latency and improve operational efficiency. Generative AI can enhance datasets using synthetic defect images, overcoming the challenge of limited real-world data. Multimodal AI, a combination of optical, thermal, and X-ray imaging, will produce better results. Integrating self-learning AI systems will lead to new defect patterns and improve inspection accuracy. These advancements will make solar wafer inspection more efficient, cost-effective, and reliable, ensuring higher-quality solar cells for the growing renewable energy sector.

REFERENCES

[1] Solarfeeds. *Solar Wafer*. 2019. url: www.solarfeeds.com/mag/wiki/solar-wafer/.

[2] Ula Hijjawi et al. "A review of automated solar photovoltaic defect detection systems: Approaches, challenges, and future orientations". In: *Solar Energy* 266 (2023), p. 112186.

[3] Chief Technology Officer at Sinovoltaics Group Niclas D. Weimar. *Solar Cell Production: From Silicon Wafer to Cell*. 2015. url: https://sinovoltaics.com/solar-basics/solar-cell-production-from-silicon-wafer-to-cell/.

[4] Communications Team. *A Detailed Guide about Solar Wafers: Application and Types*. 2022. url: https://solarsquare.in/blog/solar-wafers-application/.

[5] M Abd Al Rahman and Alireza Mousavi. "A review and analysis of automatic optical inspection and quality monitoring methods in electronics industry". In: *IEEE Access* 8 (2020), pp. 183192–183271.

[6] Great Eyes. *Wafer Inspection-Highly Sensitive Wafer Inspection System*. 2022. url: https://greateyes.de/inspection-systems/wafer-inspection/.

[7] Cristina De Luca et al. "AI in semiconductor industry". In: *Artificial Intelligence for Digitising Industry–Applications*. River Publishers, 2022, pp. 105–112.

[8] Solar Quotes. *How AI is Revolutionising the Solar Industry*. Accessed: 2024-12-10. 2024. url: www.solarquotes.com.au/blog/ai-solar-industry/.

[9] Equinox's Drones. *Automate Your Solar Panel Inspection Using AI-Powered Drones*. Accessed: 2024-12-10. 2024. url: https://equinoxsdrones.com/automateyour-solar-panel-inspection-using-ai-powered-drones/.

[10] CGI. *Using Artificial Intelligence to Inspect Photovoltaic Plants*. Accessed: 2024-12-10. 2024. url: www.cgi.com/en/article/energy-utilities/using-artificial-intelligence-inspect-photovoltaic-plants.

[11] Precedence Research. *Solar AI Market Overview and Insights*. Accessed: 2024-12-10. 2024. url: www.precedenceresearch.com/solar-ai-market.

[12] Bruce G Buchanan. "A (very) brief history of artificial intelligence". In: *Ai Magazine* 26.4 (2005), p. 53.

[13] Arthur L Samuel. "Some studies in machine learning using the game of checkers". In: *IBM Journal of Research and Development* 44.1.2 (2000), pp. 206–226.

[14] Ping Kuang, Wei-Na Cao, and Qiao Wu. "Preview on structures and algorithms of deep learning". In: *2014 11th International Computer Conference on Wavelet Actiev Media Technology and Information Processing (ICCWAMTIP)*. IEEE. 2014, pp. 176–179.

[15] Ali Thakfan, and Yasser Bin Salamah. "Artificial-intelligence-based detection of defects and faults in photovoltaic systems: A survey". In: *Energies* 17.19 (2024), p. 4807.

[16] Professional AI. *Rule-Based Systems*. Accessed: 2024-10-31. 2024. url: www.professional-ai.com/rule-based-systems.html.

[17] Geeks for Geeks. *Rule-Based System in AI*. Accessed: 2025-03-02. 2024. url: www.geeksforgeeks.org/rule-based-system-in-ai/.

[18] Han Liu, Alexander Gegov, and Frederic Stahl. "Categorization and construction of rule-based systems". In: *Engineering Applications of Neural Networks: 15th International Conference, EANN 2014, Sofia, Bulgaria, September 5–7, 2014. Proceedings 15*. Springer. 2014, pp. 183–194.

[19] Warda M Shaban. "Detection and classification of photovoltaic module defects based on artificial intelligence". In: *Neural Computing and Applications* (2024), pp. 1–28.

[20] Lixing Liu et al. "Surface defect detection of solar cells based on feature pyramid network and GA-faster-RCNN". In: *2019 2nd China Symposium on Cognitive Computing and Hybrid Intelligence (CCHI)*. IEEE. 2019, pp. 292–297.

[21] Arvind Kumar, and Sartaj Singh Sodhi. "Comparative analysis of Gaussian Filter, Median Filter and Denoise Autoenocoder". In: *2020 7th International Conference on Computing for Sustainable Global Development (INDIACom)*. 2020, pp. 45–51. https://doi.org/10.23919/INDIACom49435.2020.9083712.

[22] Radha Seelaboyina, and Rajeev Vishwakarma. "Different thresholding techniques in image processing: A review". In: *ICDSMLA 2021: Proceedings of the 3rd International Conference on Data Science, Machine Learning and Applications*. Springer. 2023, pp. 23–29.

[23] Encord. *Image Thresholding in Image Processing: Concepts, Techniques, and Applications*. Accessed: 2024-10-31. 2024. url: https://encord.com/blog/imagethresholding-image-processing/.

[24] Jyotismita Chaki, and Nilanjan Dey. *Texture Feature Extraction Techniques for Image Recognition*. Singapore: Springer, 2020. https://doi.org/10.1007/978-981-15-0853-0. url: https://link.springer.com/book/10.1007/978-981-150853-0.

[25] Semi Engineering. *Wafer Inspection*. Accessed: 2024-10-31. 2024. url: https://semiengineering.com/knowledge_centers/manufacturing/process/waferinspection/.

[26] Lee Abdullatif. *Introduction to Genetic Algorithms Including Example Code*. Accessed: 2024-10-31. 2019. url: https://towardsdatascience.com/introductionto-genetic-algorithms-including-example-code-e396e98d8bf3.

[27] Pranav Pathak. *Confusion Matrix for Machine Learning*. Accessed: 2024-10-31. 2020. url: www.analyticsvidhya.com/blog/2020/04/confusionmatrix-machine-learning/.

[28] Neeraj Tomar et al. "Role of artificial neural networks in predicting design and efficiency of dye sensitized solar cells". In: *Journal of Renewable Energy and Applied Sciences* 15.2 (2024), pp. 45–60. https://doi.org/10.1016/j.jreas.2024.02.003.

[29] IBM. *What Are Neural Networks?* Accessed: 2024-10-31. 2024. url: www.ibm.com/topics/neural-networks.

[30] Sharmarke Hassan, and Mahmoud Dhimish. "Dual spin max pooling convolutional neural network for solar cell crack detection". In: *Scientific Reports* 13 (2023), p. 11099. https://doi.org/10.1038/s41598-023-38177-8.

[31] Priyanka Kumari. *ML Beginner's Guide on Solar Panel Inspection*. Accessed: 2024-10-31. 2023. url: www.labellerr.com/blog/ml-begineers-guide-onsolar-panel-inspection/.

[32] Ibrahem Kandel, and Mauro Castelli. "The effect of batch size on the generalizability of the convolutional neural networks on a histopathology dataset". In: *Journal of Medical Imaging* 10.1 (2023), p. 015004. https://doi.org/10.1117/1.JMI.10.1. 015004. url: www.spiedigitallibrary.org/journals/journal-ofmedical-imaging/volume-10/issue-1/015004/The-effect-of-batch-sizeon-the-generalizability-of-the/10.1117/1.JMI.10.1.015004.full.

[33] Deeplizard. *Introduction to Neural Networks*. Accessed: 2024-10-31. 2024. url: https://deeplizard.com/learn/video/U4WB9p6ODjM.

[34] Huaiguang Liu et al. "Research on online defect detection method of solar cell component based on lightweight convolutional neural network". In: *International Journal of Photoenergy* 2021.1 (2021), p. 7272928.

[35] Anton Shilov. *Applied Materials' Wafer Inspection Tools Use AI and Big Data to Cut Inspection Costs by 3×*. Accessed: 2025-03-01. 2021. url: www.anandtech.com/show/16565/applied-materials-wafer-inspection-toolsuse-ai-and-big-data-to-cut-inspection-costs-by-3x.

[36] Intel Corporation. *Manufacturing Yield Analysis with AI*. White Paper. Accessed: 2025-03-01. Intel Corporation, 2023. url: www.intel.com/content/dam/www/central-libraries/us/en/documents/intel-it-manufacturingyield-analysis-with-ai-paper.pdf.

[37] Susan Rambo, and Ed Sperling. "AI in inspection, metrology, and test". In: *Semiconductor Engineering*. Accessed: 2025-03-01. Apr 2021. url: https://semiengineering.com/ai-in-inspection-metrology-and-test/.

5 Introduction to Characterization and Fabrication Techniques Employed for Advanced Solar Cell Structures

Sushmita Chaudhari, Kasturi Ghosh, and Hafizur Rahaman

5.1 INTRODUCTION

The increasing demand for renewable energy has led to the rapid development of solar energy technology. The paradigm of photovoltaic (PV) energy generation has changed from the traditional first-generation crystalline silicon solar cells to more complex and efficient new generations such as perovskite solar cells (PSCs), organic solar cells (OSCs), multi-junction solar cells, and tandem cells. To meet the increasing demand for efficiency, economy, and productivity of solar energy, the improvement of fabrication and characterization techniques has become important. This improved technology has a direct impact on the performance, durability, and commercial potential of PV devices, making it ready to solve the problems of transition from standard measurement to commercial use. The chapter of this book aims to provide knowledge of all probable fabrication and characterization techniques used in the design and improvement of advanced solar cell structures. It aims to bridge the gap between theoretical and practical knowledge and provide researchers and engineers with the tools they need to innovate in the field of PV devices.

The primary goal of solar cell research is to enhance power conversion efficiency (PCE). Advanced manufacturing processes such as chemical vapour deposition (CVD), molecular beam epitaxy (MBE), solution processing, and inkjet printing allow the control of waste, layer thickness, and fine operation. This level of control is necessary to achieve the optimum light absorption, generation of charge carriers, and collection required for high-performance solar panels. For example, a defect-free, high-purity, multi-junction solar cell is fabricated using MBE. These devices, made up of layers of different semiconductor materials,

DOI: 10.1201/9781003515784-5

can achieve over 40% efficiency over silicon-based devices. Similarly, PSCs have been fabricated using solution-based technology, which is a new emerging technology, and showed high-performance and low-cost mass production potential. These fabrication techniques provide flexibility to tune the material properties such as bandgap of the materials and carrier mobility of charge carriers, leading to significant improvement in PCE.

Reduction of the cost of solar cell production while maintaining efficiency is very central to getting solar energy to be closer to fossil energy sources. In this regard, highly advanced techniques such as roll-to-roll (R2R) processing [1] and inkjet printing [2] are extremely promising. Using these methods, it becomes possible to attain uniform deposition of solar cells on low-cost, flexible substrates and at a high rate and therefore the commercialization of thin, flexible, wearable PV technology.

This scalable production makes it possible for researchers and manufacturers to develop large-area solar panels at a one-tenth fraction of the cost of standard crystalline silicon cells. In, especially, low-temperature deposition, less material waste can be attributed to sol–gel method, which also reduces costs in manufacturing. This strategy is relevant to the commodification of perovskite and OSCs, which are cheap to produce and highly effective.

Characterization methods are also critical to the development of efficient solar cells. The optical, electrical, and structural qualities of solar cell material can be analysed by techniques including photoluminescence (PL) spectroscopy, X-ray diffraction (XRD), and scanning electron microscopy (SEM). These methodologies are essential for defect recognition and material quality control, as well as for understanding the relationship between fabrication processes and the behaviour of the devices.

For instance, details on surface morphology and grain boundaries of thin films that can inform possible sources of recombination losses can be seen in the SEM. The presence of different phases and crystallinity of the materials can be determined through XRD, and these are factors that impact the charge transport characteristics in the near neighbourhood. Scientists can save time for the discovery of new materials by improving the device layout and fabrication process since the material characteristics of solar cells can be characterized at different stages of the fabrication process.

The authors of this book chapter have attempted to educate the readers on the diverse fabrication techniques, coupled with characterization methods that are fundamental to the development of new PV solar cells. The solar PV area is changing quickly from crystalline silicon bulk to thin-film perovskite, organic, multi-junction, and tandem devices. This shift makes it evident how imperative it is to understand the essential process of fabrication as well as characterization to increase efficiency and production. Thus, reviewing the basic principles of solar cell operation at the physical level explaining the processes of light absorption, charge separation, and carrier collection, as well as highlighting the importance of accurate techniques for solar cell manufacturing like CVD, MBE, solution processing, and inkjet printing, this chapter tried to fill this knowledge gap.

Finally, the chapter also looks into some characterization techniques necessary when analysing material characteristics as well as device performance as illustrated by quantum efficiency (QE) measurements, XRD, SEM, and Raman spectroscopy. The purpose of the chapter is to illustrate how these approaches assist in developing new, efficient, and fairly inexpensive solar systems through their implementation. For scientists and engineers in the field of solar energy, it provides an understanding of theoretical fundamentals and practical applications.

5.2 FUNDAMENTAL PRINCIPLES OF SOLAR CELLS

Solar PV: A French physicist A.-E. Becquerel in 1839, while experimenting, noticed the PV effect and found that a weak electric current could be produced when some material was exposed to the sun. This is a direct conversion of light energy form into the electrical form of energy. When the material is exposed to sunlight, photons may reflect, absorb, or pass through the material. The energy possessed by a photon depends on the colour or frequency of the light. When light interacts with a material, then the energy from the photon is imparted to an electron and moves the electron onto an upper energy level in the material, and the hole is left behind. Since the lifetime of the electrons in the excited state is so short, the electrons will return to their basic state very fast, thus recombining with the hole. In the case of PV devices, the built-in potential at the depletion region forces the excited electrons to an external circuit before recombining [3].

Solar PV Cell: It is an electronic device having PV property by which light can directly be converted into electricity. For the exploitation of solar light into electrical power, it is required to have light-absorbing material, commonly known as the PV absorber material, which generates the electrons and holes as the charge carrier and is typically available as output terminals without much loss. The solar cell is a two-terminal device that converts sunlight into photovoltage. It consists of two types of semiconductor materials usually n- and p-type semiconductors. A p–n junction is formed whenever two different semiconductor materials are joined together, in which one of them includes a higher density of holes while the other contains a higher density of electrons. Free charge carriers move uniformly from one region to another at the contact region; that is, free electrons move from the n-region to the p-region and collapse with holes in the p-region. Therefore, there is a scarcity of free charge carriers at the junction called the p–n junction, also called the depletion region, where there are no free moving charge carriers [4].

When the incident light is falling on the p–n junction of a solar cell and the energy of the falling photon is equal to or more than the band gap value of that material, then electron–hole pairs are created at the junction. By this process, photogenerated charge carriers diffuse across the depletion region and are separated by the built-in electric field at the junction of the P & N regions. The photo-induced current generated by these processes is thereafter taken through the ohmic contacts, hence producing the electrical output of the cell. Details on the working of PV cells are presented in the following figure.

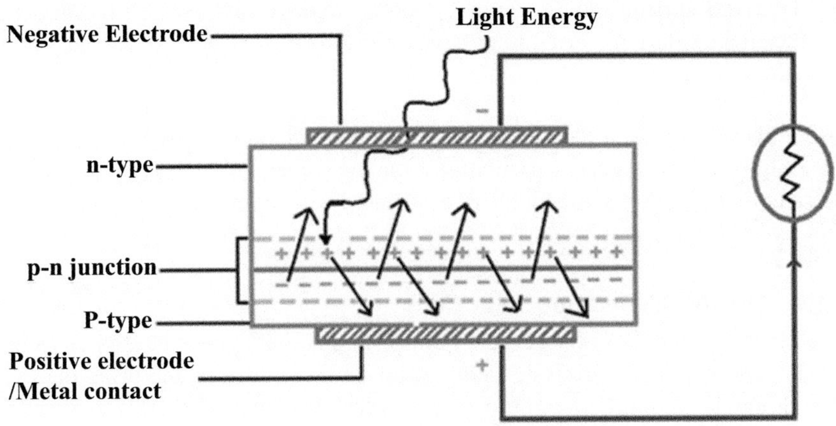

FIGURE 5.1 Working principle of photovoltaic cell [4].

5.3 FABRICATION TECHNIQUES FOR ADVANCED SOLAR CELLS

In particular, various fabrication methods are driving the advancement of solar cell technology. With solar cells evolving from first-generation crystalline silicon cells to second- and third-generation devices, the progress in fabrication methods has played a major role in improving their performance, reducing their costs, and facilitating mass production. Following is a brief description of the novel fabrication methods used in various generations of solar cells.

5.3.1 Fabrication Evolution

5.3.1.1 First-Generation Solar Cells

First-generation or amorphous silicon solar cells, which are mainly constructed from monocrystalline and polycrystalline silicon, are developed from the conventional semiconductor industry technology. These cells have laid down the preliminary structures for solar energy technologies basing its credibility on efficiency and reliability. The main fabrication techniques used for these solar cells include:

- **Wafer Production:** Crystalline silicon wafers, which act as the foundation for developing solar cells, are prepared with the Czochralski process for monocrystalline silicon and casting methods for polycrystalline silicon. These wafers are processed by slicing the silicon ingots and then polished to produce a smooth surface.
- **Doping:** To introduce the impurity and to create p-type and n-type regions diffusion or ion implantation techniques are used.
- **Screen Printing and Metallization:** These contacts include the front and back contact of the PV modules where screen printing is used to deposit metal pastes, which form the contacts after firing.

- **Thermal Oxidation:** To minimize the recombination losses of the surface, thermal oxidation is used to passivate the silicon surface with the help of a thin oxide layer.

These processes are well defined but require a huge amount of energy and thus produce a high cost of material, particularly for high-purity silicon wafers. The main constraint in scalability is the high cost associated with silicon and the energy needed for its processing.

5.3.1.2 Second-Generation Solar Cells

In the second generation of solar cells, thin-film technology was developed to minimize the amounts of materials needed to produce these devices and bring their costs down, although it compromised efficiency, to a certain extent. Thin-film solar cells are fabricated by depositing a thin semiconductor layer on a base substrate such as glass, plastic, or metal. Mostly used fabrication techniques for thin-film solar cells include:

- **CVD:** It is a widely used method to fabricate thin-film solar cells. As the name suggested, CVD takes place via the chemical vapourization of gaseous reagents and deposits the thin layers of semiconductor materials such as Cadmium Telluride (CdTe), Amorphous Silicon (a-Si), or Copper Indium Gallium Selenide (CIGS). Types of CVD, which include plasma-enhanced CVD or PECVD, are used in low-temperature deposition of amorphous silicon [5].
- **Sputtering:** In sputtering, which is one of the physical vapour deposition techniques, thin films are produced by ejecting the atoms from a target material and deposited on the substrate [6].
- **Electrodeposition:** In some of the thin-film materials, the electrodeposition technique is employed to deposit semiconductor layers. The outcome of this technique is cost-effective and can be used on a large scale, especially for CIGS- and Copper Zin Tin Sulphide (CZTS) based solar cells [7].

These fabrication processes enable mass production on large areas of flexible film, and therefore thin-film technologies are preferred for applications where the use of rigid silicon plates may not be appropriate. The diminished use of materials and the possibility of low-cost, high-throughput are the reasons for the overall adoption of this technology, although the efficiency compared to first-generation technologies is on the lower side.

5.3.1.3 Third-Generation Solar Cells: Advanced Technology and Emerging Technology

The third generation of solar cell products is intended to eliminate the efficiency and cost issues of the first and second generations. They very much contain new materials and new concepts of devices and therefore new cell structures, which, in turn, demand new fabrication methods that in the long run will offer ways of low-cost manufacturing of thin-film silicon cells with very high efficiencies. Based on third-generation research and development, the following are the more advanced technologies of solar cells: PSCs, OSCs, dye-sensitized solar cells, and multi-junction tandem cells. Commonly used fabrication methods for third-generation solar cells include:

- **Solution Processing:** In PSCs and dye-sensitized solar cells, spin coating [8], blade coating [9], ink jetting [10], and dip coating [11] techniques are employed for the deposition of perovskite layers onto a substrate. These techniques are low-temperature processing and can be extended in large-area production. The solution deposition method makes PSCs suitable for large-scale production at a reduced cost due to its simplicity.
- **R2R Processing:** This technique is very essential in the manufacturing of organic photovoltaic solar cells (OPVs) and perovskite cells. Active layers can be printed on flexible substrates using high-speed R2R processing [8]. This process makes it possible to repeatedly manufacture solar cells on flexible substrates, expanding the cost-effective, scalable, and adaptable applications of PVs in day-to-day products.
- **MBE:** For complex, multiple-junction solar cells, MBE is applied to form ultra-low defect density semiconductor layers with precise control over layer thickness and material composition. This is important for the realization of high-efficiency cells that can able to capture various portions of the solar spectrum. However, the high cost and technical complexities associated with MBE deployment restrict the usage of this technique to specific areas such as space and high-density concentrated solar power systems [12].
- **Atomic Layer Deposition (ALD):** ALD is employed for the deposition of ultra-thin films with atomic layer resolution in both third-generation and some second-generation solar cells. This is particularly important for creating passivation layers and all interfaces where layer uniformity and interface control are critical in multi-junction and tandem devices [13].
- **Inkjet Printing:** Another novel technique that is a part of the solution-based deposition is inkjet printing, which enables precise deposition of the material at the desired pattern, thus increasing efficiency and upscaling of the advanced solar cell. This method is particularly appropriate for flexible perovskite and OSCs [14].
- **Hot-Carrier and Quantum Dot Solar Cells:** Third-generation devices such as quantum dots and hot-carrier cells demand sophisticated forms of nanofabrication [15]. Such methods are the colloidal synthesis of quantum dots and quantum confinement structures that increase the absorption of light and collection of the charges.

5.3.2 Selection Criteria of Fabrication Techniques

The selection of the fabrication approach of a certain solar cell directly impacts the performance, cost, and reproducibility of the end device. Even though great progress is being made in solar cell technology, the choice of adequate fabrication methods has to be adjusted according to the proper materials used, the target conversion efficiency, and the requirements of mass production while keeping the costs and negative impact on the environment at a minimum. The following are the key criteria for selecting suitable fabrication methods.

5.3.2.1 Material Compatibility

Material compatibility is the compatibility of the chosen fabrication technique with the materials to be used in the solar cell. Solar cell materials have various physical and chemical characteristics, and the fabrication procedure must maintain or even improve such characteristics. For instance, the properties of perovskites and organic compound materials change when they are exposed to high temperatures and will degrade. These materials should be deposited at low temperatures such as by solution processing, and for materials such as silicon or cadmium telluride (CdTe), high-temperature processing is required such as CVD. It is necessary to mention that thin-film solar cells CIGS and a-Si have a complicated structure where the thickness and homogeneity of layers are critical parameters. Sputtering or evaporation is used for such applications because one is able to deposit very thin, uniform layers. From multi-junction solar cells, system structures are made using the deposition technology of atomic layer control such as MBE.

5.3.2.2 Scalability

The main challenge of technology commercialization is scalability for use in the mass production of PV solar cells. This means that the fabrication method must accommodate large-area production for all the devices while keeping the quality and efficiency as high as possible. It should also be generic to other types of substrates, rigid or flexible, and should cater to different scales of production. But, besides, the fabrication method must enable the production of a large quantity of parts to fulfil high-throughput production needs. For example, R2R processing is well suitable for OPVs and perovskites because the fabrication of solar cells in the form of a thin film on a flexible substrate is scalable and cost-efficient.

Fabrication method cost is highly influential to the market-related cost of solar cells. Certainly, high efficiency is important, but it does no good to have it if solar cells cannot be made cheap enough for widespread use and to be competitive with other energy sources. In evaluating techniques in the fabrication design, there are several factors that define cost-effectiveness including material use. Such techniques as solution processing and, in particular, the inkjet counterpart allow avoiding unnecessary material usage by depositing it selectively in the required areas. This is especially labour saving and hence advantageous for costly materials like perovskites or quantum dot inks. However, methods, such as sputtering, can lead to extravagant use of materials and hence endless costs. Another requirement which can pose a threat to cost is if the machine is used to consume a lot of energy. Processes like the CVD or the MBE are high-temperature processes, and these take much energy input, energy that is costly especially where large quantities of the semiconductor material is being produced. On the other hand, low energy methods such as R2R printing or solution processing are more suitable for large-scale production. In addition, some fabrication techniques for microsystems like MBE and CVD are capital-intensive techniques and involve expensive sophisticated tools. These techniques can also give high efficiency to cells; however, the cost may become somewhat expensive for volume production.

While electrodeposition and screen printing have a relatively low initial cost, they are used and will continue to be used for crystalline silicon as well as thin-film solar cells.

5.3.2.3 Culture and Gender Issues

However, as the technology of solar cells is being developed for large applications, the environmental and safety issues involved in fabricating the cells become more relevant. Indeed, the fabrication method has to be environmentally friendly and compliant with safety requirements. Some of the raw materials utilized in the production of solar cells including cadmium of CdTe cells are dangerous to the environment and humans. Those fabrication methods that would not expose the workers to the toxic materials or those that would use less of the toxic materials are preferred. CZTS and CIGS are considered to be more environmentally friendly than CdTe solar cells due to the lesser toxicity of the material composition. Another important point is minimal waste. This is because methods such as solution processing minimize wastage since they require only a small amount of materials, while others such as sputtering lead to increased wastage and hence a high disposal fee.

5.3.2.4 Efficiency Gains

The fabrication method selected must enhance high-efficiency solar cells through the structural and electrical quality of the device. It should enable direct control over the characteristics of materials and devices and their arrangement to achieve the optimum result. Techniques such as MBE or ALD provide a high level of control of layer thickness and composition and of the interface, which is crucial for efficient multi-junction and tandem cells. On the other hand, traditional techniques such as screen printing are sufficient for the fabrication of normal silicon solar cells but fail to provide the accuracy necessary for next-generation devices. This means that the interaction that occurs at the boundaries of the various materials is important in defining the overall efficiency. Techniques that enable the optimization of these interfaces include ALD and sputtering, with a view to optimizing or enhancing the charge transfer and reducing the recombination of charges.

5.3.3 Chemical Vapour Deposition

It is a versatile technique used to fabricate thin films and coatings by the reaction of gaseous precursor and form the solid material as a product on a substrate. It happens in a controlled environment where one or several volatile reagents are placed in a reaction chamber. These precursors react or decompose on the heated surface of a substrate and form a thin solid layer.

The CVD process typically consists of several key stages: The steps involved are (i) movement of precursor to the reaction chamber, (ii) attachment of precursor molecules to the heated film-forming substrate, (iii) chemical activity at the substrate surface that results in film growth, and (iv) evacuation of the reaction by-products from the chamber.

CVD provides a high level of control over thickness, composition, and homogeneity to contribute to the creation of high-value semiconductors, insulating and conductive films for solar cells, electronics, and optics industries. A versatile application of CVD is silicon-based solar cells, thin-film technology, as well as creating advanced

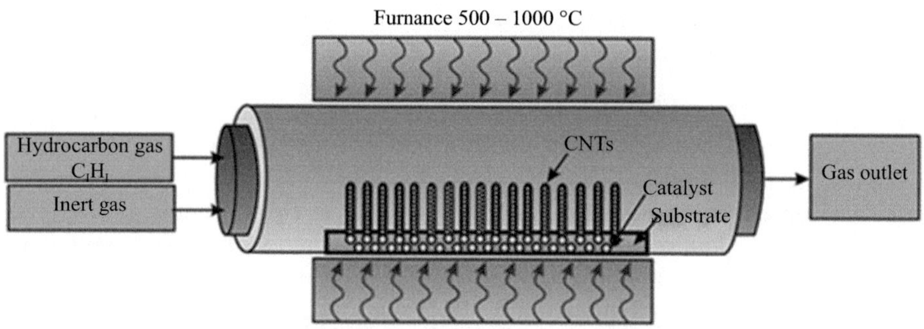

FIGURE 5.2 Vacuum chamber of chemical vapour deposition [16].

electronic and optoelectronic appliances because of the formation of conformal coatings and good material qualities [16].

Different types of CVD include plasma-enhanced CVD (PECVD), low-pressure CVD (LPCVD), and metal-organic CVD (MOCVD) that enables the deposition of materials under different deposition parameters, providing increased versatility. Different types of CVD techniques include:

1. Atmospheric Pressure CVD carries out the deposition at atmospheric pressure and has high deposition rates. Mostly it is used to form oxide layers.
2. LPCVD uses low pressure to ensure that it gives very good uniformity and step coverage, which makes it suitable for high-quality film in microelectronics applications.
3. PECVD introduces plasma to reduce the temperature for deposition, while rendering improved film properties on temperature-sensitive substrates.
4. MOCVD uses metal-organic precursors mainly for the synthesis of compound semiconductors such as GaN and InP in LEDs and solar cell applications.
5. Hot-Wire CVD provides low substrate temperature deposition with a high deposition rate due to the use of a heated filament that decomposes the gas precursors.

5.3.3.1 Challenges and Advancements in CVD for Large-Scale Production

However, there are several challenges that have been faced by CVD when scaled up for large-scale production, but this is being addressed by the latest development: A few examples follow:

Challenges

1. **Uniformity across Large Substrates**: Uniformity in film thickness and composition on large wafers or substrates is a challenging task due to dynamic gas flow inside the chamber, temperature gradient, and pressure variable. These non-uniformities can end up with poor deposition and as a result poor performance.

2. **High-Temperature Requirements**: Conventional CVD processes run typically at high temperatures, thus restricting the type of material and substrate that can be used. These may cause stress or harm to heat-sensitive substrates such as polymers.
3. **Cost and Throughput**: The scaling up of CVD for bulk production is very costly in terms of the cost of the design of the reactor, energy consumption, and supply of precursors inside the CVD chamber. The other trade-off is getting the right throughput while maintaining the quality of the film.
4. **Precursor Availability and Waste**: Certain high-quality films need special or expensive precursor materials. Furthermore, the reaction can give rise to potentially toxic by-products or unused precursor waste, which is environmentally and, probably, safety-sensitive.
5. **Complexity of Multilayer Deposition**: For multi-junction solar cells or advanced devices like OLEDs, multilayer structure formation is needed and requires precise control during deposition. The scale of overall repeatability and accuracy might be technically challenging.

Advancements

1. **Plasma-Enhanced CVD (PECVD)**: It provides deposition at lower temperatures and is more suitable for temperature-sensitive substrates. Additionally, it enhances the film uniformity and rate of deposition for large-scale production.
2. **R2R CVD**: To apply CVD for flexible electronics and solar cells, R2R CVD systems have been developed to allow deposition on flexible substrates. This, in turn, enhances throughput and helps in reducing manufacturing expenses.
3. **Modelling and Simulation**: The high level of simulation tools enables enhanced prediction of the flow of the gas, thermal gradient of the reactor, and behaviour of the precursors and improves the reactor design and process optimization for bulk fabrication.
4. **Advanced Reactor Designs**: Introducing new reactor geometry like multizone heating reactors has led to a reduction of the non-uniformity of deposition problem by controlling the effect of temperature and pressure gradients.
5. **Green CVD Processes**: Recent efforts have been made to synthesize CVD processes that use less toxic precursors and reduce the quantity of waste by-products formed.

In doing so, these improvements are gradually eliminating the challenges in CVD and inducing larger-scale production across industries such as semiconductors, solar energy, and flexible electronics.

5.3.4 Molecular Beam Epitaxy

MBE is an ultra-high vacuum (UHV) technique, which is used to deposit thin films of crystalline materials. In MBE, source material which has to be of high purity (metals or semiconductors) is heated in separate effusion cells, and a beam of atoms or molecules is directed towards the substrate kept at a high temperature. These beams are focused on the heated substrate, where the atoms condense and then grow epitaxial layers, which implies that the grown layers will have a crystal structure similar to the substrate.

Key components of an MBE system [17] are as follows:

- **UHV Chamber:** Crucial to eliminate contamination and promote the formation of high-quality and high-purity thin films rapidly. The vacuum also eliminates contact between the evaporated atoms and the residual gases at the surface of the substrate.
- **Effusion Cells:** These are the crucibles used to keep the source materials and can withstand very high temperatures. When the material is heated, there is the formation of a beam of atoms or molecules.
- **Substrate Heater:** The substrate is placed in a holder and raised to the desired temperature suitable for epitaxial growth.
- **Shutters:** These are used to control the deposition by opening and closing in front of the effusion cells to measure the rate of deposition.
- **Reflection High-Energy Electron Diffraction (RHEED):** employed to monitor the growth and surface structure of the film in real time.
- **Cooling and Ion Pumps:** Cooling and ion pumps are used to keep the UHV environment by removing the residual gases and keeping contamination to the lowest level possible.

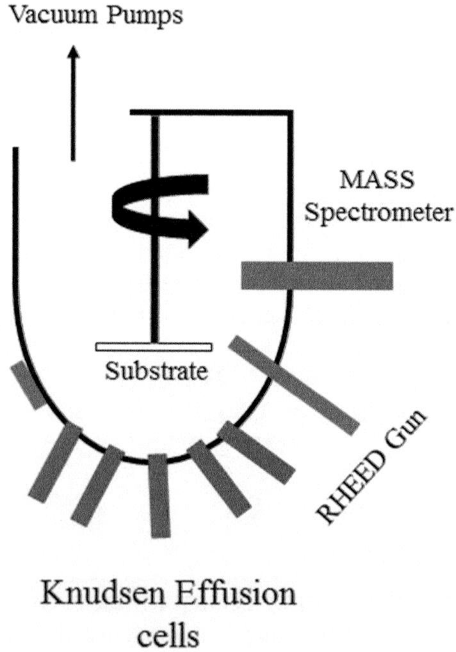

FIGURE 5.3 Schematic diagram of molecular beam epitaxy [17].

5.3.4.1 Benefits of MBE in Growing High-Quality and Defect-Less Films

MBE provides several advantages for depositing films with maximum purity and nominal density of defects. By using MBE, the deposition of the materials can be controlled at the atomic level. The great advantage of the effusion cells is that, by changing the fluxes, one can accurately determine the thickness and composition of the layers required. A further benefit of MBE is that deposition rates are generally less than 1 μm per hour, enabling great control over the growth process, which is suitable for building complicated layering structures.

Besides, due to the low growth rates and the accuracy of the technique, MBE can be used for the fabrication of complex heterostructure, which is necessary for numerous electronic- and opto-electronic applications.

The UHV conditions reduce the chances of contamination, hence obtaining very high-purity film. In-situ characterization methods such as RHEED enable the monitoring of the process of crystal growth, thus offering instant feedback on surface smoothness and layer quality. MBE is suitable for depositing various kinds of materials, for example, semiconductors, oxides, and metals; this means that MBE could be useful for a variety of research and technological fields.

Specific Applications

- **III–V Multi-Junction Solar Cells:** These cells consist of layers of III–V semiconductor materials piled in sequence; each layer is designed to absorb a segment of the solar spectrum. The MBE enables the growth of such multilayered structures without considerable defects and ensures very high efficiency.
- **Tandem Solar Cells:** In these cells, different semiconductor materials with different bandgaps are incorporated to capture more sunlight resultantly, achieving efficiencies over 40%. Sharp interfaces between different materials could be deposited by MBE and are critical in these cells.
- **Space Applications:** MBE growth multi-junction solar cells possess high efficiency and reliability, making them ideal for use in space applications. Due to the high purity and lack of defects inherent in the MBE process, these films are in high demand for space programmes that require stability and long-life product performance.
- **Concentrated Photovoltaics (CPV):** In the CPV systems, the light is focused on tiny, high-efficiency solar cells using III–V materials grown using MBE techniques. Such systems have to use thin films to prevent the degradation of performance on the high current density of light.

Thus MBE is a fabrication technique used to deposit highly efficient solar cells, which can be used in advanced techniques such as space power solar cells and concentrating solar cells.

5.3.5 Solution Processing Techniques

Solution-based deposition methods are employed in the deposition of thin films and coating of materials majorly in PV, electronics, and nanotechnology fields. These methods include the deposition of a material on a substrate and then subjecting the formed layers to drying or heat treatment processes. Solution-based methods are used because they are simpler, can be adapted for mass production, and are significantly cheaper than other deposition methods such as CVD or PVD. Spin coating and dip coating are two examples of solution-based deposition technique facilities most commonly utilized.

5.3.5.1 Spin Coating

Spin coating is a method in which a small quantity of a liquid solution is put in the middle of a substrate, and the substrate is rotated at high speeds (1,000 to 6,000 RPM). The centrifugal force disperses the solution over the surface, hence creating a thin film. The dependence of film thickness is mostly dependent on the viscosity of the solution used, the speed of spinning, and the time taken for spinning [18].

Advantages

- **Uniformity:** Spin coating proved to be effective in forming thin films of good uniformity over large areas of the substrate.
- **Control over thickness:** Other parameters like the spin speed of the total solution concentration can be varied to ensure that the exact thickness of the film is obtained.
- **Versatility:** Spin coating allows users to coat thin films with a large number of substances: polymers, semiconductors, and organic compounds.

Applications

- OSCs
- Photolithography processes used in the manufacturing of semiconductor devices
- Thin-film transistors and light-emitting diodes, commonly referred to as LEDs.

5.3.5.2 Dip Coating

Dip coating is one of the processes where the substrate is dipped into a solution, after which it is withdrawn at a specific rate so that the solution covers the surface. The thickness of the deposited film depends on the speed of its withdrawal, the viscosity of the solution, and the rate of solvent evaporation. The dip coating process usually includes multiple stages: they are immersion, dwell time, withdrawal, and drying or curing [19].

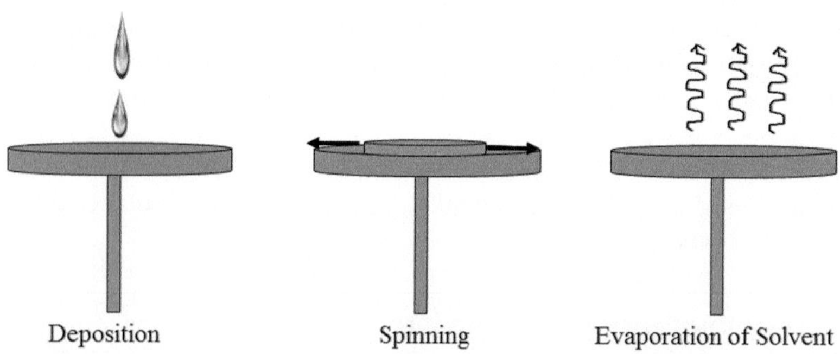

FIGURE 5.4 Different fabrication steps of spin-coating deposition [18].

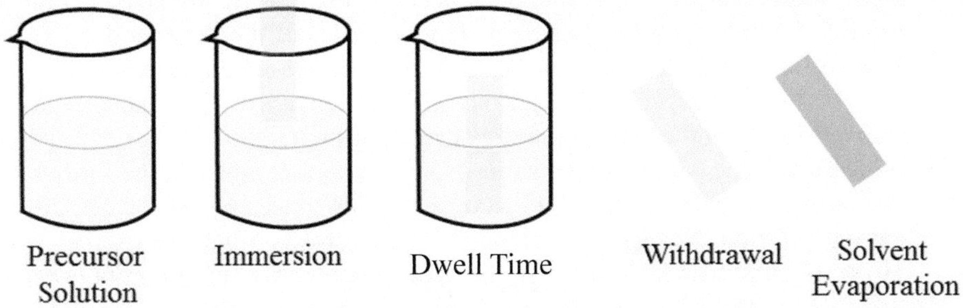

FIGURE 5.5 Schematic representation of the dip coating method [19].

- **Scalability:** Dip coating is ideal for mass manufacture and is versatile on the type of shapes the substrate can bear.
- **Simple Setup:** Compared to other methods of deposition, the equipment needed for dip coating is often not competitively developed.
- **Multilayer Capability:** The process can be repeated to deposit several layers, and this is useful more in multilayer structures.

Applications

- Dip coating has applications in anti-reflecting and protective layers on lenses, glass, and any optical uses.
- Dip coating is used in coating medical devices such as implants and stents since the coatings used are biocompatible and antimicrobial in nature.
- It is used to deposit a variety of materials including metal oxides or perovskites in the application of improving the efficiency of solar cells.
- Dip coating is particularly used in the protection of materials that are sensitive to corrosion through the use of protective polymer or metal oxide layers.
- The technique is used to deposit nanoparticles or form nanostructured films for a number of advanced uses and sizes.

5.3.5.2.1 Effects of Solvent, Thickness, and Annealing on the Performance of the Fabricated Devices

These three factors are closely related—the choice of solvent, thickness, and the annealing process and determine the overall quality and characteristics of solution-deposited films

- **Solvent Choice:** The choice of solvent in solution-based deposition method has an especially important role in the resulting deposited film quality. The solvent impacts solubility, evaporation rate, and surface tension that participate in determining the film structures. Proper solvent selection means that there are no defects on the surface of films, which is critical for device performance like solar cells or transistors.

- **Film Thickness:** The amount of deposited film has an impact on the optical, electrical, and mechanical characteristics of the deposited layer. Deposition parameters such as spin speed or solution concentration can control the thickness of the layer, and thus the light absorption and charge transportation and can improve the device parameters.
- **Annealing:** The process of annealing has an important effect on the further crystalline structure of the film by correcting any traces of the solvent left after deposition. This process enhances the microstructure characteristics of the material as related to crystallite size, phase purity, and grain boundaries. Since the film may degrade while undergoing annealing, the time and temperature for annealing should be varied depending on the material being used. Appropriate annealing enhances the device efficiency, stability, and overall performance for applications in thin-film transistors, PVs, and sensors.

5.3.5.2.2 Challenges in Solution Processing for Industrial Application

Adapting solution processing techniques such as spin coating and dip coating for mass production requires the following challenges addressed: quality, productivity, and affordability.

- Uniformity of films formed across large areas is less easily accomplished at an industrial scale than at a lab scale and results in non-uniform coatings, which in turn reduce the yield of the devices that are manufactured.
- With techniques like spin coating, it is common that a large amount of the solution is spun off during the process, and, as the scale of production increases, this waste becomes more costly and harmful to the environment.
- Bulk solution processing involves large quantities of solvents, which could be toxic or flammable. Concerns such as environmental control, safety of the workers, regulation of solvent emissions, and waste management make scaling to be a challenging process.
- Industrial production requires high throughput, but solution-based approaches may take time, especially during drying and annealing time. Another major problem is to raise the speed of deposition and at the same time to keep the quality of films at the same level.
- Solution processing methods have to be companionable with current industrial manufacturing techniques and layouts. Adapting new deposition technologies with R2R processing, automation, and quality control often poses significant engineering challenges, which translate into considerable cost and time implications.

Therefore, despite the potential of solution-based deposition methods to be used in industry, there are general issues related to uniformity, material loss, eco-friendliness, speed, and so on of technology that still need to be addressed on a large scale.

5.3.6 Inkjet Printing

Inkjet printing is one of the additive manufacturing processes where the desired material is deposited in a stipulated pattern on the substrate and is widely used in electronic, PV, and biological applications. This technology employs liquid inks, which include functional materials that are either dissolved or in suspension form in a solvent, and precisely sprays small droplets of the inks through nozzles on a surface.

Requisites:

- **Printhead:** The printhead has several tiny nozzles, which spray droplets of ink onto the substrate material. Most inkjet printers employ either the thermal or piezoelectric printing mechanism. Operation of thermal printheads includes applying heat to vapourize and form a bubble that in turn expels the liquid through the nozzle. In piezoelectric printheads, the piezoelectric element changes its shape in response to the voltage applied, giving a mechanical displacement to the ink, and it comes out of the nozzle.
- **Inks:** Ink is a functional material, such as polymers, nanoparticles, or biomolecules, in a liquid solution. The ink properties including viscosity, surface tension, and particle size of the aerosolizing fluid are required to be gradually controlled to allow the formation of the droplets and their deposition on the substrate.
- **Substrate:** The substrate is also flexible, and, based on the application, it may be glass, plastic, or other flexible material. Liquids need to be wet and spread properly on the substrate to achieve the desired properties.

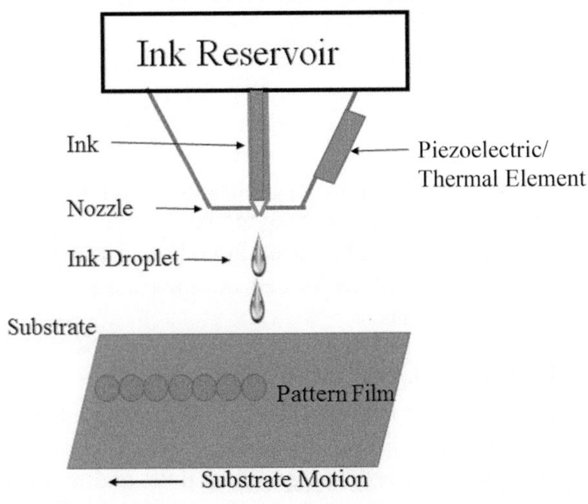

FIGURE 5.6 Schematic diagram of inkjet printing [20].

- **Droplet Formation:** It is necessary to accurately control droplet size and its ejection frequency. The volume of the droplets normally varies between picolitres and nanolitres, enabling high-resolution pattern development.

Advantages

- Inkjet printing deposits the material where it is required and not just throughout the range of the thick film breadth, thus minimizing the waste of source material.
- It can generate very thin patterns with high accuracy.
- Compatible with an extensive range of functional inks, including conductive ink, semiconductor ink, and biological ink.
- The inkjet printing technology can be particularly applied for prototyping as well as for mass production, where quick turnaround times are needed.

Applications

Inkjet printing has evolved as a potential fabrication strategy in the fabrication of organic and PSCs because it is technologically accurate and cost effective, especially for the deposition of thin films and architecture of devices. In particular, its precision, material saving, and ability to work with various forms of flexible bases make it ideal for future generations of PV devices.

- **OSCs:** OSCs require carbon-based polymers or small molecules as absorbing materials. The inkjet method is widely used for the fabrication of active layer deposition or electrodes, which results in transparent and flexible devices. Intermediate layers that enhance the collection of charges can also be deposited with high accuracy to increase the overall performance of the device.
- **PSCs:** PSCs that have shown high efficiency can also be deposited from inkjet printing. The conventional one-step deposition for the fabrication of the perovskite layer can be replaced by inkjet printing to achieve high crystallinity and low pinhole density. Piezoelectric inkjet printing for the fabrication of hole/electron transport layers has been employed to improve the charge collection capability and reduce recombination losses. Furthermore, inkjet printing offers fabrication over flexible substrates, and the lightweight, portable concepts of OPVs can be integrated into wearable electronics or building materials.

5.3.6.1 Future Trends in Inkjet Printing Industry

Predictive advancements in the field of inkjet printing technology for flexible and wearable PV applications are shifting towards the synthesis of novel inks including organic semiconductors, perovskites, quantum dots, and more. These materials enhance the efficiency, conductivity, and absorption of light. Researchers are also developing environment and health-friendly solvents.

Improvement of the printhead and its ability to control droplets refine the process of inkjet printing and enable the generation of more fine structures. This shall improve the efficiency of miniaturized PV systems.

The flexible PVs will shift attention towards extremely thin lightweight, stretchable materials that would conform to clothing or even skin interface for applications in everyday wearable devices such as smartwatches and fitness trackers. These will spur new innovations in self-charging wearable devices and portable energy-related technologies for the construction of energy harvesting systems.

5.3.7 Other Emerging Fabrication Techniques

ALD can be defined as a UHV-based thin-film formation method that enables the deposition of layers at the atomic scale. It uses processes of controlled and self-limiting chemisorption to add one atomic layer at a time. Thus, high-density conformal coatings are achieved; such is applicable to rather complex or even porous surfaces. Microelectronics, energy devices, and nanotechnology markets are key application segments for ALD primarily due to the outstanding control of film thickness and the ability to deposit multiple precise layers that are needed, for example, in the manufacturing of semiconductors, high-energy-density batteries, or barriers for flexible electronics.

R2R processing is a well-established technique for the continuous production of flexible materials like film, foil, and printed electronics for large volumes. It includes the formation of functional layers on a flexible and continuously moving substrate, making it possible to produce nanocomposite materials on a large scale and at a relatively low cost. R2R is suitable for those uses in which scale-up is imperative including solar cells, printed transistors, and detectors. It can perform several steps like coating, printing, and annealing in a single operation, thus cutting down continuous production time and even costs. Each of them, ALD and R2R processing, are known to promote the development of advanced materials and devices by enhancing accuracy, reproducibility, and effectiveness at a reduced cost.

3D Printing in Fabrication of Solar Cells: By providing opportunities for complex shapes and multi-material interfaces, 3D printing has a vast potential to facilitate solar cell fabrication. It makes the deposition of functional materials including conductive polymers, perovskites, and metals for the fabrication of light and thin PV films possible. It can also result in the designing and manufacturing of a product with considerable time saving, less material wastage and better efficiency to support the idea stated earlier. Besides, the method of 3D printing is suitable for creating prototypes and thus ideal for implementing new solar technologies. Eventually, the technology may be used for high-throughput, low-cost production of highly efficient and flexible third-generation PVs over a wide range of applications.

5.3.7.1 Hybrid Fabrication Approaches and Their Applications

Nowadays to improve the manufacturing of complex devices hybrid fabrication approach is used, which is nothing but a combination of more than one manufacturing technique such as inkjet, 3D printing, and ALD. Hybrid techniques are used to fine-control the deposition, the performance, the scalability, as well as the costs in PVs. They are applied in the formation of the multilayer structures of organic, perovskite, and flexible PV devices. Furthermore, the hybrid technique is used in other applications such as sensors and energy storage devices.

5.4 CHARACTERIZATION TECHNIQUES FOR SOLAR CELLS

The rationale for characterization is particularly critical in solar cell research because it can offer detailed comprehension of material, device, and failure characteristics. Spectra, micrographs, and I–V curves are used to characterize the chemical composition, physical structure, and nonlinear conductivity of solar cell materials. These evaluations are very important for matching important parameters, which include layer thickness, degree of crystallinity, and interfaces between the layers, as well as impact on the efficiency and durability of solar cells.

By means of characterization, structural issues and performance degradation can be discovered and addressed to modify design and material properties for better performance. For example, knowledge of optical behaviour provides insights into improved light absorption, and, on the other hand, electrical characterization provides insights into how the charge can be transported efficiently.

Furthermore, the exact specifications collected from characterization provide research professionals with the opportunity to fine-tune designs along with choosing the right materials ideal for specific applications. Through this repeated cycle, more efficient and reliable technologies can be discovered for large-scale solar applications, thus enhancing the contribution to the advancement of the technology.

The identification of suitable characterization techniques for the research of solar cells depends mostly on the material and the researched characteristics. Depending on the material of interest such as silicon, perovskites, or organic semiconductors, a different characterization approach is required. For example, crystalline materials like silicon and non-destructive characterization tools such as XRD are appropriate to determine the crystal structure and phase purity. On the other hand, organic or perovskite materials use UV–Vis spectroscopy and PL for optical properties analysis, morphology, and thin-film uniformity from SEM. Another parameter for characterizing the right characterization technique is the property of interest. As indicated earlier the desired property, whether electrical, optical, or structural, determines the mode of characterization.

Electrical properties such as conductivity and carrier mobility require Hall effect measurements or current–voltage measurements (I–V). For optical absorption, the UV–Vis technique or even PL spectroscopy can be used; in case of surface roughness and morphology, atomic force microscopy (AFM) or SEM can be employed. The choice is also dependent on the required resolution and sensitivity. For instance, HRTEM imaging may be required to study a defect or an atomic interface level in a material. Therefore, the proper choice of characterization methods is significant in obtaining desirable results.

5.4.1 Scanning Electron Microscopy

SEM stands for scanning electron microscope, which operates by focusing a high-energy electron beam on the sample being analysed. As the electrons pass through the sample, they generate signals such as secondary electrons and backscattered electrons. These signals convey information about the surface morphology and elemental composition. These signals are detected, and the information is then used to develop an image with a much higher resolution. That gives the possibility to study the surface topography of the materials down to a nanometre scale and can also be used to determine the elemental composition of the samples if the SEM is equipped with the energy dispersive X-ray spectroscopy (EDS) detector. The SEM can provide near 3D-like images at the micrometre to nanometre resolution levels and is therefore widely used for microstructure analysis of metals, thin films, coatings, and composites.

In material science, the applications of SEM include examinations of the crack sites, orientation of crystals in polycrystalline materials, and phases in composite and heterogeneous microstructures. Furthermore, there are several specialized imaging techniques including electron backscatter diffraction (EBSD) and in-situ SEM, which give precise information about the crystallographic orientation and behaviour of the material in realistic operating conditions. EBSD is a more advanced form of SEM, which is utilized to collect crystallographic data, which is central to solar cell efficiency improvements. It provides information such as grain orientations, phases, and the strain within the material. This information is vital in the characterization of polycrystalline solar cells such as silicon, perovskites, and thin films. EBSD enables control of the grain boundaries, minimization of the defects, and enhancement of the efficiency and stability of the solar cells.

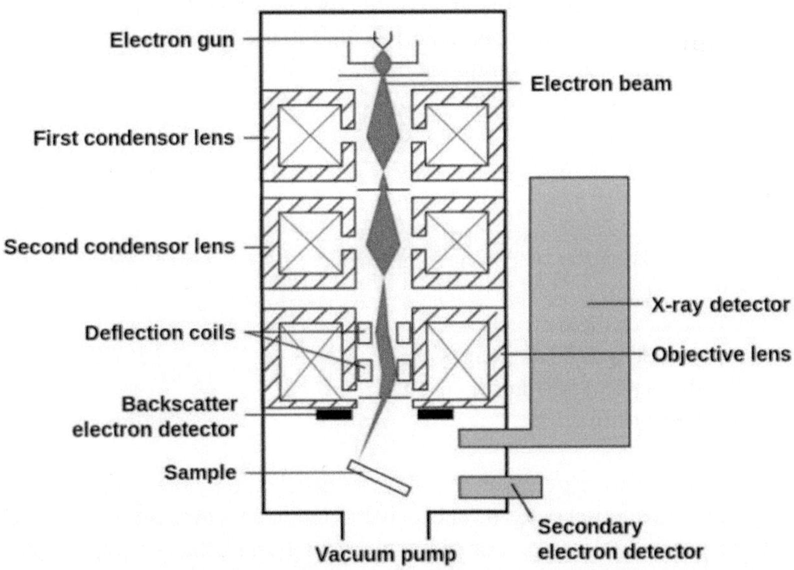

FIGURE 5.7 Schematic diagram of scanning electron microscope [21].

In-situ SEM can be used to investigate solar cell materials when subjected to mechanical stress, temperature, or electrical bias. Specialized application equipment installed in an SEM chamber allows researchers to investigate processes such as crack initiation, phase transformation, or degradation in solar cells in real time. Due to its capability of investigating the behaviour of solar cell materials in real operation conditions, the in-situ SEM is crucial for enhancing the stability, efficiency, and durability of materials utilized in perovskite and thin-film PV systems. This technique enables researchers to build more durable and effective solar cells for practical applications.

5.4.1.1 Challenges in SEM Analysis and Sample Preparation

There are numerous problems with SEM analysis, with the main issues identified in sample preparation. Samples must be conductive, which can be hard for non-metallic or organic samples and often have to be coated in a conductive layer such as gold or carbon, which may change the surface topography. Moreover, sample preparation should not cause sample contamination or sample damage because this degrades image quality. In addition, achieving appropriate vacuum pressures without compromising the shape and form of delicate materials is also very difficult. Due to the surface charging effect in the case of nonconductive samples, SEM shows low imaging resolution and therefore artefacts. Also, the preparation of cross-sectional samples or thin samples for high-resolution images requires expertise.

5.4.2 X-Ray Diffraction

Characterization of crystalline and polycrystalline materials and their phase identification are done using XRD. It is an analysis tool and a widely used instrument to derive structural information about materials, and it also gives the details of the unit cell of the materials. XRD technique is based on Bragg's law. It describes the connection between the angle of incidence of X-rays to the plane of the crystal lattice of atoms and the wavelength of the incident X-ray beam and states that an integer multiple of the wavelength of the incident X-ray beam is equal to the difference in path lengths of two diffracted X-ray beams.

$$\text{Bragg's law can be expressed as: } n\lambda = 2d \sin\theta \tag{5.1}$$

where n = order of the diffraction,
λ = wavelength of incident X-ray,
d = inter-atomic distance,
θ = angle of incident.

When X-rays are focused on a sample, they are diffracted by the atoms within a lattice and yield a diffraction pattern. These diffracted beams give information relating to the atomic positions, interplanar spacing, orientation of the crystal, and intensity of the beam. The diffraction pattern evoked by the interaction of the X-ray beam, and the crystal lattice allows for knowing the atomic structure, unit cell dimension,

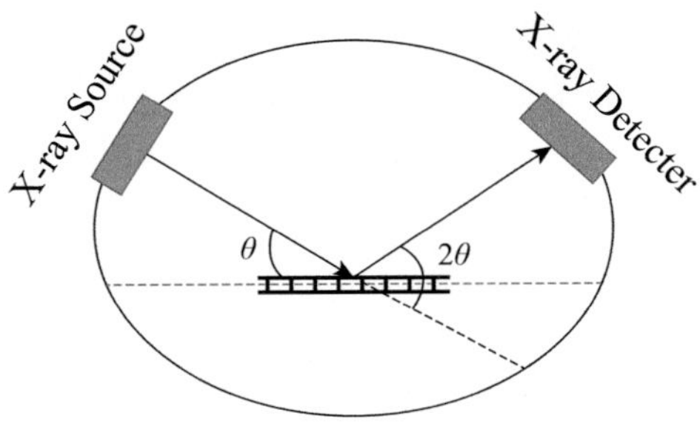

FIGURE 5.8 Schematic diagram of XRD [22].

symmetry, and purity, as well as the presence and types of defects. XRD can also determine phases present in polycrystalline materials, making it very useful in phase transformation, alloy composition, and purity analysis [22]. The quality of semiconductors such as perovskites, chalcogenide semiconductors, and silicon is evaluated using XRD, with the aim of optimizing performance through improved crystallinity and phase stability. For thin-film samples, the appropriate technique is the grazing incidence X-ray in which the angle between the incident beam and the target surface typically ranges between 0.5° and 1°. Such configuration was established to reduce the impact of the substrate and amplify the signal from the film. By using this technique, researchers are able to determine the crystallographic characteristics of the films, their strain, as well as their thickness, making them suitable for solar cells, coatings, and nanomaterials.

However, in-situ XRD permits observation of the structural evolution of materials during processes such as heating, cooling, or applying stress. It enables the monitoring of phase transformations, crystal formation, and behaviour upon use, which are essential for enhancing the reliability and efficiency of materials used in solar panels and energy storage systems, respectively. This technique is useful in case of determining material behaviour during fabrication or when it is subjected to some loads.

5.4.2.1 Challenges in Interpreting XRD Data

The analysis of computerized XRD data can be complex because diffraction patterns are themselves complex following the nature of the sample, which may be multi-phase or nanostructured. Due to the phase overlapping, peaks from different phases, or from different phases of the same material with similar crystal structures, can often be hardly resolved. Moreover, the problems arising from small crystallite size, strain, and defects may lead to peak broadening and hence pose a challenge. Thus, accurate calibration and complex mathematical algorithms are used in fitting the curves. Some general problems concern sample preparation; for example, poor crystallinity and uneven sample surfaces may result in weak or even distorted signals that diminish the reliability of data interpretation to a certain extent.

5.4.3 Raman Spectroscopy

Raman spectroscopy is also a non-destructive characterization technique employed to extract the data concerning molecular vibrations of the samples and used as a signature for defining samples and for the structural analysis of samples. The sample can be in liquid form, solid form, or gaseous form. Raman spectroscopy depends upon Raman scattering of a monochromatic light. However, inelastic scattering or Raman scattering is defined as the change in the photon's frequency owing to interaction with the sample. The changing frequency of incident photons may be positive or negative known as the Raman Effect. It involves the use of laser light interaction with the sample and identification of in-elastically scattered light [23].

Raman spectra of different materials assist in ascertaining the chemical composition of the film like perovskites, CdTe, or CIGS. The technique can differentiate between different phases present in the film and can provide the information regarding quality and homogeneity of the film. Also, Raman spectroscopy facilitates the analysis of crystallinity depending on the peak position and width and can provide data concerning the strain, defects, and grain size.

Surface-enhanced Raman spectroscopy (SERS) and Raman mapping significantly improve the performance of conventional Raman spectroscopy in thin-film solar cell studies. SERS is a characterization tool where the Raman scattering from molecules adsorbed on the surface of metallic nanostructures is enhanced. SERS improves the Raman signal and offers better sensitivity for identifying surface properties and the chemical reactivity of material at the nanoscale.

However, Raman mapping requires focusing the Raman laser beam on the sample to generate maps of elemental composition and crystal structure across the sample surface. This technique produces plane as well as 3D maps representing variations in elemental composition across the film. Raman mapping is practically irreplaceable

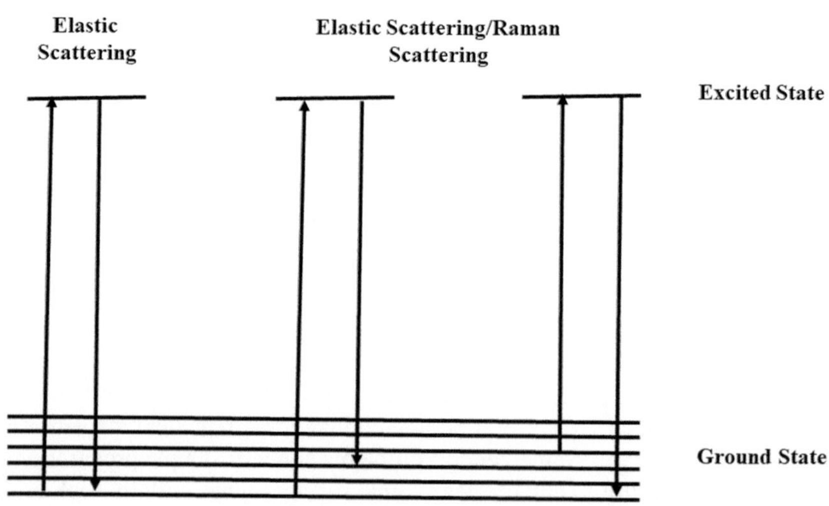

FIGURE 5.9 Elastic and inelastic scattering of light [23].

when it comes to evaluating uniformity, defects present, and spatial distribution of the phases between the layers of thin-film solar cells.

These new advanced Raman techniques thus provide better analytical tools that enable optimization of the solar cell material and its performance, thereby enhancing the efficiency and stability of solar cells.

5.4.3.1 Challenges in Interpreting Raman Spectra

In the present study, there are several issues with interpreting Raman spectra due to the complexity of the data acquired. When a structure has different phases or different types of materials, the signals may coincide since the components vibrating may be the same. Furthermore, the fluorescence of some of the materials that are being studied causes high interference and therefore low sensitivity and obscured information. Sample preparation methods add variability, which could eradicate the spectra reproducibility. Moreover, the identification of spectra sometimes involves knowledge about molecular vibrations and may sometimes involve complex computational analysis for spectrum identification. Such factors require precise tuning and analysis to gain valuable information from Raman data in thin-film solar cells.

5.4.4 PL Spectroscopy

PL is defined as an optical technique that can examine the type of electronic and optical behaviour existing in materials without destroying them. When a material is excited by light (e.g., laser), the photons are absorbed by the valance band electrons and jump into the conduction band. When these enthusiastic electrons "relax" back to this regular state, light is emitted (PL), and this is what is measured. The energy, intensity, and spectral distribution of this emitted light carry significant insights into the band structure, defects, and recombination processes of the material.

The variation in intensity and wavelength of emitted light allows PL to identify non-radiative recombination centres and impurities in solar cell materials. Solar cells suffer from defects or discontinuities as they can behave as trapping agents to the charge carriers. Therefore, PL assists in defect identification. The intensity of emitted photons in PL is also proportional to the bandgap of the material, which is critical for

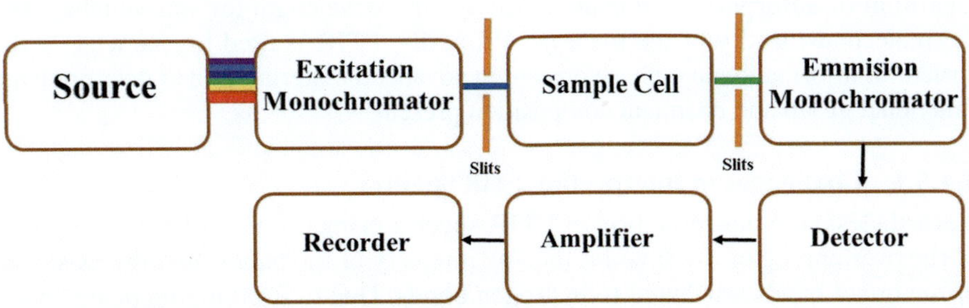

FIGURE 5.10 Schematic diagram of photoluminescence [24].

the optimization of the efficiency of the absorption of solar cells. Besides, it aids in determining carrier lifetime and recombination processes, with the help of how long photogenerated carriers (electrons and holes) take for recombination and emission of light. A longer carrier lifetime suggests a better prospective for PV conversion efficiency due to a longer time for charge collection. In conclusion, PL is a valuable characterization method for enhancing the performance and reliability of solar cell materials through the understanding of their electronic and optical nature.

5.4.5 Fourier Transform Infrared (FTIR) Spectroscopy

FTIR spectroscopy is a well-established non-destructive technique of analytical chemistry used for the identification and analysis of chemical compounds and materials. It works by determining the absorption of IR light of a particular wavelength and provides information on the structure, molecular arrangement, and also interaction of the material. FTIR is suitable for solid, liquid, and gaseous states, and hence it is perfect for all areas like material science, engineering, and chemistry.

FTIR spectroscopy works on the molecular absorption of a particular frequency of IR radiation corresponding to molecular vibrations. These frequencies are essentially equal to the frequency of vibrations of chemical bonds within the material. Similar to fingerprints, no two distinct chemical bonds/molecular combinations generate the same infrared spectrum. FTIR spectroscopy works on the principle of Beer–Lambert law, which gives the relationship between incident light intensity (I_0), transmitted light intensity (I), and absorbance (A), which again depend on the concentration of the absorbing species (c) and the path length of absorbing medium (l).

Beer–Lambert Law can be expressed as $A = \log_{10}(I_0/I) = \varepsilon.\ c.\ l$, where ε is molar absorptivity.

In FTIR spectroscopy, first, the IR light interacts with the sample and certain frequencies are absorbed while others are transmitted. The information from every wavelength is then concurrently present in the resultant signal. This data is then taken through a Fourier Transform algorithm to get the spectrum. This spectrum shows the logarithm of absorption or transmittance against wavelength (or wavenumber) and the molecular identity of the material. Primarily, FTIR is used to characterize the structural information and phase present in advanced materials and to determine the functional group and chemical composition present.

5.4.5.1 Challenges in Interpreting FTIR Spectra

Nevertheless, the interpretation of FTIR spectra brings some difficulties because of the overlapping of the spectra, the low intensity of the bands, and the existence of the broad bands attributed to hydrogen bonds. Due to overlapping peaks in the region (400–1500 cm^{-1}), it is difficult to identify individual functional groups, especially in complex compounds. Signals can be obscured by water vapour, CO_2,

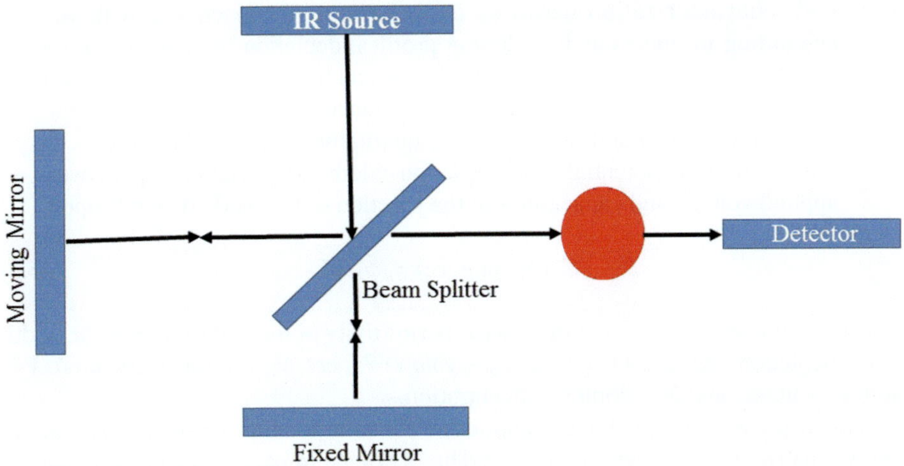

FIGURE 5.11 Schematic diagram of FTIR spectroscopy [25].

and solvents. A number of issues such as poor sample preparation, baseline drift, and even instrumental noise make the analysis challenging. They may be structural isomers, or similar compounds, which may result in near-identical spectra, which only add confusion. They elaborate additional challenges, pointing to the lack of reference data for novel compounds as one of those. To counter these problems, the right sample pre-treatment and optimization and the use of additional characterization tools, for instance, NMR or Raman spectroscopy are often required.

5.4.6 I–V/C–V Characterization

I–V and C–V characterization are common electrical methods used for the measurement of the performance and properties of a device such as semiconductor devices, PV cells, diodes, and capacitors.

- **I–V characterization** quantifies current response with respect to voltage (using the idea diode equation), offering information on charge carriers, electrical conductivity, and device performance.

$$I = I_0(e^{qV/\eta kBT} - 1) \tag{5.2}$$

where I = current, I_0 = reverse saturation current, q = electron charge, V = applied voltage, η = ideality factor, k_B = Boltzmann constant, and T = temperature.

The voltage applied across the device under test current is measured, and current–voltage characteristic is plotted and analysed to extract series resistance, shunt resistance, diode reverse saturation current, ideality factor, rectification characteristics, and so on.

- **C–V characterization** describes the behaviour of capacitance with voltage aiding in understanding doping profiles, depletion regions, as well as an interface by inspecting semiconductor junctions. A little AC voltage is added to a large DC voltage, and the capacitance measured is used to deduce the charge distribution plot to get the qualitative detail of doping concentration, built-in potential, and depletion width. The relationship between applied voltage and capacitance at the junction of the diode depends upon

$$C\ (\text{capacitance}) = \varepsilon A/W \tag{5.3}$$

where A is the area of the junction, ε is the permittivity of the material, W is the width of the depletion region, and $W = (2\varepsilon(V_{bi} - V)/qN)^{1/2}$; here, V_{bi} is built-in potential, V is applied voltage, and N is doping concentration.

This expression is specifically applicable for p–n junction solar cells having an abrupt change in the depletion region and has a crucial impact on the performance of solar cells, for example, crystalline Si solar cells and thin-film solar cells but does not accurately define capacitance, for example, PSCs where the charge transport occurs due to ionic movement rather than traditional depletion region behaviour.

5.4.7 Electrochemical Impedance Spectroscopy

Electrochemical impedance spectroscopy (EIS) is an essential non-destructive technique employed in determining the electrical characteristics of materials, interfaces, and electrochemical systems. It is used to determine the measure of impedance in a circuit at different frequencies of AC. In EIS small AC voltage (or current) is applied to the electrochemical system across a frequency range, and the corresponding AC current (or voltage) is recorded. Therefore the impedance is derived from the phase shift as well as the amplitude of the response. The data is in the form of a Nyquist or Bode plot and gives details of such processes including charge transfer resistance, double-layer capacitance, diffusion, and reaction kinetics in the system. Especially in PV cells, EIS evaluates the interface properties and optimizes materials to enhance the efficiency of PV cells. EIS is also used to calculate electron lifetime analysis, which helps in reducing recombination losses.

In solar cell research, on the other hand, a variety of optical characterization techniques is used side by side to ensure a detailed description of the properties of materials, including bandgap, carrier dynamics, and defects. UV–Vis spectroscopy technique deals with the absorption and transmittance of the sample in the UV–visible region to find the optical gap and absorption coefficient of the thin film. It is useful to determine how effectively materials absorb sunlight, which is also an important parameter from the efficiency of solar cells' point of view.

Ellipsometry involves a change in polarization of the light reflected on a material's surface. It gives strong information about thin-film thickness, refractive index, and optical constants, which is also very useful for controlling the thickness of layers in multilayer solar cell structures. These techniques when used simultaneously provide a 360-degree solution to boosting the performance of the solar cells.

5.4.8 QE Measurements

QE is a characterization technique applied to photodetectors and PV structures, denoting the efficiency of conversion of absorbed photons to charge carriers. QE is denoted by the ratio of the number of charges produced by a device to the number of photons being incident on the device. It is given in per cent and depends on the wavelength of the light. For solar cells, it is very important to control QE as this characteristic defines how efficient is the device.

There are two types of QE, namely, external quantum efficiency (EQE) and internal quantum efficiency (IQE). EQE assumes all the incident photons on the surface, regardless of whether they are absorbed or not, while IQE considers only those photons that get absorbed by the absorber layer. Thus, the significance of QE is to provide the ability to evaluate the efficiency of the light-sensitive material in conditions close to the operating environment. For instance, one might get a high QE in solar cells, meaning that it will be able to absorb more light creating greater energy conversion efficiency. Hence, QE is employed as an important characterization in the development and advancement of optoelectronic devices applied in energy, communication, and imaging sectors.

A monochromatic light source is used to measure the EQE of the device when the device is illuminated at different wavelengths. The photocurrent produced is measured, and EQE is determined by dividing charge carrier production by the photon flux at individual wavelengths. For IQE, EQE data is fitted with reflectance and transmittance values. As contrasted with EQE, IQE does not undertake allowance for reflection and transmission and is thus capable of estimating the actual efficacy of a device to convert the demanded sort of light into electric charges.

QE is employed extensively to characterize light absorption and carrier collection efficiency in PV devices and photodetectors. EQE enables the investigator to determine the effectiveness that a device has on light absorption within a certain range of wavelengths. Thus, a higher EQE means improved collection and conversion of photons into charge-carrying entities. IQE plays a vital role in the internalization of collection efficiency by deciding the extent of how the generated carriers are collected at the electrodes. This understanding is of paramount importance for designing the right device, reducing recombination losses, and enhancing the efficiency of solar cells, LEDs and photodetectors.

5.4.8.1 Challenges in QE Measurements and the Impact of Spectral Response

Factors which could complicate QE measurements include fluctuations in light sources, how best to ensure uniform light across the surface being measured, and how to deal with reflections or scattering losses. There are other losses in the device that make increasing the evaluation of precise QE difficult, recombination, and also poor carrier collection.

A variation in the device's spectral response with wavelength affects the QE measurements. Device response may vary at different wavelengths due to the properties of the materials or the thickness of layers. Designing and optimizing a device for a particular spectral range is possible after gathering knowledge of the spectral response in the application areas, such as solar cells.

5.5 CONCLUSION

The current chapter has given an overview of fabrication and characterization methods that are relevant to improving the designs of solar cell structures. The structures narrowed the focus on fabrication processes that include CVD, solution processing, and inkjet printing, hence highlighting the importance of these processes in evolving better PV technology. Such methods allow precise control over the properties of the materials used, which is crucial when it comes to achieving the best result. Picking the right fabrication technique that fits the material type, fabrication cost, and upscaling are the most important strategies that can enhance device performance.

Instruments like SEM, AFM, XRD, and QE measurements are also vital in studying the characteristics of any material and in determining the performance characteristics of a device and the potential areas where improvement could be made. These techniques assist the investigators to optimally control the structure and performance of solar cells, including their efficiency and reliability.

Thus, the combination of innovative material synthesis and characterization is very essential for the development of advanced solar technologies. Future advancements in science and technology will contribute to increased efficiency and cost reduction of the PV systems, thus leading to adequate renewable energy solutions, which will define a sustainable world.

5.6 FUTURE TRENDS OF SOLAR CELLS

Solar cell development for the future aims for improved performance combined with a larger manufacturing scale and reduced production expenses. Solar cells based on tandem and multi-junction technology have reached 32.5% efficiency levels, thus making them ready for commercial deployment. Quantum dots along with hot-carrier cells hold the potential to achieve over 50% efficiency because they maintain better performance with high-energy photons. Artificial intelligence helps discover new materials, which improve performance levels, while minimizing costs. Relevant applications of printed solar cells include implementations into construction materials and mobile objects such as vehicles in addition to wearable devices. The efficiency increases from various light-trapping methods, anti-reflection treatments, and passivation procedures. The combination of these innovative technologies will enhance solar power efficiency while lowering its operational costs, making it more accessible, which will speed up the adoption of sustainable energy systems.

REFERENCES

[1] H. C. Weerasinghe, N. Macadam, J.-E. Kim, L. J. Sutherland, D. Angmo, L. W. T. Ng, A. D. Scully, F. Glenn, R. Chantler, N. L. Chang, M. Dehghanimadvar, L. Shi, A. W. Y. Ho-Baillie, R. Egan, A. S. R. Chesman, M. Gao, J. J. Jasieniak, T. Hasan, and D. Vak, "The First Demonstration of Entirely Roll-to-Roll Fabricated Perovskite Solar Cell Modules Under Ambient Room Conditions," in Nature Communications, vol. 15, p. 1656, 2024, https://doi.org/10.1038/s41467-024-46016-1.

[2] S. K. Karunakaran, G. M. Arumugam, W. Yang, S. Ge, S. N. Khan, X. Lin, and G. Yang, "Recent Progress in Inkjet-Printed Solar Cells," in Journal of Materials Chemistry A, vol. 7, pp. 13873–13902, 2019, https://doi.org/10.1039/C9TA03155C.

[3] Jenny Nelson. The Physics of Solar Cells. 1st edition. Imperial College Press London UK, 2003.

[4] R. Boylestad and L. Nashelsky. Electronic Devices and Circuit Theory. 7th edition. Prentice Hall, 1998.

[5] T. L. Chu, "Thin Film Cadmium Telluride Solar Cells by Two Chemical Vapor Deposition Techniques," in Solar Cells, vol. 23, issues 1–2, pp. 31–48, 1988, https://doi.org/10.1016/0379-6787(88)90005-1.

[6] T. P. Dhakal, C.-Y. Peng, R. R. Tobias, R. Dasharathy, and C. R. Westgate, "Characterization of a CZTS Thin Film Solar Cell Grown by Sputtering Method," in Solar Energy, vol. 100, pp. 23–30, 2014, https://doi.org/10.1016/j.solener.2013.11.035.

[7] S. Chaudhari, S. Palli, K. P. K., and S. R. Dey, "Pulsed Electrodeposition of Cu_2ZnSnS_4 Absorber Layer Precursor for Photovoltaic Application," vol. 600, pp. 169–174, 2016. https://doi.org/10.1016/j.tsf.2016.01.021

[8] L. Tzounis, T. Stergiopoulos, A. Zachariadis, C. Gravalidis, A. Laskarakis, and S. Logothetidis, "Perovskite Solar Cells from Small Scale Spin Coating Process Towards Roll-to-Roll Printing: Optical and Morphological Studies," in Materials Today: Proceedings, vol. 4, issue 4, part B, pp. 5082–5089, 2017, https://doi.org/10.1016/j.matpr.2017.04.117.

[9] A. S. Marques, R. M. Faria, J. N. Freitas, and A. F. Nogueira, "Low-Temperature Blade-Coated Perovskite Solar Cells," in Industrial & Engineering Chemistry Research, vol. 60, no. 19, pp. 7145–7154, 2021, https://doi.org/10.1021/acs.iecr.1c00789.

[10] S. Ge, H. Xu, T. Wang, E. Gu, X. Lin, and G. Yang, "9.2% Efficient $Cu_2ZnSn(S,Se)_4$ Thin Film Solar Cell Based on Inkjet-Printed Absorber," in ACS Materials Letters, vol. 5, no. 6, pp. 1767–1771, 2023, https://doi.org/10.1021/acsmaterialslett.3c00300.

[11] P. A. Mithari, A. C. Mendhe, B. R. Sankapal, and S. R. Patrikar, "Process Optimization of Dip-Coated MWCNTs Thin-Films: Counter Electrode in Dye-Sensitized Solar Cells," Journal of the Indian Chemical Society, vol. 98, no. 11, p. 100195, 2021, https://doi.org/10.1016/j.jics.2021.100195.

[12] M. Levillayer, A. Arnoult, I. Massiot, S. Duzellier, T. Nuns, and C. Inguimbert, "As-Grown InGaAsN Subcells for Multijunction Solar Cells by Molecular Beam Epitaxy," in IEEE Journal of Photovoltaics, vol. 11, no. 5, pp. 1271–1277, Sept. 2021, https://doi.org/10.1109/JPHOTOV.2021.3093048.

[13] M. A. Hossain, K. T. Khoo, X. Cui, G. K. Poduval, T. Zhang, X. Li, W. M. Li, and B. Hoex, "Atomic Layer Deposition Enabling Higher Efficiency Solar Cells: A Review," in Nano Materials Science, vol. 2, no. 3, pp. 204–226, Sept. 2020, https://doi.org/10.1016/j.nanoms.2019.10.001.

[14] T. M. Eggenhuisen, Y. Galagan, A. F. K. V. Biezemans, T. M. W. L. Slaats, W. P. Voorthuijzen, S. Kommeren, S. Shanmugam, J. P. Teunissen, A. Hadipour, W. J. H. Verhees, S. C. Veenstra, M. J. J. Coenen, J. Gilot, R. Andriessen, and W. A. Groen, "High Efficiency, Fully Inkjet Printed Organic Solar Cells with Freedom of Design," in Journal of Materials Chemistry A, vol. 3, pp. 7255–7262, 2015, https://doi.org/10.1039/C5TA00540J.

[15] P. Singhal and H. N. Ghosh, "Hot Charge Carrier Extraction from Semiconductor Quantum Dots," in The Journal of Physical Chemistry C, vol. 122, no. 31, pp. 17586–17600, 2018, https://doi.org/10.1021/acs.jpcc.8b03980.

[16] O. Zaytseva and G. Neumann, "Carbon Nanomaterials: Production, Impact on Plant Development, Agricultural and Environmental Applications," in Chemical and Biological Technologies in Agriculture, vol. 3, p. 17, 2016, https://doi.org/10.1186/s40538-016-0070-8.

[17] D. Suthar, S. Chuhadiya, R. Sharma, Himanshu, and M. S. Dhaka, "An Overview on the Role of ZnTe As An Efficient Interface in CdTe Thin Film Solar Cells: A Review," in Materials Advances, vol. 3, pp. 8081–8107, 2022, https://doi.org/10.1039/d2ma00817c.

[18] J. Shojaeiarani, D. S. Bajwa, N. M. Stark, T. M. Bergholz, and A. L. Kraft, "Spin Coating Method Improved the Performance Characteristics of Films Obtained from Poly(lactic Acid) and Cellulose Nanocrystals," in Sustainable Materials and Technologies, vol. 26, p. e00212, 2020, https://doi.org/10.1016/j.susmat.2020.e00212.
[19] S. Chaudhari, P. K. Kannan, K. Madhuri, and S. R. Dey, "Formulation of Selenium-Rich $Cu_2ZnSn(S_xSe_{1-x})_4$ Film through Non-Vacuum Dip Coating Technique," in Journal Name, vol. 32, pp. 19102–19109, 2021, https://doi.org/10.1007/s10854-021-06427-y.
[20] M. J. Uddin, J. Hassan, and D. Douroumis, "Thermal Inkjet Printing: Prospects and Applications in the Development of Medicine Technologies," in Technologies, vol. 10, no. 5, p. 108, 2022, https://doi.org/10.3390/technologies10050108.
[21] O. P. Choudhary and Priyanka, "Scanning Electron Microscope: Advantages and Disadvantages in Imaging Components," in International Journal of Current Microbiology and Applied Sciences, vol. 6, no. 5, pp. 1877–1882, 2017, https://doi.org/10.20546/ijcmas.2017.605.207.
[22] H. Jung, S. Jeong, Y. Park, Y. Shin, and H. Jeong, "X-Ray Diffraction Analysis of Damaged Layer During Polishing of Silicon Carbide," in International Journal of Precision Engineering and Manufacturing, vol. 24, pp. 25–32, 2023, https://doi.org/10.1007/s12541-022-00711-5.
[23] W. J. Peveler, S. B. Jaber, and I. P. Parkin, "Nanoparticles in Explosives Detection—The State-of-the-Art and Future Directions," in Forensic Science, Medicine, and Pathology, vol. 13, pp. 490–494, 2017, https://doi.org/10.1007/s12024-017-9903-4.
[24] A. M. Alshehawy, D-Eldin A. Mansour, M. Ghali, M. Lehtonen, and M. M. F. Darwish, "Photoluminescence Spectroscopy Measurements for Effective Condition Assessment of Transformer Insulating Oil" in Processes, vol. 9, pp. 732(1–15), 2021, https://doi.org/10.3390/pr9050732.
[25] A. Munajad, C. Subroto, and Suwarno. "Fourier Transform Infrared (FTIR) Spectroscopy Analysis of Transformer Paper in Mineral Oil--Paper Composite Insulation under Accelerated Thermal Aging.-" in Energies, vol. 11, no. 2, p. 364(1–12), 2018, https://doi.org/10.3390/en11020364.

6 AI–ML Powered Systems with Computer Vision for Wafer Inspections of Solar Cells

Technological Scenario, Implementation, Advantages, and Challenges

Kasturi Ghosh and Sushmita Chaudhari

6.1 INTRODUCTION

The quality of silicon wafers plays a crucial role in the production of solar panels. The wafers account for approximately 75% of the total solar cell cost [1]. The efficiency and durability of photovoltaic modules are directly affected by the presence of defects in a wafer. Wafer defects may be generated during fabrication, transportation, installation, and regular use of the solar panel. Thus, inspection of solar cell wafers is essential in each of these phases to replace the defective ones. The basic steps of fabrication of solar cells are depicted in Figure 6.1. Wafer inspection runs parallelly with fabrication in large-scale production of solar modules according to the requirement. Testing of solar cells can be broadly classified into two categories: electrical testing techniques and image-based testing [2]. Electrical testing approaches include the I–V curve, earth capacitance measurement, time domain reflectometry, and power loss analysis. However, electrical testing of photovoltaic modules is expensive and time consuming due to the deployment of set-up and required human resources. Moreover, exact pinpointing of the defect in solar cells is difficult in electrical testing methods. Traditional methods of image-based testing of solar cells also heavily relied on human expertise which is time consuming, costly, and sometimes inadequate in their accuracy. Today, the immense development of artificial intelligence (AI) and its all-pervading effect are replacing the traditional methods in most domains. Detection of wafer defects in photovoltaic cells is one of them. Moreover, the ever-increasing demand for automated techniques to detect defects in solar wafers without human intervention and with improved accuracy and speed has stimulated the development of AI and machine learning (ML)–assisted computer vision

DOI: 10.1201/9781003515784-6

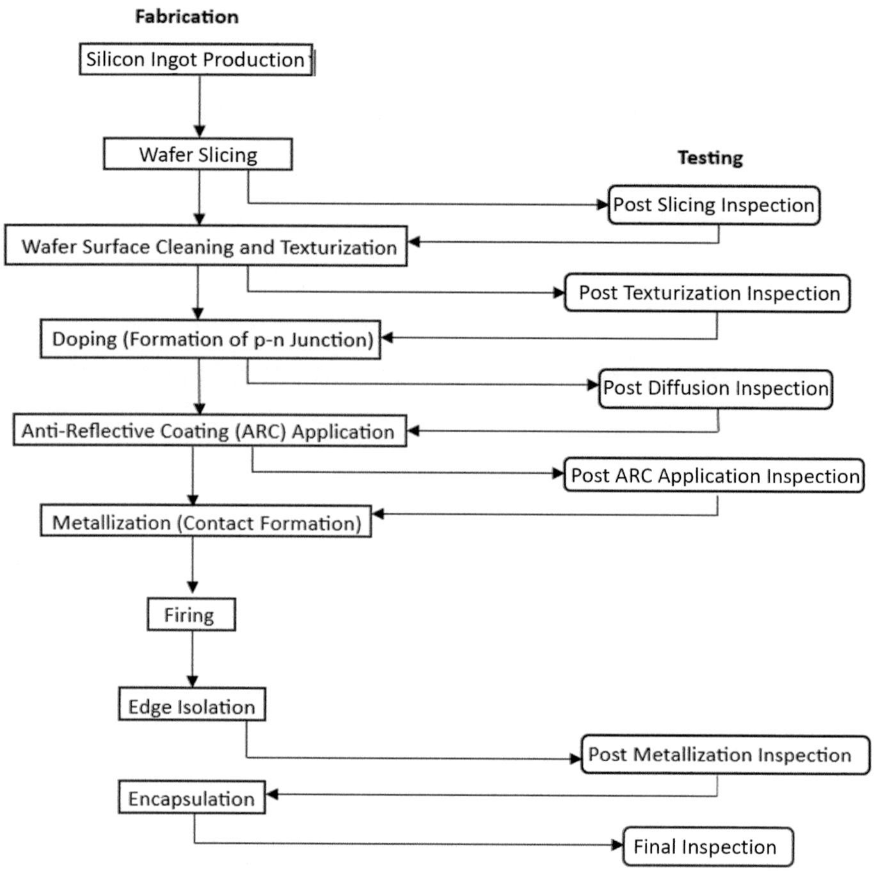

FIGURE 6.1 Flow chart of basic steps of fabrication of solar cell.

techniques. This chapter outlines effective implementation strategies for defect localization and classification in solar cells using ML-assisted computer vision techniques.

6.2 TYPE OF DEFECTS AND THEIR EFFECTS IN SOLAR CELL

Defects in solar cells can be fundamentally classified into two classes: material induced and process induced [1]. Crystallization defects, grain boundaries, impurity inclusion, and so on are material-induced defects. We have the least control over material-induced defects since their origin lies in the source of the material and impurities in the material. Process-induced defects are generated during most of the fabrication stages as well as during handling, transportation, installation, and regular use. The defects which are most commonly found in solar cells are cracks (macro and micro), surface scratches, dislocation, contamination, short circuits, saw marks, edge defects, thickness variation, uneven surface texture, colour variation, and so on.

Presently, the thickness of the wafer used in solar cells is down to 100 μm. The probability of occurring crack in the wafer is much more than any other defect, and, thus, this type of defect is most widely studied by researchers. Cracks can be generated either on the surface or underneath. Surface-generated cracks may also penetrate beneath the surface. Sawing defects may originate crack, scratch, and edge defects. Improper itching may give rise to thickness variation and uneven surface texture. Irrespective of the size, shape, and origin, wafer cracks hamper the efficiency of the solar cells. Process-generated micro-cracks increase in length under thermal stress when exposed to sunlight. Thus, the durability of the solar panel diminishes rapidly. Metal contamination may cause short circuits in solar cells.

6.3 OUTLINE OF SOLAR CELL WAFER INSPECTION USING AI AND ML-ASSISTED COMPUTER VISION

Wafer inspection/defect detection includes localization and class prediction of multiscale defects. The block diagram of the workflow for AI–ML assisted computer vision technique for wafer inspections of solar cells is shown in Figure 6.2. Each step mentioned in Figure 6.2 is illustrated in the following subsections.

6.3.1 Image Acquisition Techniques

Acquisition of images of solar cells, which can also be termed as data collection, is the first step for the detection of defects in solar cells using AI–ML-assisted computer vision techniques. The use of visible light alone is mainly capable of capturing images of surface defects, like surface scratches or surface cracks [3, 4]. In this technique, the surface defects of the wafer are identified from the change of pixel colours of the images. To capture the defects underneath the surface of the wafer, special imaging techniques are needed. Along with regular imaging techniques with visible light, other imaging techniques which are used for the detection of defects in solar cells a thermographic image, laser terahertz emission microscopy (LTEM) image, electroluminescence (EL) imaging, photoluminescence (PL) imaging, and so on. Before using the images for training ML-based models, images are properly processed to reduce the background noises.

Thermographic images can be obtained by pulse thermography, lock-in thermography, vibrothermography, and so on [5, 6]. Infrared cameras are used to capture thermographic images. In pulse thermography, the radiation of heat generated due to electromagnetic pulse-induced eddy current in solar cells is recorded with an

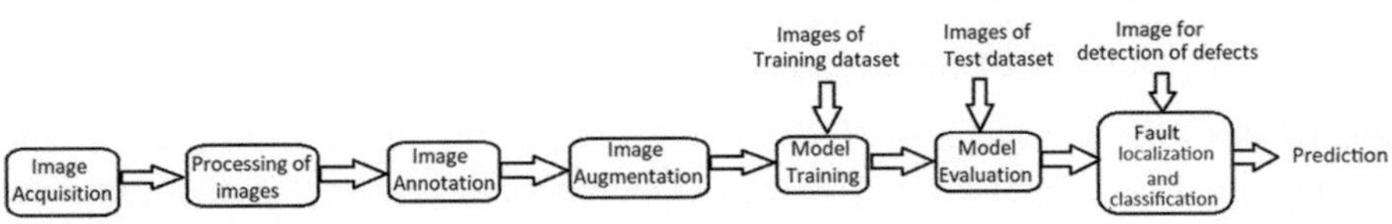

FIGURE 6.2 Flow diagram of AI–ML-assisted computer vision technique for wafer inspections of solar cells.

infrared camera. Hidden cracks are visible in pulse thermographic images since the presence of cracks affects the local current flow and hence, the resultant heat generated. In lock-in thermography, an oscillating thermal field is applied to the solar module, and resultant radiation due to generated heat in the module is captured. In vibrothermography, vibration, that is, mechanical excitation, is applied to the solar cell to generate heat. The amount of heat generated in the defect region will be different than in the defect-free region. The radiated energy is captured as images which depict the generated thermal pattern inside the solar cell. Defect regions are marked in thermographic images as different colours or varying brightness levels. LTEM images are generated by capturing the terahertz emissions from a solar cell when it is illuminated with a laser. However, LETM images have not yet been used in any ML-based automated technique for the detection of defects in solar cells.

EL is the opposite of the photovoltaic effect [7]. When PV cells are excited electrically, they convert electrical energy to photo energy, which is captured by an infrared camera in a dark room. Regions with a higher density of defects in solar cells result in lower conversion efficiency. Defects in EL images appear as dark regions. EL technique cannot be used for wafer inspection during the fabrication of solar cells since this technique needs electrical contact. However, this is a popular technique for the detection of defects in solar panels.

Photoluminescence (PL) imaging is a non-invasive technique to identify wafer defects based on the luminescence emitted from the material when exposed to light [8]. Generally, the laser is used as an external illuminator. PL imaging can be used for solar panels as well as solar cells.

6.3.2 Processing of Image Data

The input images are processed through specific techniques before applying ML algorithms to them [5]. Although the techniques depend upon the type of images used, such as EL or PL, two common steps apply to all images: resizing and noise filtering. Image resizing standardizes the dimensions of the input images, ensuring compatibility with classification algorithms. Noise filtering enhances image quality by removing unwanted background noises, improving the accuracy of subsequent processing. Greyscale conversation, histogram equalization, and so on are other preprocessing steps which are used according to the requirements. In greyscale conversation, colour images (RGB) are converted to greyscale. Histogram equalization is a contrast enhancement technique that redistributes the intensities (brightness) of pixels in an image to achieve a more uniform distribution. It improves the overall contrast, especially in low-contrast images.

6.3.3 Annotation of Images of Solar Cell

For AI–ML-based inspection of solar cells, the ML model is to be trained with labelled images of solar cells. The term “labelled images” refers to the images in which the defects are localized (using bounding boxes) and classified (assigning the type of defects). The task of localization and classification of objects (defects in this

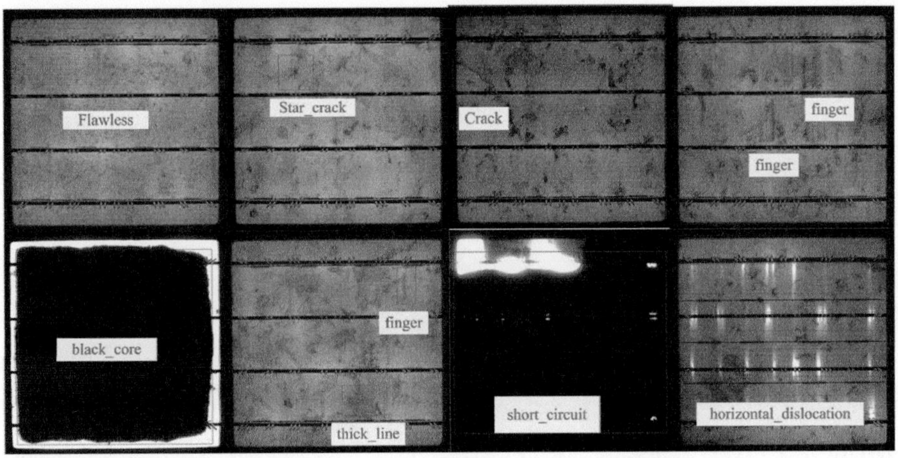

FIGURE 6.3 Labelled dataset for training of ML algorithm for solar cell wafer inspection [10].

case) in images for the generation training dataset is called annotation [9]. An annotated dataset serves as the reference or "correct" output for training, testing, and validating the ML models. Annotation of solar cell images typically involves human experts and computer vision-based techniques. Recently some AI companies have also been providing this service. Figure 6.3 represents a set of solar cell–labelled images mentioning the type of defects.

6.3.4 Image Augmentation

Augmentation is a popular method to combat the problem of inadequate training datasets. For an ML model to perform consistently well across various scenarios and conditions, it must be exposed to an extensive and varied dataset during training. By artificially generating modified versions of existing data, the size of the training dataset is increased effectively by data augmentation. Augmentation is also effective in reducing overfitting. Geometric transformation is a common method of image data augmentation which includes randomly flipping, cropping, rotating, stretching, and zooming of images. Other traditional methods of image data augmentation are colour space transformation, applying kernel filters, randomly erasing some parts of images, and mixing images. Traditional methods of image augmentation are easy to implement and fast. However, the main problem with these techniques is that the difference among the images generated with traditional methods is minimal when a large number of images are generated [11]. Image augmentation for the detection of defects in solar cells was recently addressed by creating synthetic data using ML-based models like Generative Adversarial Network (GAN), Conditional Generative Adversarial Network (cGAN), Wasserstein Generative Adversarial Network with Gradient Penalty (WGAN-GP), and so on [12]. Figure 6.4 represents the GAN architecture for solar cell image augmentation proposed by Tang et al. [11]

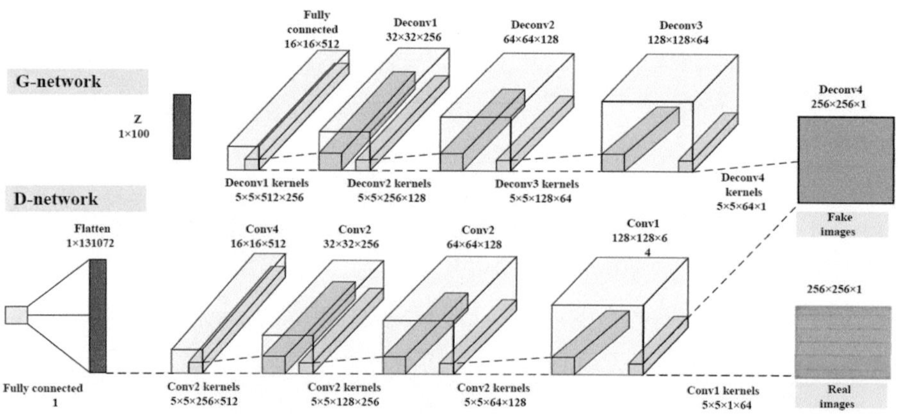

FIGURE 6.4 GAN architecture for EL image augmentation for detecting defects in PV module proposed by Tang et al. [11].

6.3.5 Training of Model

The training/learning of a model means optimizing the parameters of the model so that the model can perform a specific task with significant accuracy. To train the ML model, the processed images are first separated into two divisions among which one part (also known as the training dataset) is kept for training the model, and the other (also known as the test dataset) is for testing the trained model. The divisions are made in general in an 80:20 ratio for training and testing, respectively. A training set is a dataset of inputs along with corresponding outputs. In this case, the training set is processed labelled images of solar cells. In labelled images of solar cells, images of defective solar cells are labelled as "defective" and defect-free images are labelled as "defect free." In the case of localization and classification of defects, defects are localized with bounding boxes (also known as ground truth), and the type of defects is also mentioned. During training, a cost or loss function is determined by assessing the difference between the actual output and the model's predicted outputs for each sample within the training dataset. The model's parameters are then updated iteratively using a specific optimization algorithm to minimize the loss function. For solar cell wafer inspection, researchers primarily use two types of models: support vector machines (SVMs) and deep convolutional neural networks (DCNNs) [13]. The advancements in deep learning have revolutionized object detection in images, making it highly effective for identifying defects in solar cells. Deep learning–based defect localization in solar cells is achieved with two types of networks: two-stage network (e.g. R-CNN) [14] and one-stage network (e.g., You Only Look Once (YOLO)) [15]. Two-stage networks classify and localize defects based on proposal regions generated by the network. On the other hand, one-stage networks locate defects directly based on extracted features. Deployment of a one-stage network is easier than a two-stage network. Moreover, the performance efficiency of a two-stage network is relatively low in terms of speed and accuracy. Thus, a two-stage network is much popular for detection of defects in solar cell.

6.3.6 Model Evaluation

Trained models are tested using test data before using it for practical purposes. The results of the classification of the ML-based classifier model are generally represented with a confusion matrix. Figure 6.5 shows the structure of the binary classifier which only detects "Class 1" and "Class 0." "True Positive" (TP) represents the count of class 1 objects correctly identified as class 1. "True Negative" (TN) refers to the count of class 0 objects correctly identified as class 0. "False Negative" (FN) represents the count of class 1 objects misclassified as class 0. "False Positive" (FP) represents the count of class 0 objects misclassified as class 1. The following performance parameters of the classifier are defined from the confusion matrix.

$$\text{Accuracy} = \frac{TP + TN}{TP + TN + FP + FN} \tag{6.1}$$

$$\text{Precision} = \frac{TP}{TP + FP} \tag{6.2}$$

$$\text{Recall} = \frac{TP + TN}{TP + FN} \tag{6.3}$$

Figure 6.5(b) represents the confusion matrix solar cell defect detector classifier proposed in [5], where three classes of defects are considered.

The first row of the table (in Figure 6.5(b)) indicates that among the test data 98 "Healthy" samples are classified as "Healthy," 2 "Healthy" samples are mistakenly classified as "Non-faulty hot spot," while 2 other "Healthy" samples are incorrectly classified as "Faulty." Similarly, the second and third rows provide information regarding the classification of classes "Non-faulty hot spot" and "Faulty," respectively.

Localization and classification of multi-class defects in solar cells are determined by the prediction of the position of the bounding box and the probability of a particular class within the bounding box. Defect detection, that is, whether there is a defect within the bounding box or not is determined from *Intersection over Union (IoU).*

(a)

		Prediction	
		Class 1	Class 0
Actual	Class 1	True Positive	False Nagative
	Class 0	False Positive	True Negative

(b)

Class	Classified as		
	Healthy	Non-faulty hot spot	faulty
Healthy	98	2	2
Non-faulty hot spot	3	95	1
faulty	1	1	112

FIGURE 6.5 (a) Confusion matrix of binary classifier. (b) Example of confusion matrix of multi-class classifier for solar cell defect identification.

$$\begin{bmatrix} 0 & 0 & 0 & 0 & 0 & 0 \\ 0 & 1 & 1 & 2 & 4 & 0 \\ 0 & 5 & 6 & 7 & 8 & 0 \\ 0 & 3 & 2 & 1 & 0 & 0 \\ 0 & 1 & 2 & 3 & 4 & 0 \\ 0 & 0 & 0 & 0 & 0 & 0 \end{bmatrix}$$

A fictitious annotated solar wafer image is shown in Figure 6.6(a), where the green box denotes the actual position (also known as ground truth) of the defect, and the red box is the predicted (by ML model) position (or predicted bounding box) of the defect. In this case, IoU is approximately 40%. If the threshold of IoU is 0.5, then it will be considered that the predicted bounding box (red box) in Figure 6.6(a) does not contain the defect. If the threshold of IoU is considered 0.3, then the predicted bounding box (in Figure 6.6(a)) contains the defect. Thus, in the case of localization, the count of TP and FP depends upon the IoU threshold. In the case of Figure 6.6(b), if the IoU threshold is considered 0.3, then,

TP = 2
FP = 1
FN = 1

Similarly, if the IoU threshold is considered 0.5, then,

TP = 1
FP = 2
FN = 1

Average precession is defined as follows:

$$\text{Average Precesion} = \int_0^1 \text{Precision d}(\text{recall}) \tag{6.5}$$

For a multi-class defect classifier, mean average precision (mAP) is more significant, which is defined as follows.

$$\text{mAP} = \frac{1}{n}\sum_{i=1}^{n}(\text{AP})_i \tag{6.6}$$

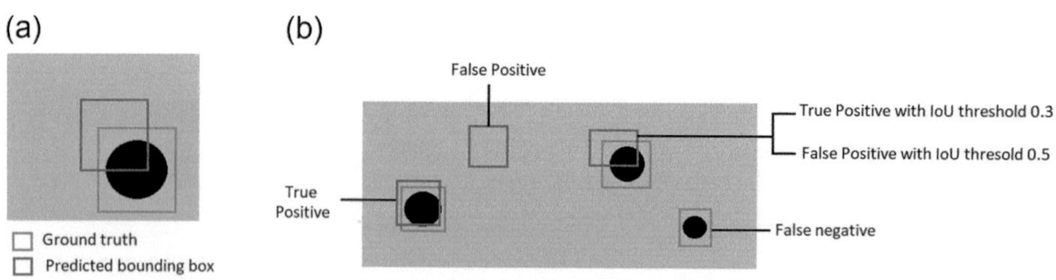

FIGURE 6.6 Fictitious annotated solar wafer image (a) showing ground truth and predicted bounding box. (b) Determination of TP, TN, FP, and FN based on IoU threshold.

where *n* represents the number of defect classes. Most of the researchers consider the IoU threshold to be equal to 0.5 for defect localization in solar cells. Correspondingly, the mAP of the ML model is denoted as mAP@0.5.

6.3.7 Defect Detection

Defect detection in solar cell wafers involves two tasks: localization of defect by bounding boxes in the images of solar cells and classification of the defect inside the predicted bounding boxes. The bounding box is generally identified by five values of which four are coordinates to locate the bounding box in the image. The fifth value will be zero or one indicating whether the box contains the defect or not. For the classification of the object, the model predicts a probability for each of the classes. The probability of the actually present defect within the bounding box should be highest.

6.4 DEFECT CLASSIFICATION USING SVM

The major steps of SVM-based defect classification in solar cell images are extraction of features from processed image data, training SVM, and class prediction. Different algorithms which are used to extract features from images are Histogram of Oriented Gradients (HOG), SIFT (Scale-Invariant Feature Transform), KAZE, Speeded Up Robust Features (SURF), Adaptive and Generic Accelerated Segment Test, Pyramid

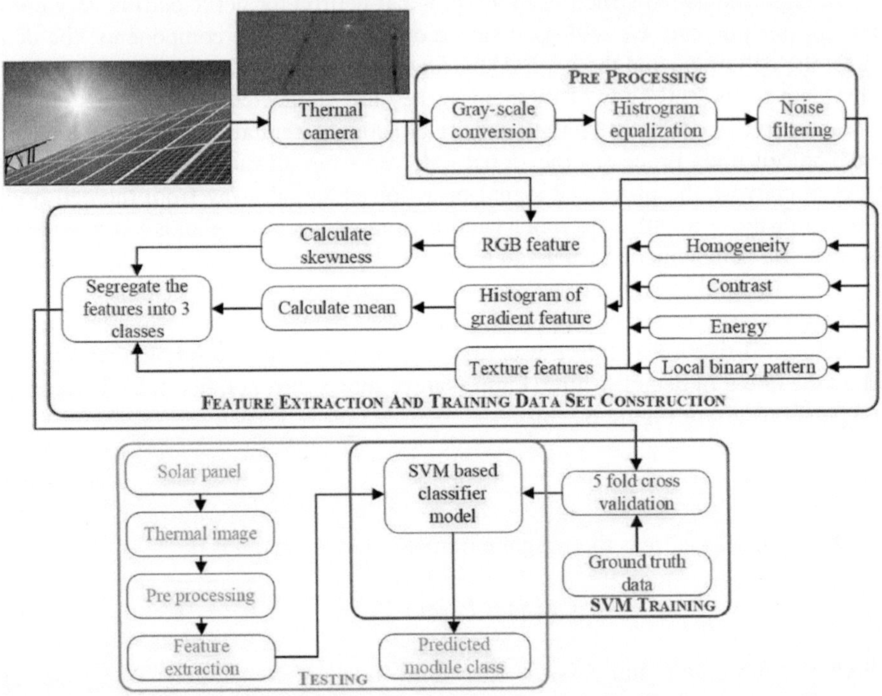

FIGURE 6.7 Flow chart of SVM-based defect classification of photovoltaic module proposed by Ali et al. [5].

Histogram of VisualWords (PHOW), VLFEAT, Visual Geometry Group (VGG), LBP feature, and so on [13].

Deitsch et al. showed SVM-based defect classification of photovoltaic modules using EL imaging where two classes are considered defective and functional [13]. The average accuracy of classification using SVM was found to be 82.44%. Ali et al. reported defect classification of photovoltaic modules using SVM. In this work, infrared thermographic images of solar cells are classified into three classes: healthy, non-faulty hotspot, and faulty. Figure 6.7 shows the flow chart of the SVM-based defect classification of the photovoltaic module proposed by Ali et al. [5].

6.5 DEFECT CLASSIFICATION USING DEEP NEURAL NETWORK

Before going into the details of deep learning–based defect identification and localization, it is necessary to understand the working principle of artificial neural networks (ANNs).

6.5.1 Background of Deep Learning

The invention of deep learning has revolutionized the concept of automation in many fields, among which wafer inspection is one [16–18]. Deep learning has emerged from the concept of ANNs, which in turn is designed as a simplified version of biological neural networks. Figure 6.8(a) depicts the structure of biological neurons.

Although the actual structure of a biological neuron (or nerve cell) is very complex, for our purpose, we will concentrate on its three prime components: the dendrites, the cell body, and the axon. Dendrites collect inputs from other neurons and relay them to the cell body of the neuron they are connected to. Dendrites play an important role in determining the intensity of action potentials produced by the neuron. The cell body processes the signal received from all the dendrites attached to it. Axon conveys the processed signal or action potential away from the cell body to transmit those impulses to other neurons, muscle cells, or glands via a synapse. The synapse is the junction where an axon of one neuron connects with a dendrite of another, allowing the signal to be passed to another neuron or target cell.

The pioneering work on ANN was the invention of perceptron (a single artificial neuron) in 1957. Perceptron was first proposed by Frank Rosenblatt who is also called the father of deep learning. Figure 6.8(b) shows the operational block diagram of a perceptron. For Figure 6.8(b),

$$n = \boldsymbol{wp} + b \tag{6.7}$$

p is the input matrix, $\boldsymbol{w}$ is the weight parameter matrix, and b is the bias.

$$a = f(n) = f(wp + b) \tag{6.8}$$

where a is the output and f is the activation function. Equation (6.7) can also be written as

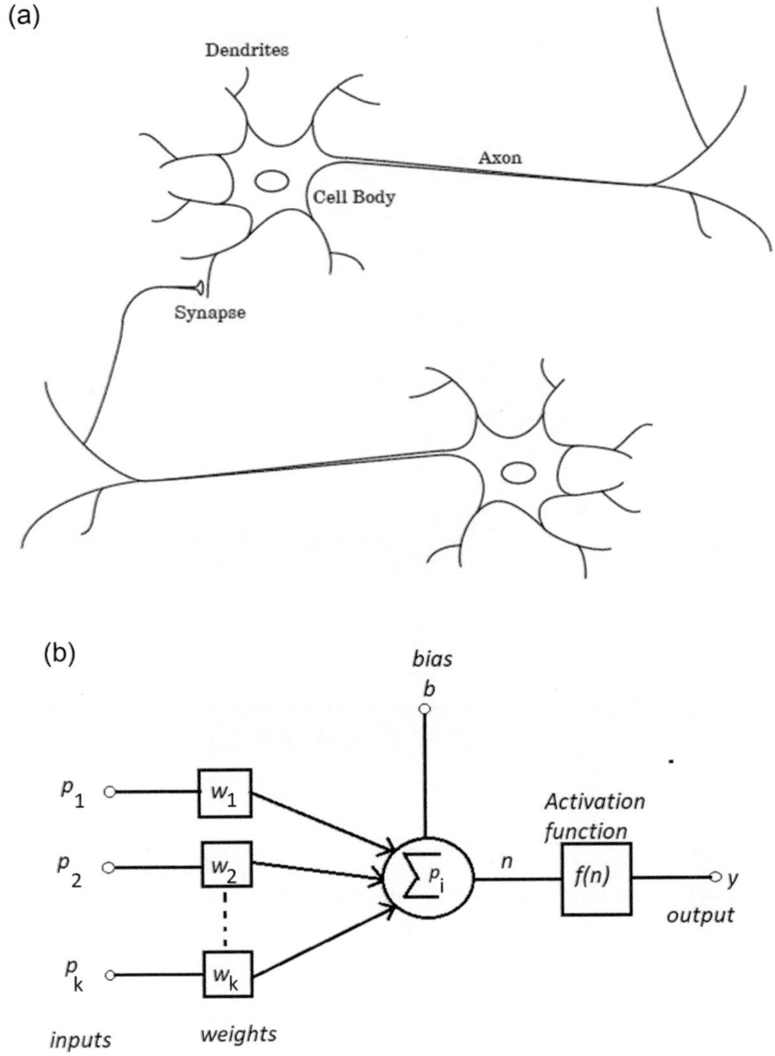

FIGURE 6.8 (a) Biological neuron [19]. (b) Structure of perceptron [19].

$$n = \begin{bmatrix} w_1 & w_2 & w_3 \end{bmatrix} \begin{bmatrix} p_1 \\ p_2 \\ p_3 \end{bmatrix} + b \tag{6.9}$$

Figure 6.9 represents a single layer of ANN, which consists of three neurons with four inputs. The working formula of the network is the same as equation (6.7). The

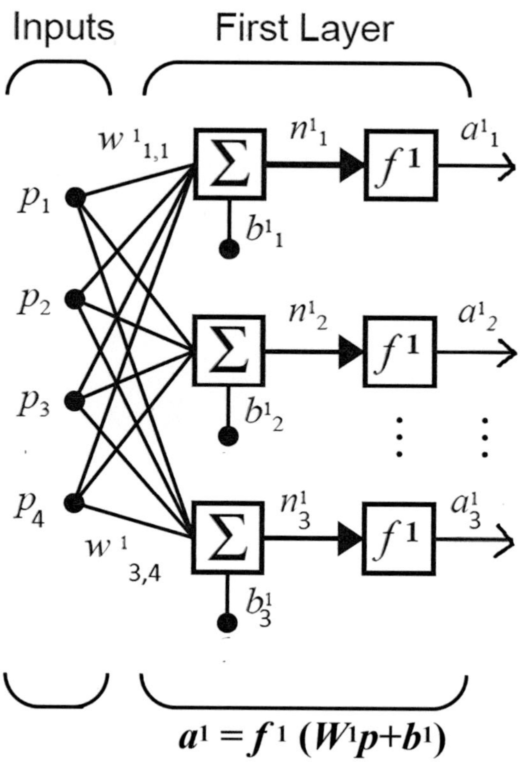

FIGURE 6.9 Multiple-input multiple-output perceptron [19].

only difference is that both, ***a*** and ***n*** are matrices here. Thus, the working formula of the architecture of Figure 6.9 is as follows:

$$\begin{bmatrix} n_1 \\ n_2 \\ n_3 \end{bmatrix} = \begin{bmatrix} w_{11} & w_{12} & w_{13} & w_{14} \\ w_{21} & w_{22} & w_{23} & w_{24} \\ w_{31} & w_{32} & w_{33} & w_{34} \end{bmatrix} \begin{bmatrix} p_1 \\ p_2 \\ p_3 \\ p_4 \end{bmatrix} + \begin{bmatrix} b_1 \\ b_2 \\ b_3 \end{bmatrix} \tag{6.10}$$

$$\text{and} \begin{bmatrix} a_1 \\ a_2 \\ a_3 \end{bmatrix} = \begin{bmatrix} f(n_1) \\ f(n_2) \\ f(n_3) \end{bmatrix} \tag{6.11}$$

$$\text{that is, } n_1 = w_{11}p_1 + w_{12}p_2 + w_{13}p_3 + w_{14}p_4 + b_1 \text{ and } a_1 = f(n_1) \tag{6.12}$$

$$n_2 = w_{21}p_1 + w_{22}p_2 + w_{23}p_3 + w_{24}p_4 + b_2 \text{ and } a_2 = f(n_2) \tag{6.13}$$

$$n_3 = w_{31}p_1 + w_{32}p_2 + w_{33}p_3 + w_{34}p_4 + b_3 \text{ and } a_3 = f(n_3) \tag{6.14}$$

where a_1, a_2, and a_3 are the outputs of the first, second, and third neurons, respectively. The weight matrix w has columns corresponding to the count of inputs to the network and rows corresponding to the number of neurons. Each element w_{ij} in the matrix represents the weight linking the j-th input to the i-th neuron (e.g., w_{32} is the weight for the third neuron and the second input). Additionally, the bias b_i is associated with the i-th neuron, so b_3 corresponds to the third neuron.

Figure 6.10 depicts a three-layer ANN where the second layer serves as the hidden layer. In a multi-layer neural network, the input of the hidden layers is the output of the previous layers. The network's operation, as shown in Figure 6.10, is governed by the following formula:

$$\begin{bmatrix} n_1^1 \\ n_2^1 \\ n_3^1 \end{bmatrix} = \begin{bmatrix} w_{11}^1 & w_{12}^1 & w_{13}^1 \\ w_{21}^1 & w_{22}^1 & w_{23}^1 \\ w_{31}^1 & w_{32}^1 & w_{33}^1 \end{bmatrix} \begin{bmatrix} p_1 \\ p_2 \\ p_3 \end{bmatrix} + \begin{bmatrix} b_1^1 \\ b_2^1 \\ b_3^1 \end{bmatrix} \tag{6.15}$$

$$\begin{bmatrix} a_1^1 \\ a_2^1 \\ a_3^1 \end{bmatrix} = \begin{bmatrix} f^1\left(n_1^1\right) \\ f^1\left(n_2^1\right) \\ f^1\left(n_3^1\right) \end{bmatrix} \tag{6.16}$$

$$\begin{bmatrix} n_1^2 \\ n_2^2 \\ n_3^2 \end{bmatrix} = \begin{bmatrix} w_{11}^2 & w_{12}^2 & w_{13}^2 \\ w_{21}^2 & w_{22}^2 & w_{23}^2 \\ w_{31}^2 & w_{32}^2 & w_{33}^2 \end{bmatrix} \begin{bmatrix} a_1^1 \\ a_2^1 \\ a_3^1 \end{bmatrix} + \begin{bmatrix} b_1^2 \\ b_2^2 \\ b_3^2 \end{bmatrix} \tag{6.17}$$

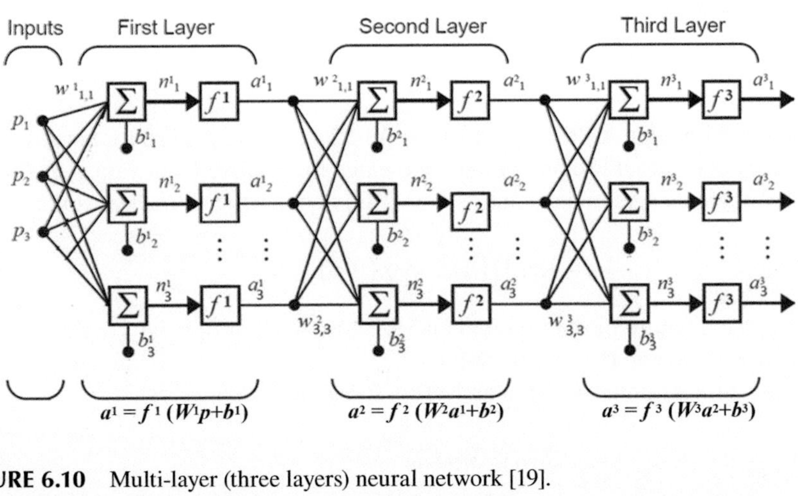

FIGURE 6.10 Multi-layer (three layers) neural network [19].

$$\begin{bmatrix} a_1^2 \\ a_2^2 \\ a_3^2 \end{bmatrix} = \begin{bmatrix} f^2\left(n_1^2\right) \\ f^2\left(n_2^2\right) \\ f^2\left(n_3^2\right) \end{bmatrix} \tag{6.18}$$

$$\begin{bmatrix} n_1^3 \\ n_2^3 \\ n_3^3 \end{bmatrix} = \begin{bmatrix} w_{11}^3 & w_{12}^3 & w_{13}^3 \\ w_{21}^3 & w_{22}^3 & w_{23}^3 \\ w_{31}^3 & w_{32}^3 & w_{33}^3 \end{bmatrix} \begin{bmatrix} a_1^2 \\ a_2^2 \\ a_3^2 \end{bmatrix} + \begin{bmatrix} b_1^3 \\ b_2^3 \\ b_3^3 \end{bmatrix} \tag{6.19}$$

The final output is

$$\begin{bmatrix} a_1^3 \\ a_2^3 \\ a_3^3 \end{bmatrix} = \begin{bmatrix} f^3\left(n_1^3\right) \\ f^3\left(n_2^3\right) \\ f^3\left(n_3^3\right) \end{bmatrix} \tag{6.20}$$

Here, w_{ij}^k and b_i^kstand for weight parameter and bias associated with k-th layer. f^k represents the activation function of the k-th layer, which may or may not be the same for all layers. The term "deep neural networks" (DNNs) denotes networks that contain a large number of hidden layers.

6.5.1.1 Activation Function

In deep learning, activation functions allow neural networks to identify and learn complex patterns. Well-known activation functions used in a DNN are defined in the following text.

Sigmoid activation function maps values to a range between 0 and 1, typically used in the output layer for binary classification.

$$f\left(n\right) = \frac{1}{1+e^{-n}} \tag{6.21}$$

For rectified linear unit (ReLU) activation function, the output is 0 for negative input values, and the output is equal to the input value for positive input values. It's widely used for hidden layers in convolutional and fully connected neural networks.

$$\text{ReLU}\left(n\right) = \max\left(0, n\right) \tag{6.22}$$

The graphical nature of the Sigmoid and ReLU activation functions are represented in Figure 6.11 (a) and (b), respectively.

Sigmoid linear units, or SiLU activation function, is defined as

$$f\left(n\right) = n\sigma\left(n\right) \tag{6.23}$$

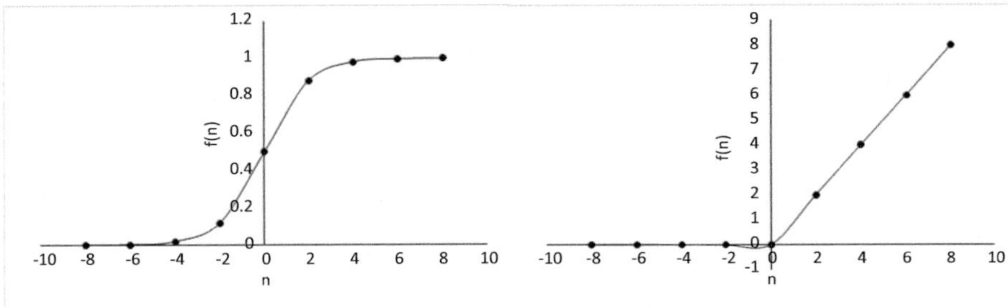

FIGURE 6.11 (a) Sigmoid activation function. (b) ReLU activation function.

The output of the SiLU activation function is the sigmoid function multiplied by its input.

The Softmax activation function can be defined as mathematically as follows.

$$f(n) = \frac{e^{n}}{\sum_{j}\left(e^{n_j}\right)} \tag{6.24}$$

The Softmax activation function converts logits (raw scores from the model, which can be any real values) into probabilities. It normalizes the logits into a probability distribution, making it useful for multi-class classification problems. Each output of the Softmax function represents the likelihood of a particular class. The graphical nature of SiLU and Softmax activation function are shown in Figure 6.12(a) and (b), respectively.

6.5.1.2 Training of ANN

Training a neural network involves adjusting its weights and biases to improve performance. The process follows these key steps:

Step 1: All weights and biases are initially assigned random values.
Step 2: The network predicts the output for a specific input in the dataset.
Step 3: The cost/loss function is computed as a function of predicted and actual output.

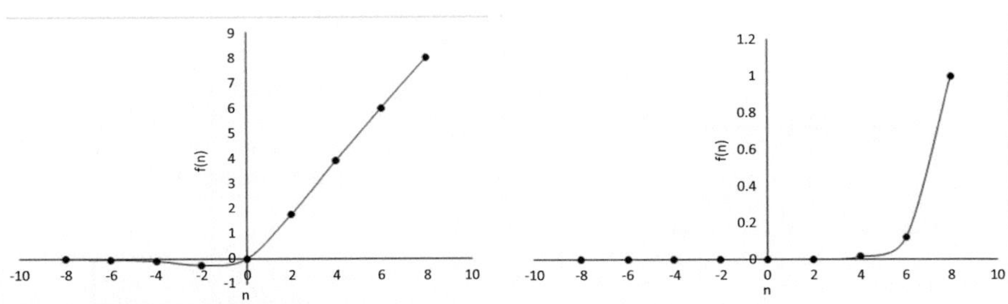

FIGURE 6.12 (a) SiLU activation function. (b) Softmax activation function.

Step 4: Compute new weights and biases using the formula of the learning algorithm.
Step 5: Repeat Steps 2–4 for each data in the dataset.
Step 6: Traverse the dataset for the given number of epochs.

The Gradient Descent Algorithm (GDA) is a commonly employed optimization method in neural networks to find local minima (of the loss function) during training. According to this algorithm, the updated value of the weight and bias are as follows:

$$w_{new} = w_{old} - \alpha \frac{dJ}{dw} \tag{6.25}$$

$$b_{new} = w_{old} - \alpha \frac{dJ}{db} \tag{6.26}$$

where w and b are the parameters to be optimized, w_{new} is the updated value of w, w_{old} is the old value of w, α is the user-defined learning rate, and $J(w)$ is the cost/loss function.

The cost/loss function is computed from the actual output (from the dataset) and predicted output. Two loss functions are involved in the detection of defects in solar cell images: one for bounding box regression and the other for classification of defects inside the bounding box. Bounding box regression is determined by the mean square error between the coordinates of predicted bounding boxes and ground truth bounding boxes. The cross entropy loss function is generally used for the classification of defects. Figure 6.13 represents the flowchart of gradient descent algorithm–based

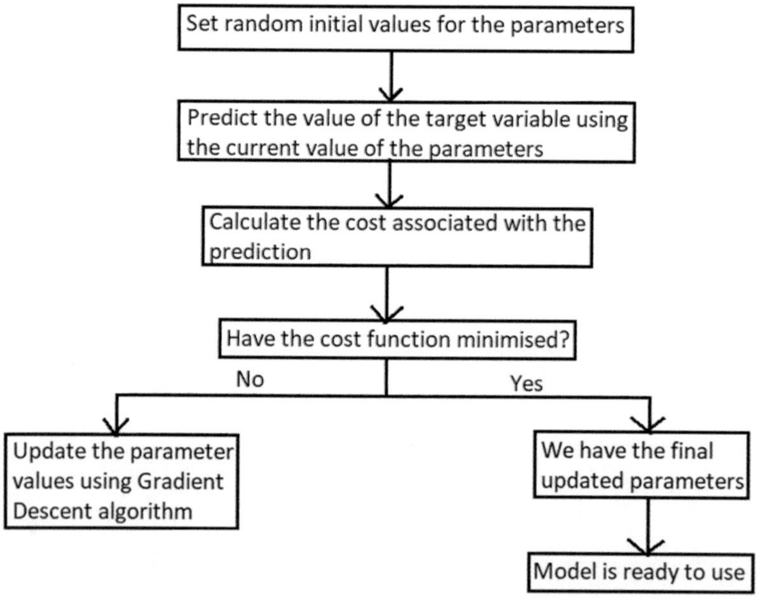

FIGURE 6.13 Flowchart of gradient descent algorithm–based optimization.

optimization technique. Adam is another popular optimization technique where the learning rate α is also determined by optimization.

6.5.2 Convolutional Neural Network

CNN is a special branch of neural work designed for images and video. A typical CNN architecture has the following layers.

Input Layer: The input layer takes the digital image.
Convolution + Activation function: This layer extracts features from the images.
Pooling: Pooling is for downsampling the feature matrix.
Fully Connected Layer: This layer combines features and makes predictions.
Output Layer: The purpose of this layer is final classification or output prediction.

CNN architecture for image classification is illustrated in Figure 6.14.

6.5.2.1 Convolution

The convolution layer applies filters (also called kernels) to the input to extract features. The filter moves across the image performing element-wise multiplication (convolution), and the output is a feature map. Stride controls how much the filter moves at each step.

Let the input (digital image) is $\begin{bmatrix} 1 & 1 & 2 & 4 \\ 5 & 6 & 7 & 8 \\ 3 & 2 & 1 & 0 \\ 1 & 2 & 3 & 4 \end{bmatrix}$. Consider the filter $\begin{bmatrix} 1 & 0 & 0 \\ 0 & 1 & 0 \\ 0 & 0 & 1 \end{bmatrix}$.

To get the output of the same dimension as the input, zero padding is necessary.

After zero padding, the input will be $\begin{bmatrix} 0 & 0 & 0 & 0 & 0 & 0 \\ 0 & 1 & 1 & 2 & 4 & 0 \\ 0 & 5 & 6 & 7 & 8 & 0 \\ 0 & 3 & 2 & 1 & 0 & 0 \\ 0 & 1 & 2 & 3 & 4 & 0 \\ 0 & 0 & 0 & 0 & 0 & 0 \end{bmatrix}$

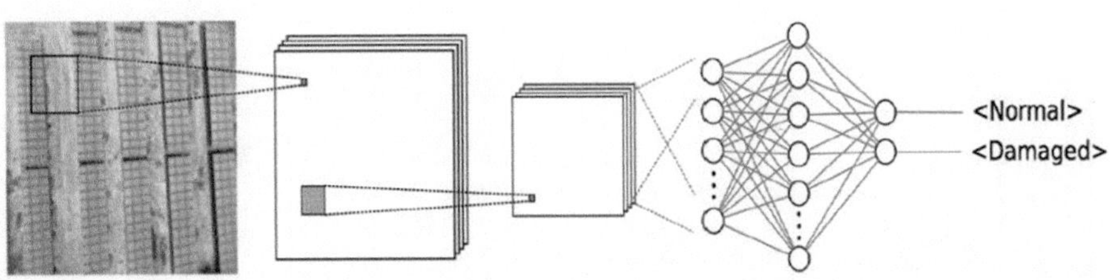

FIGURE 6.14 CNN architecture for image classification [20].

Figure 6.15 shows the movement of the filter across the image for computation of the each element of the output matrix. The encircled position is the position of the corresponding element in the output matrix. Thus, the first element of the first row of the output matrix can be computed as follows.

$$\begin{bmatrix} 0 & 0 & 0 & 0 & 0 & 0 \\ 0 & 1 & 1 & 2 & 4 & 0 \\ 0 & 5 & 6 & 7 & 8 & 0 \\ 0 & 3 & 2 & 1 & 0 & 0 \\ 0 & 1 & 2 & 3 & 4 & 0 \\ 0 & 0 & 0 & 0 & 0 & 0 \end{bmatrix} \text{ that is, } \begin{bmatrix} 0 & 0 & 0 \\ 0 & 1 & 1 \\ 0 & 5 & 6 \end{bmatrix} * \begin{bmatrix} 1 & 0 & 0 \\ 0 & 1 & 0 \\ 0 & 0 & 1 \end{bmatrix} = \begin{array}{l} (0 * 1) + (0 * 0) + \\ (0 * 0) + (0 * 0) + \\ (1 * 1) + (1 * 0) + \\ (0 * 0) + (5 * 0) + \\ (6 * 1) = 1 + 6 = 6 \end{array} \tag{6.27}$$

Similarly, the second element of the first row of the output matrix can be computed as follows.

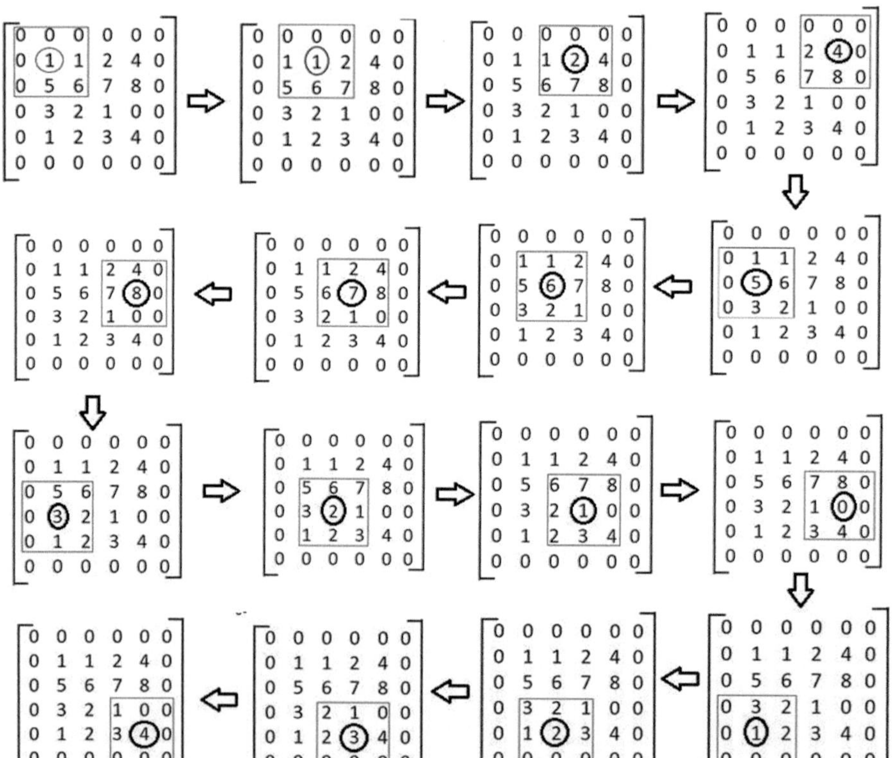

FIGURE 6.15 Filter movement across the image for convolution.

$$\begin{bmatrix} 0 & 0 & 0 & 0 & 0 & 0 \\ 0 & 1 & \textcircled{1} & 2 & 4 & 0 \\ 0 & 5 & 6 & 7 & 8 & 0 \\ 0 & 3 & 2 & 1 & 0 & 0 \\ 0 & 1 & 2 & 3 & 4 & 0 \\ 0 & 0 & 0 & 0 & 0 & 0 \end{bmatrix} \text{ that is, } \begin{bmatrix} 0 & 0 & 0 \\ 1 & 1 & 2 \\ 5 & 6 & 7 \end{bmatrix} * \begin{bmatrix} 1 & 0 & 0 \\ 0 & 1 & 0 \\ 0 & 0 & 1 \end{bmatrix} = 1+7 = 8 \tag{6.28}$$

All elements of the output matrix can be computed in the same manner and the final output is

$$\begin{bmatrix} 6 & 8 & 10 & 4 \\ 7 & 8 & 8 & 10 \\ 5 & 10 & 10 & 7 \\ 1 & 5 & 5 & 5 \end{bmatrix}$$

6.5.2.2 Activation Function

The activation function enables the CNN to capture intricate patterns. Generally, the ReLU activation function is used in the hidden layers.

Applying ReLU activation function on the final output of the convolution layer in this case, we get

$$\begin{bmatrix} 6 & 8 & 10 & 4 \\ 7 & 8 & 8 & 10 \\ 5 & 10 & 10 & 7 \\ 1 & 5 & 5 & 5 \end{bmatrix}$$

6.5.2.3 Pooling

Pooling is applied to the output feature matrix obtained from the convolution layer to reduce the spatial dimensions of the feature map (downsampling), which in turn decreases computation and control overfitting. Generally, two kinds of pooling are used in CNN: max pooling and average pooling.

Max Pooling: A window (e.g., 2 × 2 or 3 × 3) slides over the input, and, for each window, the maximum value is taken. It emphasizes the most prominent features (e.g., edges or textures) and discards less relevant information.

Input (4x4): $\begin{bmatrix} 6 & 8 & 10 & 4 \\ 7 & 8 & 8 & 10 \\ 5 & 10 & 10 & 7 \\ 1 & 5 & 5 & 5 \end{bmatrix}$

Max pooling with a 2 × 2 window with stride 2 would return: $\begin{bmatrix} 8 & 10 \\ 10 & 10 \end{bmatrix}$

Average Pooling: A window slides over the input, and, for each window, the average of the values is taken. It produces a smoother output by preserving information from the entire region, rather than focusing on the most prominent features.

Application of average pooling on the previous feature matrix with a 2 × 2 window with stride 1 would return: $\begin{bmatrix} 7.25 & 8 \\ 5.25 & 6.75 \end{bmatrix}$

6.5.2.4 Data Flattening

Data flattening means to convert the pooled data into a one-dimensional array. The flattening of greyscale pooled data is shown here.

$$\begin{bmatrix} 7.25 & 8 \\ 5.25 & 6.75 \end{bmatrix} \rightarrow \begin{matrix} 7.25 \\ 8 \\ 5.25 \\ 6.75 \end{matrix}$$

Flattened data will be fed to a fully connected layer of NN.

6.5.3 Defect Detection in Wafer of Solar Cell Using DCNN

Deitsch et al. showed defect detection in solar wafers using a DCNN [13].

The deep neural architecture proposed by Deitsch et al. is depicted in Figure 6.16 [13]. The network's input consists of processed EL images of solar cells of pixel size of 300 × 300. Since the inputs are RGB images, the input layer has three channels. A 3 × 3 filter with a stride of 1 is used, which produces an output of size 300 × 300. The input layer uses 64 filters, enabling the extraction of 64 features. The ReLU activation function is applied to the output matrix of the convolution layer. Then, a max-pooling operation with a 2 × 2 window and a stride of 2 is applied to the ReLU output, reducing its size to 150 × 150. The convolution + ReLU layers followed by max pooling are repeated six times, with a total of five max-pooling layers, ultimately producing an output of size 9 × 9. A global average pooling layer is then

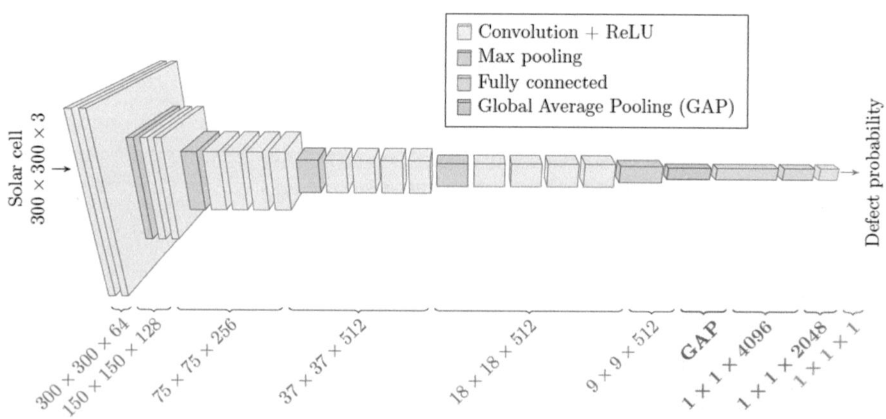

FIGURE 6.16 Architecture of deep convolutional neural network for the detection of defects in solar cell wafer proposed by Deitsch et al. [13].

applied to the 9 × 9 output, followed by two fully connected layers with 4,096 and 2,048 neurons, respectively. The final layer consists of a single neuron, which provides the classification output for the solar cell.

6.5.3.1 Two-Stage Networks Represented by Faster R-CNN

The vast potential of DCNNs for object detection in computer vision has driven researchers to develop advanced neural network architectures that prioritize efficient feature extraction, faster training, and higher detection accuracy. In 2013, Girshick et al. introduced R-CNN. The main components of R-CNN are CNN, SVM, selective search, and regression. CNN extracts the features. SVM is used for classification. Selective search is used to generate object candidate region and object localization is performed with the help of regression [21]. In 2015, a newer version of R-CNN, known as fast R-CNN was proposed since R-CNN was time consuming [22]. While fast R-CNN improved training efficiency, its detection performance remained suboptimal. To address this, fast R-CNN was introduced where region proposal networks (RPNs) are created using a shared convolutional network [23]. This RPN-based deep network significantly enhanced both the speed and accuracy of object detection.

Zhang et al. proposed a multiscale region proposal fusion network, shown in Figure 6.17, for detecting surface defects in solar cells [24]. The network comprises four key modules: feature extraction, an RPN fusion module, a region of interest (ROI) pooling layer, and a classification regression module. The feature extraction network consists of five consecutive layers, each consisting of convolution, ReLU activation, and pooling. To combat the gradient vanishing problem, cross-layer connections are introduced. Feature maps generated by these layers are fed into three separate RPN networks.

The RPN architecture, shown in Figure 6.18, includes two branches. One branch determines whether a proposed region contains an object, while the other branch identifies the location of the object within the image if detected. Since solar cell images often contain defects of varying sizes, the proposal layer discards regions that are too small. The outputs from the three proposal layers are merged in the multiscale RPN fusion layer, which ranks the region proposals by confidence score and selects the highest-ranking one. The output of feature extraction layers and RPN layer are fed into the ROI layer which converts inputs of varying shapes into fixed-length output vectors. The output of the ROI layer is the input of the classification regression module which finally predicts the class and position of the identified region using a fully connected layer and a bounding box algorithm.

6.5.3.2 One-Stage Networks Represented by Faster YOLO

Among different object detection algorithms, YOLO is most popular for its fast and reliable detection and localization of objects within images. YOLO was first proposed by Redmon in 2016 [15]. Since then, the YOLO family has grown, with each new version demonstrating improved performance over the previous one.

Figure 6.19 shows an approximate timeline of the development of the YOLO family. With the advancement of the YOLO algorithm, the architecture became more complex. From YOLOv4 onwards, all YOLO architectures are composed of three parts: backbone, neck, and head [25]. The backbone is typically composed of a

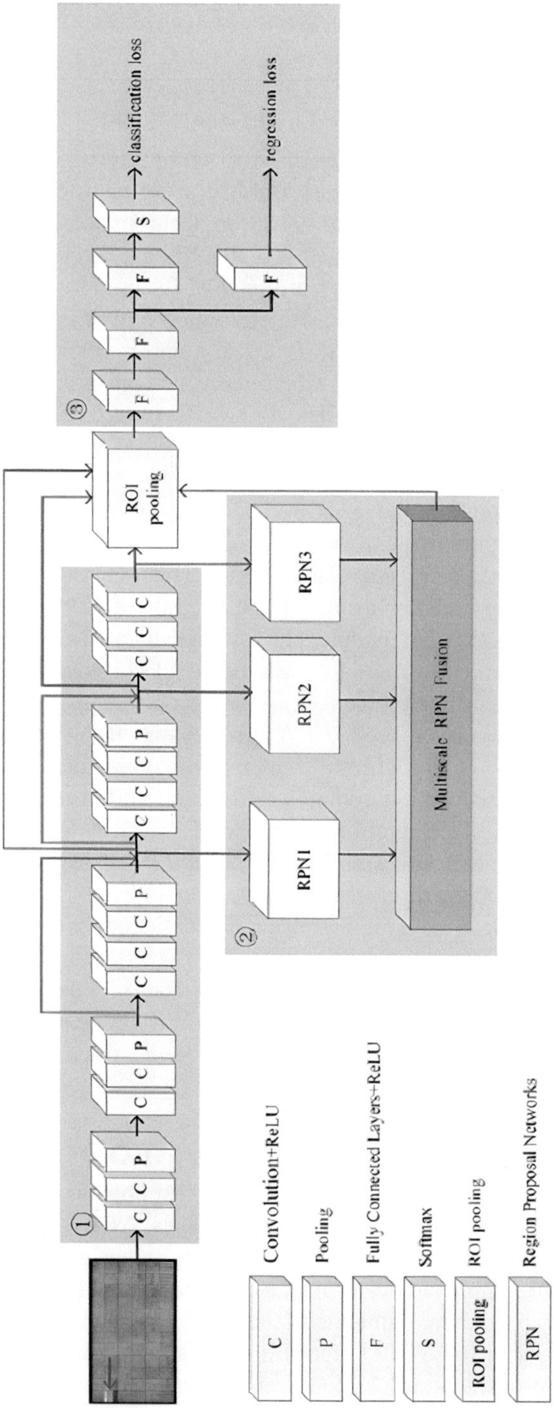

FIGURE 6.17 Network proposed by Zhang et al. for detecting surface defects in solar cells [24].

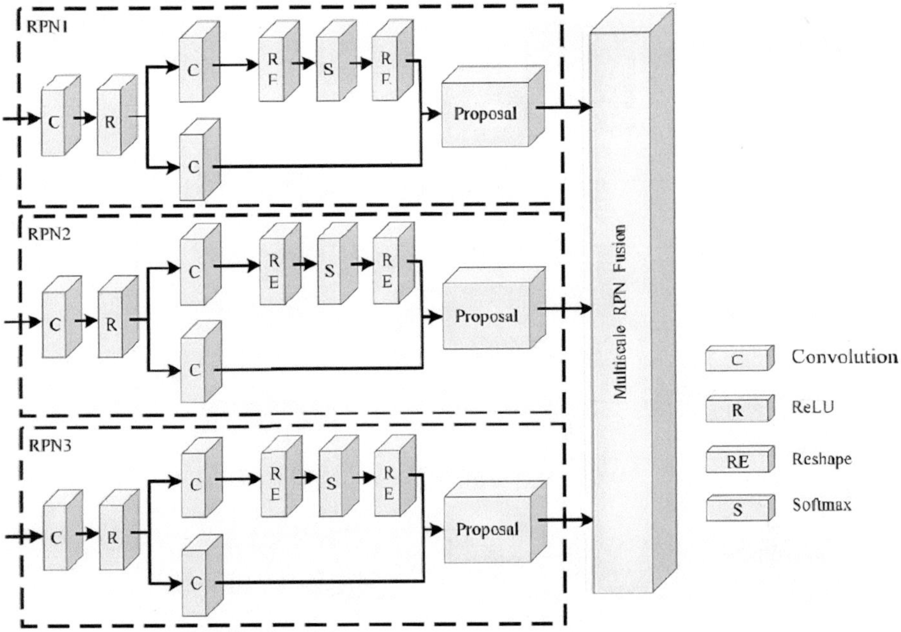

FIGURE 6.18 RPN architecture proposed by Zhang et al. [24].

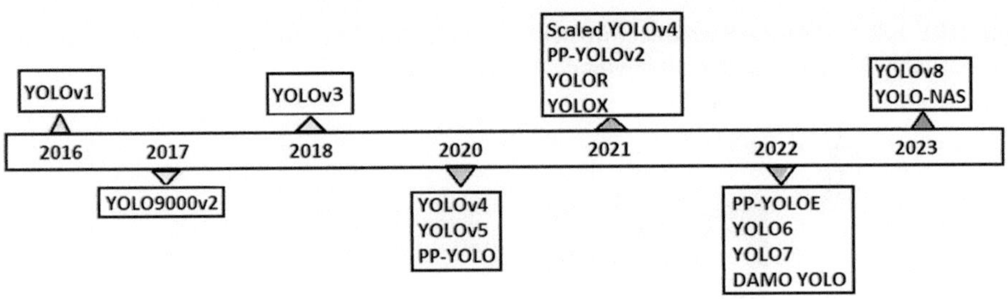

FIGURE 6.19 A timeline for the evolution of YOLO versions.

convolutional network and is responsible for extracting features. The neck is responsible for aggregating and refining the extracted features. The neck is composed of additional convolutional networks, feature pyramid networks, or any other advanced techniques to improve the feature representations. Finally, the head performs the job of classification and localization based on the obtained features from the neck.

Lang and Lv reported photovoltaic defect detection from EL images combining an attention mechanism and the transformer module with the baseline of YOLOv8 architecture [26]. The complete architecture is shown in Figure 6.20. Convolutional, C2f, SSPF, and CCT blocks (shown in Figure 6.21) constitute the backbone of the model. The convolutional block in turn comprises the fundamental convolutional layer, a BatchNorm layer, and the SiLU activation function.

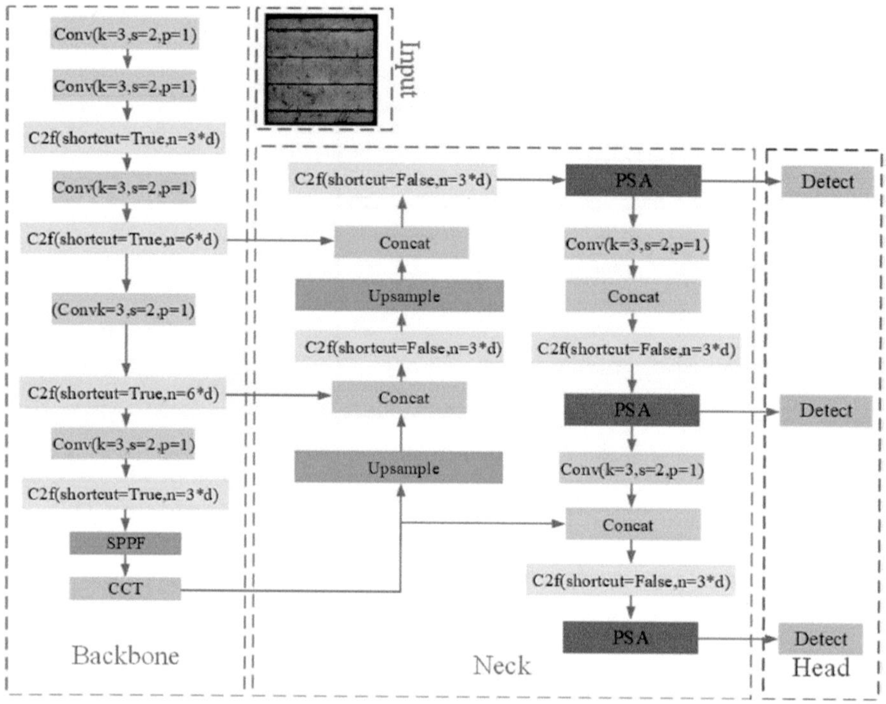

FIGURE 6.20 YOLOv8 architecture with attention mechanism and transformer module proposed by Lang and Lv for photovoltaic defect detection [26].

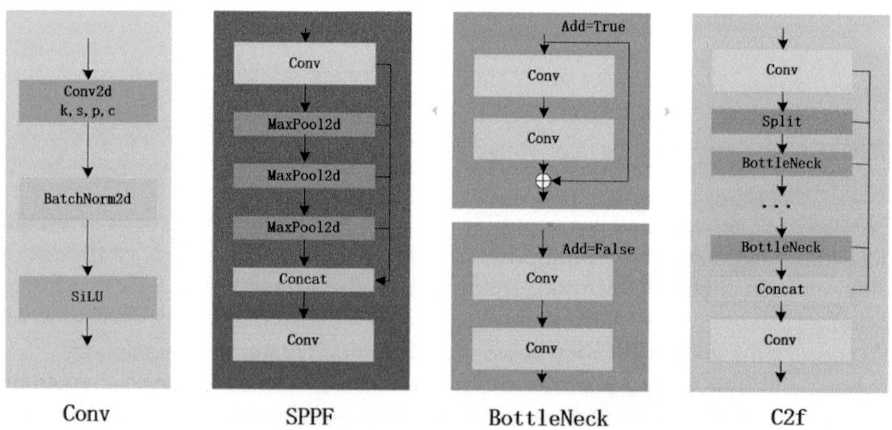

FIGURE 6.21 Modules of the YOLOv8-based proposed architecture in [26].

The Batch Normalization (BatchNorm) layer is a mechanism used in ANN to improve training speed and stability. Instead of normalizing the inputs and feeding them to the network, in batch normalization, the output of every layer is normalized. It first normalizes the data with a mean of 0 and a variance deviation of 1. Then it scales and shifts the values. The parameter for scaling and shifting is learned during training. The C2f unit is built up with Bottleneck units connected with residual network. On the other hand, the Bottleneck unit is itself a collection of convolution units interconnected with the residual network. Residual Network (ResNets) is widely used in a DNN in which alternative paths are provided for data to reach the latter parts of the neural network by skipping some layers. The special structure of the C2f unit helps the architecture to collect feature information from multiple layers. Spatial Pyramid Pooling Fusion module consists of max-pooling layer with varying kernel size.

The last module of the backbone unit is the CCT or CNN combined transformer. Transformer is a deep learning mechanism developed by the researchers of Google in 2017 [27]. The purpose of the CCT module is to combine the extraction of contextual features and global features. Traditional CNN block is responsible for contextual features, whereas transformer architecture extracts global features. The neck is an interconnection of a number of C2f blocks, up-sampling, convolutional block, and polarized self-attention (PSA) [28]. The architecture of PSA is shown in Figure 6.22. PSA allows aggregation of the important features of the neighbour of the target object, which helps to enhance detection accuracy. Figure 6.23 represents the confusion matrix of the YOLOv8-based model proposed by Lang and Lv for photovoltaic defect detection [26].

Table 6.1 represents a summary of ML-based models proposed by researchers for defect detection in solar cells.

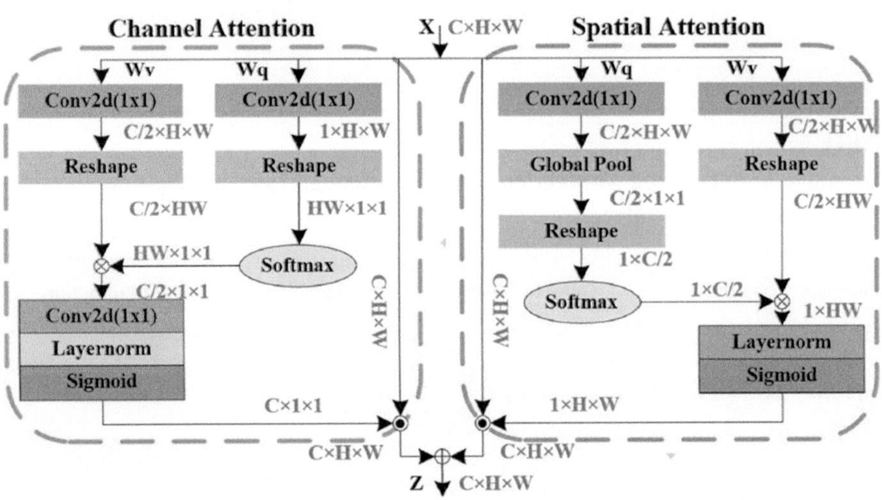

FIGURE 6.22 Polarized self-attention (PSA) modules proposed in [26].

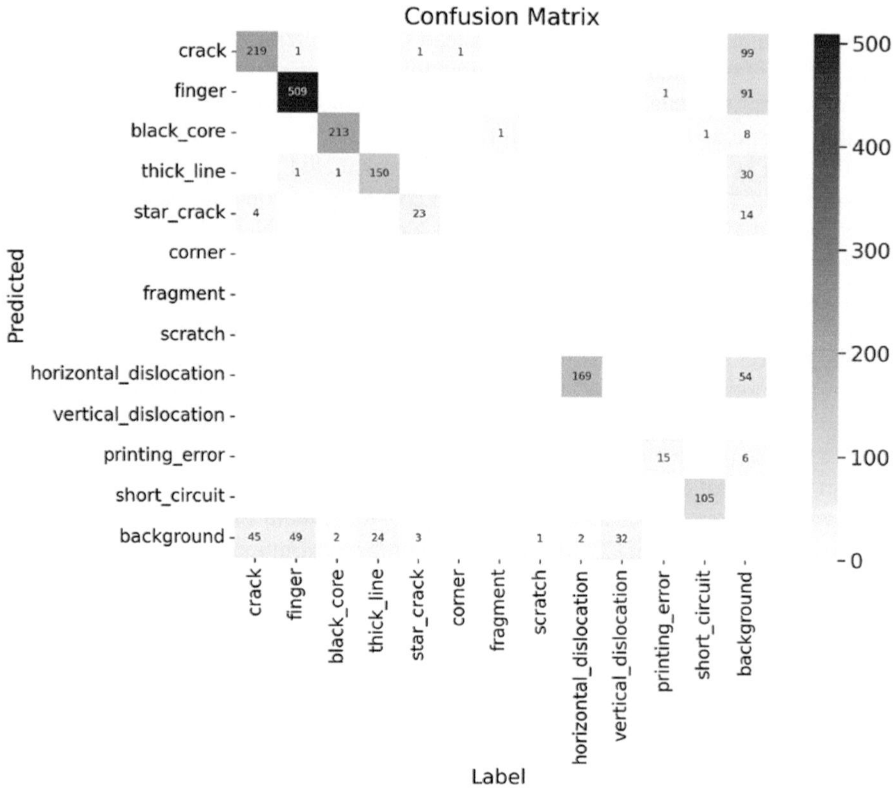

FIGURE 6.23 Confusion matrix of the YOLOv8-based model proposed in [26].

TABLE 6.1
Summary of ML-Based Models Proposed by Researchers for Defect Detection in Solar Cells

Reference	Image Type	Defect Classes	Algorithm	Performance Parameter of Prediction
[29]	EL	Cracks, black core, and finger interruption	YOLOv5	mAP: 89.64%
[30]	EL	Five categories of defects: Short circuit, short circuit cell/low-power cell, crack, micro-crack, and other errors (dark spot, scratch, and contamination)	YOLOv5, YOLOv8	mAP@0.5: 76.3%

TABLE 6.1 *(Continued)*
Summary of ML-Based Models Proposed by Researchers for Defect Detection in Solar Cells

Reference	Image Type	Defect Classes	Algorithm	Performance Parameter of Prediction
[26]	EL	Twelve categories of fault: crack, finger, black core, thick line, star crack, corner, fragment, scratch, horizontal dislocation, vertical dislocation, printing error, short circuit	YOLOv8	mAP@50: 77.9%
[10]	EL	PVEL-AD dataset [31]	VarifocalNet	mAP: 85.7%
[32]	EL	Different types of defective cells	SVM, CNN	Accuracy: CNN: 91.58% SVM-HOG: 69.95% SVM-KAZE: 71.04% SVM-SIFT: 68.90% SVM-SURF: 72.74%
[24]	Image with visible light	Surface defects	MF-RPN	mAP: 85%
[33]	IR	Battery string, hot spot, dust, fracture, plant, occlusion	PV-YOLO	mAP: 92.56%
[6]	IR	Healthy, hotspot, and faulty	SVM	Accuracy: 97%
[5]	IR	Healthy, non-faulty hotspot, and faulty	SVM	Accuracy: 92%
[7]	EL	PVEL-AD dataset [31]	ASDD-Net	mAP: 88.81%
[34]	Colour image with visible light	Surface defects	CNN	Accuracy: 94.30%
[11]	EL	Defect-free, micro-crack, finger interruption and break	CNN	Accuracy: 83%
[35]		Normal PV module, defective PV module		Accuracy: 93.02%
[15]	Colour image with Visible light	Surface defects	DCNN	Accuracy: 85.80%
[13]	EL	Functional, defective	DCNN, SVM	Average accuracy: DCNN: 88.42% SVM: 82.44%

6.6 AI-POWERED INDUSTRIES FOR WAFER INSPECTION

AI has revolutionized object detection by enabling systems to identify and classify objects in images with high accuracy. A few commercial companies also provided AI-driven solutions for the application domain. Wafer inspection is one of these domains. KLA Corporation, Empirisch Tech, Robovision, DataXquad, Faststream Technologies, VIA Wafer Inspection Solution, SolVision, and Averroes have significantly advanced wafer inspection processes. Their solutions utilize computer vision and deep learning algorithms to detect nano-scale defects, such as scratches and irregularities, with unprecedented precision. By automating inspection workflows, these companies reduce manual intervention, enhance production efficiency, and minimize defects in semiconductor manufacturing. AI-powered tools like TCS WaferWise and Prescient Technologies further contribute by integrating predictive analytics, allowing early detection of potential issues. These advancements ensure high-quality output and meet the increasing demand for reliable semiconductor devices. Table 6.2 enlists the name of a few AI-powered well-known companies that provide service for wafer inspection.

6.7 CHALLENGES OF AI-POWERED WAFER INSPECTION FOR SOLAR CELL

AI-powered wafer inspection for solar cells offers significant improvements in defect detection, class prediction, and localization, but it also faces certain challenges.

TABLE 6.2
List of Popular AI-Powered Companies for Wafer Inspection

Company	Services
KLA Corporation	AI-enhanced inspection for reticles and wafers advanced accuracy and process efficiency.
Empirisch Tech	Deep learning solutions for detecting subtle anomalies and defects in wafers.
Robovision	AI-powered automation reduces manual dependency on defect detection.
DataXquad	AI algorithms for nano-scale defect detection like scratches and stains.
Faststream Technologies	AI-driven visual inspection systems for automating wafer defect detection.
Prescient Technologies	High-accuracy AI models for detecting critical wafer defects.
TCS WaferWise	ML-based predictive analytics to identify and predict wafer defects early in production.
VIA Wafer Inspection Solution	Computer vision for high-speed wafer defect identification.
SolVision	AI solutions for surface defect detection and yield rate improvement in semiconductor manufacturing.
Averroes AI	Real-time inspection and precise defect classification using AI.

AI-powered wafer inspection using computer vision requires a large number of high-resolution labelled images for training the model. The collection of a large number of high-quality images and accurately labelling the images are a challenging task. The lack of sufficient image data is addressed by image augmentation which in turn may result in overfitting. The image library should contain images of all possible types of defects. Correct labelling of images needs human experts. ML-based models sometimes mark benign irregularities as defects, leading to unnecessary rejection of wafers. High FP rates increase costs due to higher wastage and lower yield. Advanced solar cell wafers come in different materials. AI models trained on one type of wafer material may not perform well on others without additional training or fine-tuning. High-resolution imaging systems and deep learning models require significant computational power, memory, and GPU resources.

6.8 CONCLUSION

AI–ML-powered computer vision systems represent a pivotal advancement in wafer inspection for solar cells. It increases inspection accuracy, real-time monitoring, and the ability to detect subtle defects that might be overlooked by human inspectors. The use of AI–ML in wafer inspection not only increases operational efficiency and accuracy but also contributes to more sustainable manufacturing processes. Fast identification, classification, and localization of defects during fabrication of different stages of fabrication of solar cells for commercial application helps to find the cause of the defects and improves industrial productivity of solar panels. However, in the challenge of collecting a large number of correctly annotated images for training of ML model, a combination of traditional inspection methods with AI-based techniques is necessary to leverage the strengths of both approaches. If the labelling is performed by any ML-based algorithm, cross-verification with human experts is also necessary for better results. To keep AI models up-to-date with evolving defect types and manufacturing conditions, it is necessary to introduce continuous learning frameworks in the model.

REFERENCES

[1] M. Saleem and F. G. Al-Amri, "Multi-Attribute Analysis of Micro-Defect Detection Techniques Suitable for Automated Production Line of Solar Wafers and Cells," in *IET Renewable Power Generation*, vol. 14, pp. 1413–1423, 2020, https://doi.org/10.1049/iet-rpg.2019.0904

[2] Ula Hijjawi, Subhash Lakshminarayana, Tianhua Xu, Gian Piero Malfense Fierro and Mostafizur Rahman, "A Review of Automated Solar Photovoltaic Defect Detection Systems: Approaches, Challenges, and Future Orientations," in *Solar Energy*, vol. 266, p. 112186, 2023, https://doi.org/10.1016/j.solener.2023.112186.

[3] H. Chen, Y. Pang, Q. Hu, et al., "Solar Cell Surface Defect Inspection Based on Multispectral Convolutional Neural Network," in *Journal of Intelligent Manufacturing*, vol. 31, pp. 453–468, 2020, https://doi.org/10.1007/s10845-018-1458-z.

[4] Y. S. Balcıoğlu, B. Sezen and C. Cubukcu Cerasi, "Solar Cell Busbars Surface Defect Detection Based on Deep Convolutional Neural Network," in *IEEE Latin America Transactions*, vol. 21, no. 2, pp. 242–250, Feb. 2023, https://doi.org/10.1109/TLA.2023.10015216

[5] Muhammad Umair Ali, Hafiz Farhaj Khan, Manzar Masud, Karam Dad Kallu and Amad Zafar, "A Machine Learning Framework to Identify the Hotspot in Photovoltaic Module Using Infrared Thermography," in *Solar Energy*, vol. 208, pp. 643–651, 2020, https://doi.org/10.1016/j.solener.2020.08.027.

[6] W. Ahmed, M. U. Ali, S. J. Hussain, A. Zafar and S. A. Hasani, "Visual Vocabulary Based Photovoltaic Health Monitoring System Using Infrared Thermography," in *IEEE Access*, vol. 10, pp. 14409–14417, 2022, https://doi.org/10.1109/ACCESS.2022.3148138.

[7] Yuqi Liu, Yiquan Wu, YuBin Yuan and Langyue Zhao, "Deep Learning-Based Method for Defect Detection in Electroluminescent Images of Polycrystalline Silicon Solar cells," in *Optics Express*, vol. 32, pp. 17295–17317, 2024.

[8] R. Leonard, M. Conrad, E. Van Brunt, J. Witry and E. Balkas, "Implementation of Large Scale Deep Learning Non-Destructive Methods for Characterizing 4H-SiC Materials," in *DDF*, vol. 426, pp. 3–9, 2023, https://doi.org/10.4028/p-08c7e9.

[9] U. Otamendi, I. Martinez, I. G. Olaizola and M. Quartulli, "A Scalable Framework for Annotating Photovoltaic Cell Defects in Electroluminescence Images," in *IEEE Transactions on Industrial Informatics*, vol. 19, no. 9, pp. 9361–9369, Sept. 2023, https://doi.org/10.1109/TII.2022.3228680.

[10] Y. Jia, G. Chen and L. Zhao, "Defect Detection of Photovoltaic Modules Based on Improved VarifocalNet," in *Scientific Report*, vol. 14, p. 15170, 2024, https://doi.org/10.1038/s41598-024-66234-3.

[11] Wuqin Tang, Qiang Yang, Kuixiang Xiong and Wenjun Yan, "Deep Learning Based Automatic Defect Identification of Photovoltaic Module Using Electroluminescence Images," in *Solar Energy*, vol. 201, pp. 453–460, 2020, https://doi.org/10.1016/j.solener.2020.03.049.

[12] Mustafa Yusuf Demirci, Nurettin Beşli and Abdülkadir Gümüşçü, "An Improved Hybrid Solar Cell Defect Detection Approach Using Generative Adversarial Networks and Weighted Classification," in *Expert Systems with Applications*, vol. 252, Part A, p. 124230, 2024, https://doi.org/10.1016/j.eswa.2024.124230.

[13] Sergiu Deitsch, Vincent Christlein, Stephan Berger, Claudia Buerhop-Lutz, Andreas Maier, Florian Gallwitz and Christian Riess, "Automatic Classification of Defective Photovoltaic Module Cells in Electroluminescence Images," in *Solar Energy*, vol. 185, pp 455–468, 2019, https://doi.org/10.1016/j.solener.2019.02.067.

[14] R. Pierdicca, M. Paolanti, A. Felicetti, F. Piccinini and P. Zingaretti, "Automatic Faults Detection of Photovoltaic Farms: SolAIr, a Deep Learning-Based System for Thermal Images," in *Energies*, vol. 13, p. 6496, 2020, https://doi.org/10.3390/en13246496.

[15] J. Redmon, S. Divvala, R. Girshick and A. Farhadi, "You Only Look Once: Unified, Real-Time Object Detection," *2016 IEEE Conference on Computer Vision and Pattern Recognition (CVPR)*, Las Vegas, NV, USA, 2016, pp. 779–788, https://doi.org/10.1109/CVPR.2016.91.

[16] U. Batool et al., "A Systematic Review of Deep Learning for Silicon Wafer Defect Recognition," in *IEEE Access*, 2021, https://doi.org/10.1109/ACCESS.2021.3106171.

[17] A. UL-HAQ et al., "Modeling and Fault Categorization in Thin-Film and Crystalline PV Arrays Through Multilayer Neural Network Algorithm," in *IEEE Access*, 2020, https://doi.org/10.1109/ACCESS.2020.2996969.

[18] Manas Parai, Supriyo Srimani, Kasturi Ghosh and Hafizur Rahaman, "Multi-Source Data Fusion Technique for Parametric Fault Diagnosis in Analog Circuits," Integration, vol. 84, pp. 92–101, 2022, https://doi.org/10.1016/j.vlsi.2022.01.005.

[19] Martin T. Hagan, Howard B. Demuth, Mark Hudson Beale and Orlando De Jesús, Neural Network Design, 2nd Edition.

[20] T. Hussain, M. Hussain, H. Al-Aqrabi, T. Alsboui and R. A. Hill, "A Review on Defect Detection of Electroluminescence-Based Photovoltaic Cell Surface Images Using Computer Vision," in *Energies*, vol. 16, p. 4012, 2023, https://doi.org/10.3390/en16104012.

[21] R. Girshick, J. Donahue, T. Darrell and J. Malik, "Rich Feature Hierarchies for Accurate Object Detection and Semantic Segmentation," in *Proc. IEEE Conference on Computer Vision and Pattern Recognition*, Jun. 2014, pp. 580–587.

[22] R. Girshick, "Fast R-CNN," in *Proc. IEEE International Conférence on Computer Vision*, Dec. 2015, pp. 1440–1448.

[23] S. Ren, K. He, R. Girshick and J. Sun, "Faster R-CNN: Towards Realtime Object Detection with Region Proposal Networks," in *IEEE Transactions on Pattern Analysis and Machine Intelligence*, vol. 39, no. 6, pp. 1137–1149, Jun. 2017.

[24] X. Zhang, T. Hou, Y. Hao, H. Shangguan, A. Wang and S. Peng, "Surface Defect Detection of Solar Cells Based on Multiscale Region Proposal Fusion Network," in *IEEE Access*, vol. 9, pp. 62093–62101, 2021, https://doi.org/10.1109/ACCESS.2021.3074219.

[25] J. Terven, D. M. Córdova-Esparza and J. A. Romero-González, "A Comprehensive Review of YOLO Architectures in Computer Vision: From YOLOv1 to YOLOv8 and YOLO-NAS," in *Machine Learning and Knowledge. Extraction*, vol. 5, pp. 1680–1716, https://doi.org/10.3390/make5040083.

[26] D. Lang and Z. Lv, "A PV Cell Defect Detector Combined With Transformer and Attention Mechanism," in *Scientific Report*, vol. 14, p. 20671, 2024, https://doi.org/10.1038/s41598-024-72019-5.

[27] Ashish Vaswani, Noam Shazeer, Niki Parmar, Jakob Uszkoreit, Llion Jones, Aidan N. Gomez, Łukasz Kaiser and Illia Polosukhin, "Attention is All you Need (PDF)," in *Advances in Neural Information Processing Systems*, vol. 30, 2017. Curran Associates, Inc.

[28] Huajun Liu, Fuqiang Liu, Xinyi Fan and Dong Huang, "Polarized Self-Attention: Towards High-Quality Pixel-Wise Mapping," in *Neurocomputing*, vol. 506, pp. 158–167, 2022, https://doi.org/10.1016/j.neucom.2022.07.054.

[29] M. Zhang and L. Yin, "Solar Cell Surface Defect Detection Based on Improved YOLO v5," in *IEEE Access*, 2022, https://doi.org/10.1109/ACCESS.2022.3195901.

[30] M. Zhang and L. Yin, "Solar Cell Surface Defect Detection Based on Improved YOLO v5," in *IEEE Access*, vol. 10, pp. 80804–80815, 2022, https://doi.org/10.1109/ACCESS.2022.3195901.

[31] B. Su, Z. Zhou and H. Chen, "PVEL-AD: A Large-Scale Open-World Dataset for Photovoltaic Cell Anomaly Detection," in *IEEE Transactions on Industrial Informatics*, vol. 19, no. 1, pp. 404–413, 2023.

[32] A. Ahmad, Y. Jin, C. Zhu, I. Javed, A. Maqsood and M. W. Akram, "Photovoltaic Cell Defect Classification Using Convolutional Neural Network and Support Vector Machine," in *IET Renewable Power Generaion*, vol. 14, pp. 2693–2702, 2020, https://doi.org/10.1049/iet-rpg.2019.1342.

[33] W. Yin, S. Lingxin, L. Maohuan, S. Qianlai and L. Xiaosong, "PV-YOLO: Lightweight YOLO for Photovoltaic Panel Fault Detection," in *IEEE Access*, vol. 11, pp. 10966–10976, 2023, https://doi.org/10.1109/ACCESS.2023.3240894.

[34] H. Chen, Y. Pang, Q. Hu, et al., "Solar Cell Surface Defect Inspection Based on Multispectral Convolutional Neural Network," in *Journal of Intelligent Manufacturing*, vol. 31, pp. 453–468, 2020, https://doi.org/10.1007/s10845-018-1458-z.

[35] M. Waqar Akram, Guiqiang Li, Yi Jin, Xiao Chen, Changan Zhu, Xudong Zhao, Abdul Khaliq, M. Faheem and Ashfaq Ahmad, "CNN Based Automatic Detection of Photovoltaic Cell Defects in Electroluminescence Images," in *Energy*, vol. 189, p. 116319, 2019, https://doi.org/10.1016/j.energy.2019.116319.

7 Principles, Methods, and Industrial Applications of Perovskite Silicon Tandem Solar Cell

Velpuri Leela Devi

7.1 INTRODUCTION

A more efficient and affordable photovoltaic (PV) technology has been developed in response to the increasing demand for renewable energy on a worldwide scale. One of the most abundant and environmentally friendly power sources, solar energy is generated by these systems by converting solar radiation into electricity through the PV effect. Solar cells have come a long way since 1954, when Bell Labs unveiled the first silicon (Si) p–n junction solar cell. A power conversion efficiency (PCE) of 6% was the first performance of this cell [1]. With advancements in optical design, surface treatment techniques, device architecture, and Si defect passivation, Si-based PV has come a long way since then [2, 3]. The most recent efficiency table from the National Renewable Energy Laboratory (NREL) shows that the highest reported efficiencies are 26.7% for Si heterojunction cells, roughly 26.1% for single-crystal Si cells, and about 23.3% for multi-crystalline Si cells. Nevertheless, there are growing constraints on future advancements as single-junction solar cells (SJSCs) get closer to the theoretical efficiency of 29% for Shockley–Queisser (S–Q)[4, 5].

Recently, a promising PV technology based on organic–inorganic hybrid metallic halide perovskite (PVK) materials, such as $CH_3NH_3PbI_3$ (commonly known as $MAPbI_3$), has demonstrated significant potential to achieve higher PCE than SJSC [6]. According to Figure 7.1, perovskite solar cells (PSCs) have shown a substantially faster efficiency growth rate compared to Si-based PV technology. This is a significant development in the field of PV, which is considered to be a breakthrough. The optoelectronic properties of metallic halide PVKs make them promising materials for use in solar energy systems of the future. Long carrier diffusion length and lifespan with a low exciton binding energy are some of these features [7]. There has been a significant advancement in PSC technology since 2009 when the first PVK-based dye-sensitized solar cell reported a 3.8% of PCE [8], PVK PV technology has attracted significant global attention, leading to rapid advancements. Over a decade of development, its efficiency has been steadily improved, reaching a certified value of 25.5% [9]. Major advances in optimizing device structures, modifying PVK compositions,

 DOI: 10.1201/9781003515784-7

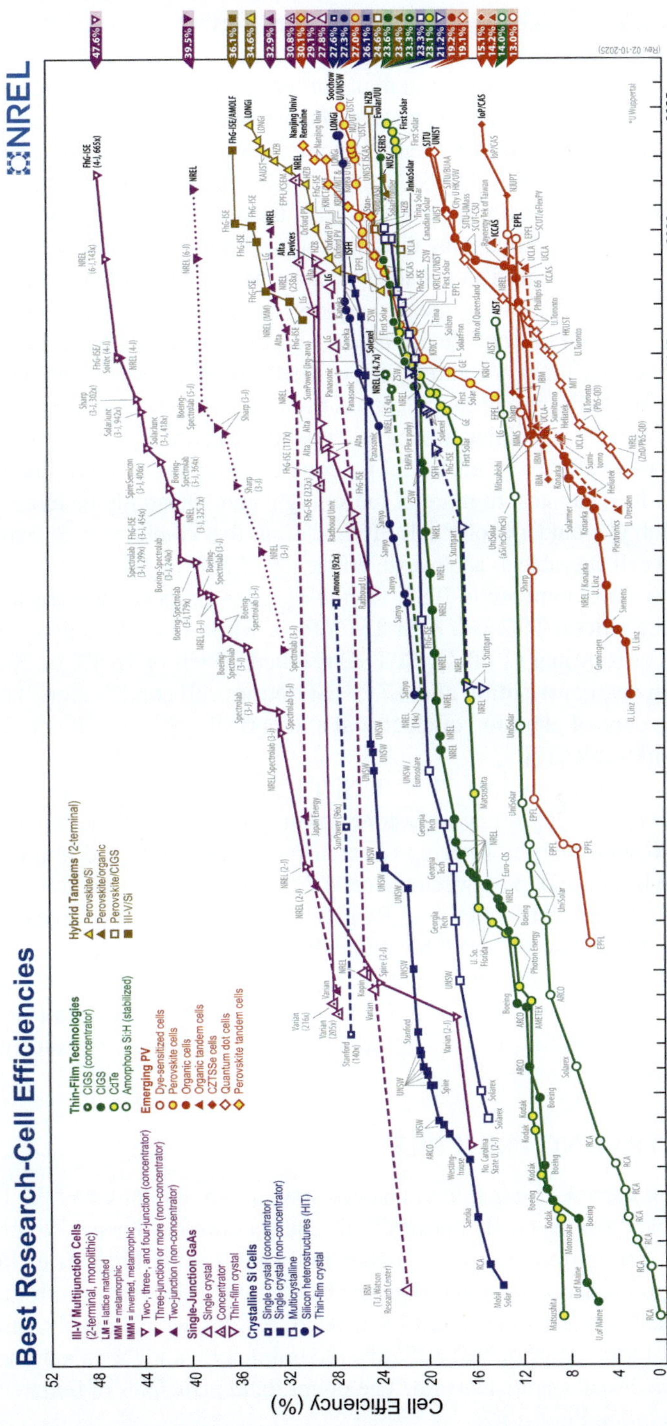

FIGURE 7.1 1975–2025 best research-cell PCE for different solar technologies [4].

and accurately controlling crystal formation are responsible for this breakthrough [10, 11]. The efficiency and stability of PSC have been further improved by efforts to passivate PVK bulk and defects in the surface, optimize charge-transporting layers (CTLs), engineer device interfaces, and make progress in related fields [12, 13].

Recent advancements in PSC efficiency have been remarkable, as shown in Figure 7.1. On the other hand, improvements are happening at a slower pace, which means that the PCE of PSC is getting close to the S–Q limit. Under perfect circumstances, free of angular constraints, theoretical models predict that the efficiency limit of a PVK-based SJSC can exceed 30% [14]. Enhancing the efficiency of PVK-based SJSC beyond that of Si-based SJSC remains a significant challenge, as surpassing the S–Q limit is unlikely. While PVK PV technology has progressed rapidly, its practical application is still hindered by limitations such as small device areas and insufficient long-term stability [15]. Traditional Si-based technology controls 95% of the PV market share, with thin-film PV technology making up the remaining 5%. It has been suggested to combine a Si sub-cell with a semiconductor material that has a greater bandgap in order to make Si-based PV converters even more efficient. This technique can be used to make multi-junction solar cells (MJSCs), which can absorb more of the sunlight and potentially be more efficient than SJSC. Notably, a tandem solar cell (TSC) design that combines a Si bottom cell with a III–V top cell was able to achieve an efficiency of 33% [16].

According to developments in TSC technology, researchers have been investigating perovskite silicon TSC (PVK/Si TSC) for a while now. An initial proof-of-concept for a four-terminal (T) PVK/Si TSC reached a PCE of 13.4% in 2014, with a 6.2% PVK top cell contribution and a 7.2% Si bottom cell contribution. The study found that, with careful planning of electrical and optical systems, efficiencies of up to 31.6% are achievable [17].

A year later, in 2015, the first 2T PVK/Si TSC was reported, reaching an overall PCE of 14.3%, marking a key milestone in integrated TSC PV technology [18]. Simulations indicated that, by filtering the transparent electrode and improving the interconnecting layers (ICLs) between the cells, PCE of up to 32% could be achieved. From 2017 to 2020, the certified PCE of PVK/Si TSCs increased dramatically, going from 23.5% to 29.5%, which is higher than the S–Q limit for SJSC as shown in Figure 7.1. However, PCE levels have not yet reached their upper limit [19]. Research in optics and electronics suggests that PVK/Si TSCs hold the potential to exceed 40% efficiency, highlighting promising opportunities for further advancements in TSC performance.

7.2 TSC OPERATING PRINCIPLE

The top-junction narrow-bandgap cell absorbs low-energy photons and the bottom-junction wide-bandgap (E_g) cell captures high-energy photons during the formation of a TSC. This configuration enables higher PCE, minimizes hot carrier thermalization losses, and broadens the wavelength absorption range. Henry introduced an intuitive graphical method for determining TSC PCEs [20], which visually quantifies two fundamental loss mechanisms: radiative recombination and the misalignment of an SJSC with the broad solar spectrum. The operational principles of both SJSC and MJSC are depicted in Figure 7.2.

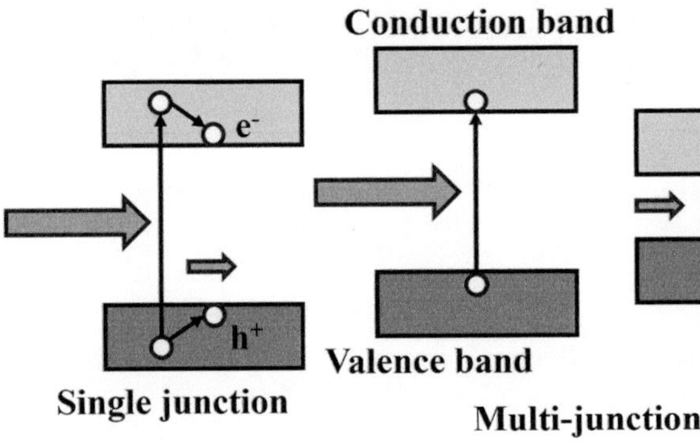

FIGURE 7.2 SJSC and MJSC working principle.

The standard test situations are 25°C, an AM 1.5G solar spectrum is used, and the irradiance is 100 mW cm^{-2}. This approach takes a working temperature (T) of 298 K for all sub-cells into consideration. The following is the process to calculate the integrated photon flux, which is given in photons m^{-2} s^{-1} [21]:

$$n_{ph}\left(E_g\right) = \int_{E_g}^{\infty} \frac{dn_{ph}}{dh\nu} dh\nu$$

$$= \sum_{i=E_g}^{\infty} \left(h\nu_{i+1} - h\nu_i\right) \frac{1}{2} \left(\frac{dn_{ph}}{dh\nu}\left(h\nu_{i+1}\right) + \frac{dn_{ph}}{dh\nu}\left(h\nu_i\right) \right) \tag{7.1}$$

Here, h stands for Planck's constant, which is 6.626×10^{-34} J s. The relationship between the speed of light (c) and the frequency of light (ν) can be expressed as $\nu = c/\lambda$, where ν is the wavelength in nanometers and 2.998×10^8 m s^{-1} is the value of c. Consequently, photon flux is used to express the AM 1.5G spectrum irradiance. You can find the load's current density, $J = I/A$, by using this formula:

$$J = J_{ph} + J_{th} + J_{rad} \tag{7.2}$$

The current density that is linked to thermal radiation is represented as J_{th}, while that generated by absorbed solar energy is denoted as J_{ph}, Additionally, radiation that is either absorbed by the substrate or emitted by the upper surface is also known as J_{rad}.

According to their E_g values, semiconductors act as opaque materials when the photon energy is higher than them and as transparent when the photon energy is lower. When the E_g value is 0.30 eV or greater, J_{ph} and J_t become insignificant for semiconductors. Hence, this term can be removed to simplify the analysis, resulting in the following equation:

$$J = en_{ph} - \frac{2\pi \exp\left(n^2+1\right) E_g^2 KT}{h^3 c^2} \left(\frac{eV - E_g}{KT} \right) \tag{7.3}$$

To find the open-circuit voltage (V_{oc}), we need to set $J = 0$ using the following equation:

$$eV_{oc} = E_g - KTIn\left(\frac{\frac{2\pi \exp\left(n^2+1\right)E_g^2 KT}{h^3c^2}}{en_{ph}}\right) \tag{7.4}$$

By putting the derivative dJV/dV equal to zero, we can get the maximum power point (MPP), referred to as J_{MPP} and V_{MPP}.

$$eV_{MPP} = eV_{oc} - KTIn\left(1+\frac{eV_{MPP}}{KT}\right) \tag{7.5}$$

$$J_{MPP} = \frac{en_{ph}}{1+KT/eV_{MPP}} \tag{7.6}$$

Combining all (7.4),(7.5), and (7.6),

$$W_{MPP} = E_g - KT\left(In\left(\frac{\frac{2\pi \exp\left(n^2+1\right)E_g^2 KT}{h^3c^2}}{en_{ph}}\right) + In\left(1+\frac{eV_{MPP}}{KT}\right)+1\right) \tag{7.7}$$

To find the maximum PCE, Henry's graphical quantum efficiency study takes into account the two main intrinsic losses. The greatest area contained is reached by adjusting ΔE_g until this is accomplished [22].

7.3 ARCHITECTONICS OF THE TSC

The comparable circuit in Figure 7.3 shows that TSCs are mostly 2T monolithic or 4T. Semi-transparent wide-bandgap (E_g) top cells are physically layered above narrow E_g bottom cells in a 4T configuration. The 4T cell, with electrically divided upper and lower cells, allows sub-cell optimization. Thus, each cell's power output is retrieved separately. In contrast, a 2T builds upper and lower sub-cells immediately on top of each other. Accurate fabrication and matching of currents between the two series-connected sub-cells are prerequisites for this design. Since fewer transparent electrodes are used in a 2T cell as opposed to a 4T cell, reflection and parasitic absorption losses are reduced and efficiency is raised [23, 24]. In a 2T, the series connection of sub-cells requires constant synchronization of upper and bottom cell currents. It's vital to avoid the sub-cell, with the lower current becoming a device current bottleneck. In contrast, 4T TSC are immune to current matching limitations

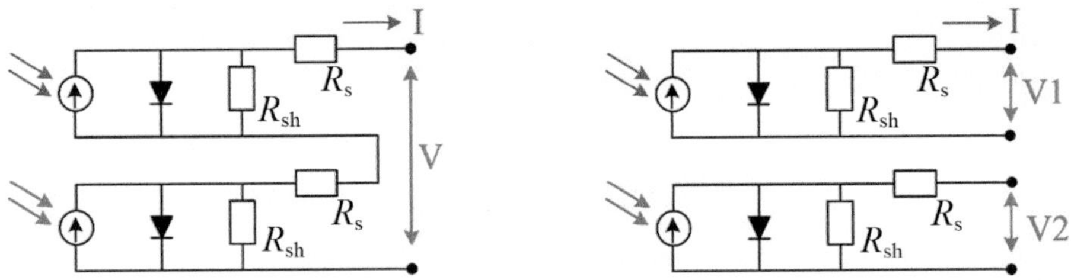

FIGURE 7.3 Equivalent circuit representations of 4T and 2T PVK/Si TSC.

due to electrical separation between top and bottom cells. This independence lets each sub-cell be optimized separately, improving performance by increasing top cell efficiency or bottom cell energy conversion.

Organic–inorganic hybrid PVK materials are considered highly efficient semiconductors for TSC due to their tunable E_g, which can be adjusted between 1.3 and over 2 eV through compositional engineering [25]. Earlier computations indicated that, for a TSC to attain its maximum PCE, the optimal top cell for integration should have an E_g value ranging from 1.6 to 1.8 eV, with a bottom cell composed of crystalline Si having an E_g value of 1.1 eV [26]. The PVK retains great transparency for photon energies below its E_g due to ultralow sub-E_g absorption. This reduces bottom cell optical losses, improving TSC efficiency. It is possible to directly integrate a PVK onto a Si by producing the PVK layer using either solution processing or vacuum thermal evaporation. The development of 2T TSCs with improved PCE and scalability is made possible by this [27]. Table 7.1 presents a comparative analysis of 2T and 4T PVK/Si TSC, highlighting their respective advantages and disadvantages.

7.3.1 PVK/Si TSC

Even though PVK/Si TSCs have been around only for a little over a decade, research on them has been expanding at a steady rate. Figure 7.4 shows the PVK/Si 2T TSC, which was initially proposed in 2014 by Loper et al. [28]. To assess the limiting PCE of a PVK/Si 2T TSC, researchers employed an optical model. Under ideal conditions, defined by the absence of parasitic absorption and 100% external quantum efficiency (EQE), the maximum theoretical PCE was determined to be 35.67%. Subsequently, the same research group conducted an experimental demonstration of a 4T PVK/Si TSC, achieving 13.4% PCE. Löper et al. projected that, by minimizing optical losses, particularly through the use of high-transparency contacts, the 4T TSC could potentially attain a PCE exceeding 30% [17].

After that, Bailie and colleagues showed how to make a PVK top cell with a transparent electrode made of silver nanowires. A comparable 4T PVK/Si TSC attained a PCE of 17% through using this method [29]. With a consistent PCE of 13.7%, the first proof-of-concept for PVK/Si TSC was shown in early 2015. Connecting the top and bottom cells using an ICL designed of thin Si improved charge transmission and allowed for better device integration [18]. Instead of using the traditional spiro-OMeTAD layer, Futscher and Ehrler suggested using a transparent hole-transporting

TABLE 7.1
Evaluation of PVK/Si 2T and 4T TSCs

Feature	2T TSC	4T TSC
Structure	Monolithic, which is series-connected	Mechanically stacked
Electrical configuration	Shared current as in current-matched	Independent operation of both cells
Current matching	Required because both sub-cells must have matching currents	Not required as each sub-cell operates separately
Power extraction	Constrained with the lower current output	Optimal maximum power point (MPP)
Fabrication complexity	High as monolithic integration is needed	Lower due to separate fabrication
Optical losses	Lower in fewer interfaces, which makes it better light absorption	Higher as extra layers increase parasitic absorption
Efficiency (lab scale)	~32.9% which is current best	>36%, which is higher potential efficiency
Scalability	High as large-scale manufacturing is easier	Moderate as integrating into PV systems is more challenging
Manufacturing cost	Lower as fewer external components are needed	Higher as it requires additional processing steps
Stability considerations	Requires stable interconnection layers	More resilient to degradation due to independent sub-cells
Application suitability	Ideal for commercial solar modules	Suitable for research and in high-performance PV applications

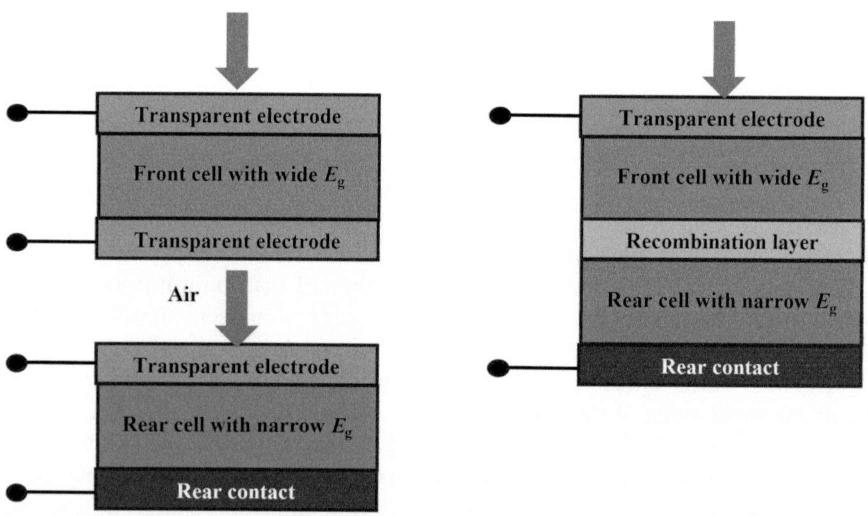

FIGURE 7.4 Layer architecture of 4T vs. 2T perovskite/Si TSC.

layer (HTL) to improve 2T TSC PCE. However, their findings indicated that using PVK as the top cell significantly restricted the overall PCE, highlighting the need for further optimization in material selection and device architecture [27]. Better performance in PVK/Si TSCs was achieved in early 2016 as a result of composition engineering improvements to a PVK material with an E_g of 1.74 eV. An impressively efficient 4T TSC PCE was 25.2% with this configuration [30].

The 4T PVK/Si TSC was designed using a transparent electrode for the PVK top cell, resulting in a total PCE of 23% [31]. Due in large part to their ease of manufacture, 4T PVK/Si TSC were the subject of considerable research prior to 2017. A semi-transparent PVK was set above a Si to execute this approach, which enabled straightforward manufacture. Improvements in ICLs, PVK composition, and top transparent electrodes resulted from researchers' focus shifting to 2T PVK/Si TSCs after 2017. Qiu et al. explored 2T PVK/Si TSCs with varying E_g energies and PVK compositions. Notably, a 2T TSC achieved a 20% PCE integration of a PVK with a Si [32].

In the following years, extensive investigation of PVK/Si TSCs has been conducted, with a strong focus on improving transparent top electrodes, optimizing PVK and Si cells, advancing interface engineering, and innovating ICLs. These efforts have contributed to significant advancements in efficiency, stability, and overall device performance [33]. In 2020, significant advancements were achieved in the PCE of both 4T and 2T PVK/Si TSCs. Early that year, Helmholtz Zentrum Berlin (HZB) successfully developed a 2T PVK/Si TSC that attained a certified PCE of 29.15%, marking a major milestone in TSC technology [33]. When it came to certified PCE for a 2T PVK/Si TSC, Oxford PV achieved a record-breaking 29.5% by the end of 2020. Nevertheless, information about the design of the devices and their PV performance measures is still limited. Additionally, 4T TSC achieved 28.2%, which is a verified world-record PCE through improvements in the overall device performance, optimization of PVK E_g, and the making of a high-transparency top electrode [34]. The PCE of PVK/Si TSCs is expected to exceed 30% in the near future, as indicated in Table 7.2.

7.3.2 Perovskite–Perovskite (all-PVK) TSC

7.3.2.1 4T all-PVK TSC

Figure 7.5 shows that the development of all-PVK TSCs has become a prominent priority in the PVK research field, due to the adjustable E_g of PVK materials. However, the use of highly polar solvents in PVK film deposition poses challenges for the fabrication of monolithic TSCs. The most basic method uses a 4T PVK TSC to physically stack two PVK cells, one with a wide E_g and one with a narrow E_g. One of the most effective and widely used strategies for reducing the E_g in PVK materials is the replacement of lead with tin [35]. To create high-performance 4T all-PVK TSCs, Wu and Wang [36] used a Pb–Sn binary PSC, which showed high PCE and stability.

Challenges arise when fabricating an Sn-based PVK film that is smooth and pinhole-free on a flat substrate [37, 38].

Because their carrier lifetimes are so short, Pb–Sn-based PSCs show a decrease in light sensitivity at wavelengths exceeding 700 nm [39]. Zhao et al. [40] state that Pb–Sn PVK films' carrier lifetimes are significantly affected by their crystallinity and

TABLE 7.2
Important Timeline Achieved for PVK/Si TSC Development

Year	Key Milestone
2009	First PSC by Kojima et al., who introduced PVK-based PV, achieving an efficiency of around 3.8%.
2013	First PVK/Si TSC concept proposed by researchers, who integrated PVKs with Si to surpass the S–Q limit.
2014	Theoretical 35.67% PCE limit calculated for an ideal 2T PVK/Si TSCs. And 4T PVK/Si TSCs achieved 13.4% efficiency.
2015	First efficient 4T PVK/Si TSCs by Oxford PV and other researchers, who demonstrated a 4T PVK/Si TSC with around 13.7% efficiency. And 2T PVK/Si TSCs achieved 13.7% efficiency.
2016	First 2T monolithic PVK/Si TSCs by École Polytechnique Fédérale de Lausanne (EPFL), which achieved 18.1% efficiency, and 4T PVK/Si TSCs with 1.74 eV E_g PVK, which achieved 25.2% PCE.
2017	2T PVK/Si TSCs reached 20.57% efficiency with 1.69 eV E_v PVK. 4T PVK/Si TSCs with ultrathin metal electrodes reached 23.0% PCE.
2018	Efficiency surpasses 25% by a research group from Oxford PV, KAUST, and EPFL. Large-area 16 cm² 2T PVK/Si TSCs developed without an indium tin oxide (ITO) layer between sub-cells. 2T PVK/Si TSCs reached 25.5% efficiency using a light management strategy.
2019	25.4% 2T PVK/Si TSCs achieved by controlling PVK grain growth. Mechanically stacked 2T PVK/Si TSCs reached 25.9% efficiency.
2020	Oxford PV reported a record efficiency of 29.5% for a 2T PVK/Si TSC by the end of the year. In parallel, Helmholtz-Zentrum Berlin (HZB) achieved a 29.1% PCE with their own 2T PVK/Si TSCs. Additionally, 4T PVK/Si TSC devices reached an efficiency of 28.2%, highlighting significant progress across different tandem architectures.
2021	Commercial scaling begins by an Oxford PV, which announced plans to commercialize 28–30% efficient PVK/Si TSC modules.
2022	Record efficiency of 31.3% by HZB, which set a new record with a 31.3% efficiency for 2T PVK/Si TSCs.
2023	According to records from the National Renewable Energy Laboratory (NREL), PVK/Si TSCs have surpassed 32.9% efficiency. In laboratory conditions, 4T configurations have achieved efficiencies exceeding 36%, marking a significant milestone in photovoltaic performance.
2024	Industrial development advances more companies, including LONGi and JinkoSolar, investing in scaling PVK/Si TSCs technology for mass production.
2025 and beyond	Targeting 35%+ efficiency and long-term stability. Focusing on improving stability, scalability, and commercial viability for large-scale deployment.

grain size. In a PSC, they produced a PCE of 17.6% using a $(FASnI_3)_{0.6}(MAPbI_3)_{0.4}$ film that was 620 nm thick and had extended carrier lifetimes of 255 ns. Additionally, the mechanical stacking of a semi-transparent $FA_{0.3}MA_{0.7}PbI_3$ (with an E_g of 1.58 eV) as the top cell resulted in a PCE of 21.2%. An effective semi-transparent $FA_{0.8}Cs_{0.2}Pb(I_{0.7}Br_{0.3})_3$ with an E_g of around 1.75 eV was also developed as the top cell,

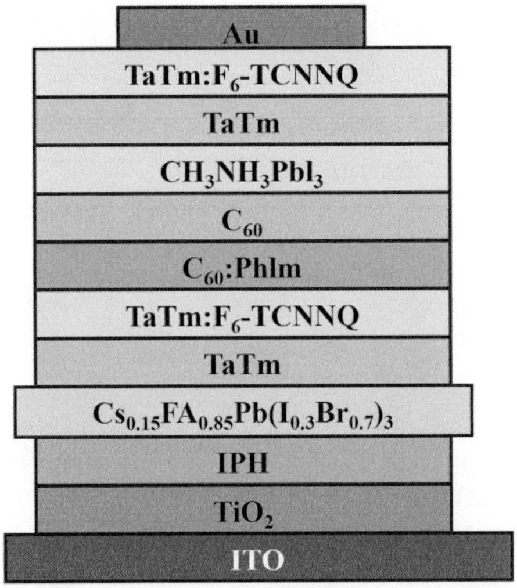

FIGURE 7.5 Schematic of 2T PVK TSCs.

achieving a PCE of 23.1% in the optimal 4T PVK TSC. The top and bottom cells were optically coupled using paraffin oil to reduce optical losses [41]. The optimal 4T of PVK TSCs achieved PCE was around 23.1%. Recently, Tong et al. [42] developed a film with an E_g of 1.25 eV, made of Pb–Sn-based PVK, and showed electron–hole recombination time surpassing 1 μs by adding guanidinium thiocyanate (GuSCN) to the Pb–Sn PVK film. The combination of this with a top cell made of $Cs_{0.05}FA_{0.8}MA_{0.15}PbI_{2.55}Br_{0.45}$ and an E_g of 1.63 eV allowed for the achievement of a 25.4% PCE by a 4T PVK TSC [43]. These developments demonstrate the ongoing progress in Pb–Sn PVK TSCs, with various strategies such as grain size control, carrier lifetime optimization, and interface engineering, leading to improved PCEs and enhanced device stability. The integration of advanced techniques like optical coupling and vacuum-assisted growth is pushing the limits of TSC performance.

7.3.2.2 2T all-PVK TSCs

When working with 2T all-PVK TSCs, one of the primary challenges is the selection of ICLs, which need to be wisely chosen and adequately protected to prevent damage to the underlying layers. Using polymer ICLs, Jiang et al. [44] obtained 2T all-PVK TSC. To improve the TSC device's performance and reliability, this method employs protective layers to prevent solvent degradation while deposition is taking place. This keeps the underlying PVK film stable. Figure 7.6 shows that the ICLs in TSCs serve both as charge recombination layers and as a protective barrier for the primary PVK film when the top cell PVK layer is being deposited. A TSC using the same $MAPbI_3$ as both cells managed only a PCE of 7%, even with these improvements.

Heo and Im [45] made 2T PVK TSC by layering $MAPbBr_3$ with a $MAPbI_3$ as the top cell and bottom cell. The HTL comprised P3HT doped with Li-TFSI, serving as the ICL in the top cell. This tandem configuration achieved a PCE of 10.4%, with a high V_{oc}

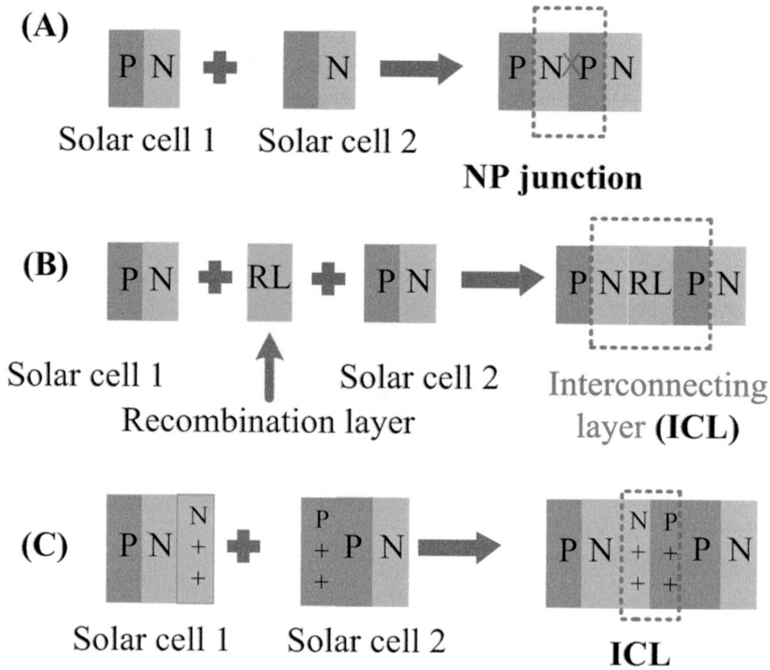

FIGURE 7.6 Illustrations of ICL.

of 2.25 V. To fabricate the bottom cell without damaging the underlying layer, Bolink et al. used a vacuum deposition process. They employed doped organic semiconductors for the charge transfer and interconnection layers. This approach led to a PCE of 15.6% in a TSC comprising a $MAPbI_3$ bottommost cell and a $Cs_{0.15}FA_{0.85}Pb(I_{0.3}Br_{0.7})_3$ top cell.

Furthermore, an optical model was utilized by Bolink et al. to optimize a vacuum-deposited $MAPbI_t$/$MAPbI_t$ TSC. The optimized device achieved an impressive V_{oc} of 2.3 V, which is quite near the theoretical limit [46]. In an unlike approach, Rajagopal et al. [47] built a 2T all-PVK TSC using a sputtered ITO recombination layer and combining E_g of 1.82 eV material and $MAPb_{0.5}Sn_{0.5}I_3$. This TSC achieved a PCE of 18.5% after interfacial and optical engineering. With all of the absorber layers (ALs) working together, V_{oc} reached 1.98 V or around 80% of the voltage limit [48], as shown in Table 7.3. These studies highlight the ongoing advancements in TSC technology, with the optimization of ICL, optical properties, and interface engineering, leading to substantial improvements in PCE and stability for all-PVK and hybrid PVK/organic TSC.

Zhao et al. [49] designed a 2T TSC using vacuum deposition techniques to apply ultrathin layers of Ag, MoO_x, and ITO as the ICLs where, Ag serves as the recombination center, while the MoO_x layer confirms a robust connection between Ag and ITO, protecting the underlying layers from potential damage during ITO sputtering. By introducing chlorine (Cl), a Pb–Sn PVK film with an E_g of 1.25 eV was developed. This configuration led to a PCE of approximately 21% with the front cell. The

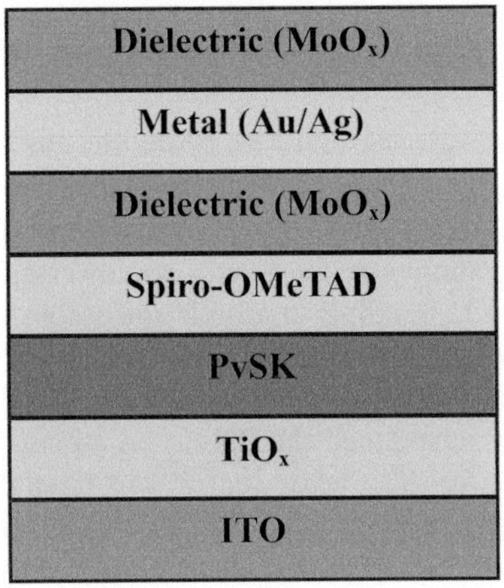

FIGURE 7.7 The configuration of a PSC featuring a multilayer transparent top electrode.

TABLE 7.3
All PVKs TSCs With Respect to Overall Performance

Type	PVK	Transparent Electrodes	ICL	V_{oc} (V)	J_{sc} (mA cm^{-2})	FF (%)	PCE (%)	Ref.
2T	$Cs_{0.05}FA_{0.8}MA_{0.15}PbI_{2.55}Br_{0.45}$ $(FASnI_3)_{0.6}(MAPbI_3)_{0.4}$	ITO	$Ag/MoO_x/ITO$	1.942	15.01	80.3	23.4	[43]
2T	$DMA_{0.1}Cs_{0.3}FA_{0.6}PbI_{2.4}Br_{0.6}$ $Cs_{0.25}FA_{0.75}Sn_{0.5}Pb_{0.5}I_3$	ITO	AZO/IZO	1.88	16.0	77.0	23.1	[50]
2T	$Cs_{0.4}FA_{0.6}PbI_{1.95}Br_{1.05}$ $Cs_{0.05}FA_{0.5}MA_{0.45}Sn_{0.5}Pb_{0.5}I_3$	ITO	$C_{60}/SnO_{1.76}$	2.03	15.2	79.7	24.4	[51]
2T	$FA_{0.6}Cs_{0.4}Pb(I_{0.65}Br_{0.35})_3$ $Cs_{0.05}FA_{0.5}MA_{0.45}Sn_{0.5}Pb_{0.5}I_3$	ITO	SnO_2/ITO	1.99	15.1	0.77	22.7	[52]
2T	$FA_{0.6}Cs_{0.4}Pb(I_{0.65}Br_{0.35})_3$ $Cs_{0.05}FA_{0.5}MA_{0.45}Sn_{0.5}Pb_{0.5}I_3$	ITO	SnO_2/ITO	1.99	15.1	0.77	22.7	[53]
2T	$FA_{0.8}Cs_{0.2}Pb(I_{0.6}Br_{0.4})_3$ $FA_{0.7}MA_{0.3}Sn_{0.5}Pb_{0.5}I_3$	ITO	SnO_2/Au	1.965	15.6	81.0	24.8	[54]
2T	$FA_{0.8}Cs_{0.2}Pb(I_{0.6}Br_{0.4})_3$ $FA_{0.7}MA_{0.3}Sn_{0.5}Pb_{0.5}I_3$	ITO	SnO_2/Au	2.013	16.0	79.8	25.6	[54]
4T	$FA_{0.83}Cs_{0.17}Pb(I_{0.5}Br_{0.5})_3$ $FA_{0.75}Cs_{0.25}Sn_{0.5}Pb_{0.5}I_3$	$SnO_2/ZTO/ITO$		0.97 0.74	20.3 7.9	79.0 73.0	20.3	[38]
4T	$FA_{0.3}MA_{0.7}PbI_3$ $(FASnI_3)_{0.6}(MAPbI_3)_{0.4}$	$MoO_x/Au/$ MoO_x		1.141 0.808	20.1 4.8	80.0 74.4	21.2	[40]
4T	$FA_{0.8}Cs_{0.2}Pb(I_{0.7}Br_{0.3})_3$ $(FASnI_3)_{0.6}(MAPbI_3)_{0.4}$	MoO_x/ITO		1.204 0.81	17.6 12.1	74.6 75.3	23.1	[41]
4T	$Cs_{0.05}FA_{0.8}MA_{0.15}PbI_{2.55}Br_{0.45}$ $(FASnI_3)_{0.6}(MAPbI_3)_{0.4}$	$SnO_2/ZTO/IZO$		1.124 0.806	20.8 10.5	79.3 80.6	25.4	[42]

PCE for the 2T all-polymer TSC was 23%, utilizing a GuaSCN-modified Pb–Sn PVK [42]. Palmstrom et al. [50] documented the use of a polyethylenimine ethoxylated (PEIE) induced the densification of a conformal and dense aluminum-doped zinc oxide (AZO) layer. The back cell was processed on a thin indium zinc oxide (IZO) layer, with AZO/IZO acting as an effective ICL, significantly mitigating solvent damage. To broaden the E_g of PVK, they employed a compensatory A-site-based technique, mixing both cations on the A-site. By lowering bromine concentration, they achieved wide E_g PVK, leading to a high and stable V_{oc}, which helped inhibit phase separation. The team produced 2T TSCs on rigid and flexible substrates, achieving a PCE of 23.1%.

Yu et al. [51] basically introduced the ICL as an atomic layer deposition (ALD)-deposited SnO_{2-x} layer above a C_{60} layer. Iodine ions from the PVK are accidentally n-doped with the C_{60} layer, making it an effective electron transport layer (ETL). The partial oxidation of the SnO_{2-x}, regulated by pulse duration, reaction time, and deposition temperature, led to a high density of Sn^{2+} ions. This allowed for ambipolar carrier transport characteristics in the $SnO_{1.76}$ layer, establishing an ohmic contact between C_{60} and $SnO_{1.76}$, facilitating charge carrier recombination between the two sub-cells. In their study, both large-area (1.15 cm²) and small-area (0.059 cm²) devices demonstrated high PCEs of 22.2% and 24.4%, respectively, for the 2T TSC based on $C_{60}/SnO_{1.76}$ ICL. These developments demonstrate the significant progress made in 2T TSCs, with ongoing innovations in ICLs and tandem cell configurations driving higher efficiencies and improved device performance.

A significant problem with Pb–Sn binary PVK is the oxidation of Sn^{2+} to Sn^{4+}, which leads to higher defect densities, shorter carrier lifetimes, and inadequate stability. Studies aiming at enhancing the stability and performance of Pb–Sn-based PSCs have been prompted by this oxidation process. Using a FA0.75Cs0.25Sn0.5Pb0.5I3 front cell as a source, Leijtens et al. [55] showed that PVK grain sizes greater than a micron were produced by post-processing FA0.75Cs0.25Sn0.5Pb0.5I3 films with MACl vapor. This greatly improved the operational, thermal, and atmospheric stability of the cells, leading to a PCE of 19.1%. The charge-collecting efficiency was limited in tin-based PSCs due to the short electron diffusion length, as noted by Yang et al. [52]. The addition of CdI_2 to the tin-based PVK led to an electron diffusion length of 2.72 ± 0.15 μm. With this upgrade, the 2T all-PVK TSC's PCE remained at 22.7%.

Lin et al. [55] incorporated metallic tin, which improved the carrier diffusion length to 3 μm. This modification allowed for a solar cell with an E_g of 1.22 eV to achieve a PCE of 21.1% for $MA_{0.3}FA_{0.7}Pb_{0.5}Sn_{0.5}I_3$. The resulting 2T TSC demonstrated verified PCEs of 24.8% for an area of 0.049 cm² and 22.1% for an area of 1.05 cm² when paired with a wide E_g PVK with an E_g of 1.77 eV. These results demonstrate that the PCE of all-PVK TSCs exceeds that of single-junction PSCs [54], further cementing their potential as high-efficiency devices.

7.3.3 Perovskite–Organic TSCs (PVK/Org TSC)

The versatility, small size, and ease of solution processing of organic solar cells (OSCs) have made them a hot topic [56, 57]. One notable development in enhancing the photo response of PSCs is the creation of a combined PVK/organic TSC, which extends the photo response of the PSC. Chen et al. [58] accomplished a 2T with PCE of 10.23%, demonstrating the potential of integrating $MAPbI_3$ front cells with PCE-10:PC71BM bottom cells. The PVK film thickness was limited to 90 nm to ensure sufficient photon absorption by the back cell. To optimize electron–hole recombination, the ICL was implemented with a silver thin film, achieving a PCE of 16% for the 2T TSC [59].

Li et al. [60] further advanced 2T PVK/Org TSCs, integrating a bulk heterojunction layer (HJL) composed of PBDB-T:SN6IC-4F and a surface-passivated layer of phenmethylammonium bromide (PMABr) with a $Cs_{0.1}(FA_{0.6}MA_{0.4})_{0.9}Pb(I_{0.6}Br_{0.4})$ composition. This tandem achieved a PCE of 15.13% and a V_{oc} of 1.85 V. Additionally, they developed a PV-driven electrolysis system integrating the TSC with water-splitting electrocatalysis, demonstrating the advantages of the high V_{oc} and achieving a solar-to-hydrogen PCE of 12.30%.

In another significant advancement, Chen et al. [61] introduced a highly 2T PVK/Org TSC configuration utilizing $FA_{0.8}MA_{0.02}Cs_{0.18}PbI_{1.8}Br_{1.2}$ for the top cell and PBDBT-2F:Y6:PC71BM (or knowingly called as PM6:Y6:PC71BM, full ternary blend) for the bottom cell. This TSC accomplished a PCE of 20.6%, along with a high V_{oc} of 1.902 V, using an all-thermally evaporated interlayer configuration. The top PVK cell functions reduce UV sensitivity and enhance the photostability of the OSC, thus improving the overall performance of the TSC. Due to their remarkable thermal stability [62, 63], all-inorganic PVK materials offer an excellent means of shielding OSCs from UV light. The top cell was CsPbBr3, while the bottom cell was PBDB-T-SF:IT-4F in a 4T inorganic PVK/org TSC constructed by Chen et al. [64], as shown in Figure 7.8.

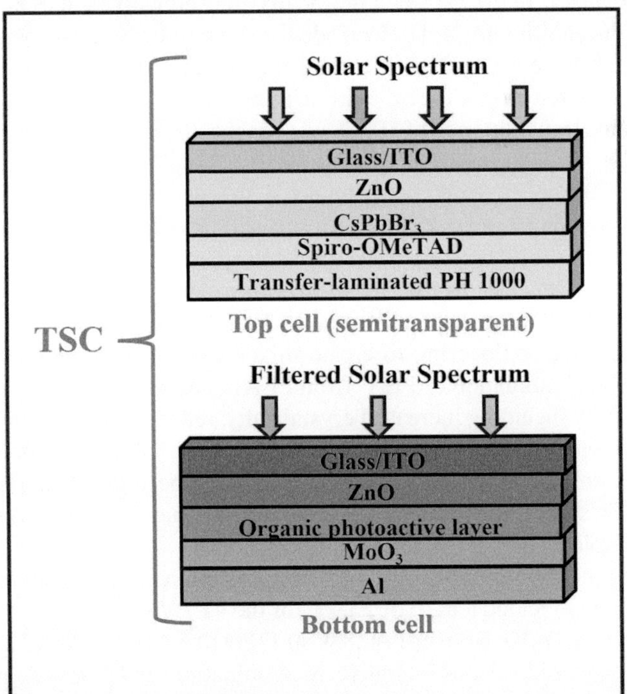

FIGURE 7.8 The structure of 2T PVK/Org TSC.

One potential method to enhance the PCE of all-inorganic PSCs is to combine OSC and PSC devices into a TSC. Xie et al. [65] demonstrated a 2T TSC using $CsPbI_2Br$ as the top cell, paired with a bulk HJL-made multilayer as the ICL. They found that UV light harvesting by $CsPbBr_3$ enhanced the photostability of the bottom cell, resulting in a PCE of 14.03% for the 4T TSC. Additionally, they designed a 2T TSC with a ternary PTB7-Th:COi8DFIC:PC_{71}BM rear cell and a $CsPbI_2Br$ front cell, achieving a PCE of 15% [66]. By optimizing device fabrication and light illumination, TSCs have demonstrated PCEs above 18%. Additionally, with optimal sub-cell E_g, the potential PCE of PVK/org TSCs is predicted to surpass 31%, which is a major advancement in the quest for high-efficiency TSCs.

7.3.4 Perovskite–Cu(In, Ga)Se_2TSCs

7.3.4.1 PVK/CIGS (Copper Indium Gallium Selenide) TSC

CIGSs are a feasible PV option with advantages over c-Si technology. CIGS bottom cell E_g is usually set at 1.0 eV for tandem designs. The possibility for great efficiency and design flexibility has made PVK/CIGS TSCs popular in recent years. MAPbI3 top cells use Ag nanowires as transparent electrodes, enabling the early creation of a 4T PVK/CIGS TSC. When combined with a CIGS bottom cell, this arrangement had 18.6% PCE [67]. In another advancement, Fu et al. [68] utilized In_2O_3:H as a transparent electrode, which led to the development of a semi-transparent PSC with a PCE of 14.2% and a typical transmittance in the range of 800 to 1150 nm. The 4T TSC achieved a PCE of 20.5% when physically stacked with a CIGS bottom cell. By using the high-mobility In_2O_3:H electrode, an inverted semi-transparent PSC was successfully fabricated. Combining a semi-transparent MAPbI3 device with CIGS bottom cells resulted in a 4T TSC with a PCE of 22.1% and 20.9%, respectively, while an average transmittance of 80.4% was reached by a semi-transparent MAPbI3 device [69].

Shen et al. [70] achieved high performance with low hysteresis using a CsRbFAMA PVK material with an E_g of 1.75 eV. Both sub-cells incorporated antireflection coatings to enhance transmittance, with the semi-transparent PSCs demonstrating a usual transmittance of over 80% in the 700–1200 nm range. The optimal 4T PVK/CIGS TSC, using a CIGS bottom cell, achieved a PCE of 23.9%. Kim et al. [71] used bimolecular and additive engineering to create mixed-dimensional 3D–2D–2D PSC for the top cell. The combination of PEA^+ from PEAI and SCN^- from $Pb(SCN)_2$ boosted the film quality by including increased crystallinity, reduced PbI_2 production, reduced defect density, enhanced carrier mobility, and lifetime. Table 7.4 shows that these improvements enhanced performance significantly. These developments show that PVK/CIGS TSCs can achieve high PCE, adaptability, and cost-effectiveness, making them a formidable challenger in the solar market.

When combined with a 1.12 eV CIGS bottom cell, a semi-transparent PSC with an IZO top contact yielded a PCE of 25.9% for the 4T PVK/CIGS TSC. Gharibzadeh et al. [75] deposited n-BABr over FA0.83Cs0.17Pb(I0.6Br0.4) to build a 2D/3D PVK heterostructure to reduce non-radiative recombination and improve performance. This reduction in recombination increased V_{oc} and PCE, stabilizing the 4T PVK/CIGS TSC PCE at 25%. Feurer et al. [74] passivated the CIGS surface post-deposition, reducing defect density.

TABLE 7.4
PVK/CIGS TSC With Respect to Overall Performance

Type	PVK Front Cells	Transparent Electrodes	Recombination Layers	V_{oc} (V)	J_{sc} (mA cm^{-2})	FF (%)	PCE (%)	Ref.
2T	$Cs_{0.09}FA_{0.77}MA_{0.14}Pb(I_{0.86}Br_{0.14})_3$	ITO/MgF_2	BZO/ITO	1.774	17.3	73.1	22.4	[72]
2T	$Cs_{0.05}(MA_{0.17}FA_{0.83})_{0.95}Pb(_{I0.83}Br_{0.17})_3$	SnO_2/IZO	AZO/IZO	1.68	19.17	71.9	23.26	[73]
4T	$MAPbI_3$	IZO		1.034	20.79	79.8	25.0	[74]
				0.565	19.4	74.2		
4T	$MAPbI_3$	$MoO_x/In_2O_3:H$		1.104	17.4	73.6	20.5	[68]
				0.667	12.7	74.9		
4T	$MAPbI_3$	$In_2O_3:H$		1.116	19.1	75.4	22.1	[69]
				0.669	12.1	73.6		
4T	$Cs_{0.1}Rb_{0.05}FA_{0.75}MA_{0.15}PbI_{1.8}Br_{1.2}$	$MoO_x/IZO/MgF_2$		1.14	21.8	74.0	23.9	[70]
				0.62	13.0	72.0		
4T	$FA_{0.65}MA_{0.2}Cs_{0.15}Pb(I_{0.8}Br_{0.2})_3$	IZO		1.137	19.6	76.8	25.9	[71]
				0.715	15.6	79.2		
4T	$FA_{0.83}Cs_{0.17}Pb(I_{0.6}Br_{0.4})_3$	$MoO_x/ITO/MgF_2$		1.16	19.7	78.7	25.0	[75]
				0.71	13.6	78.1		

7.3.4.2 2T PVK/CIGS TSC

Todorov et al. [76] first noticed TSC development in 2015. They adjusted the E_g of MAPbX3 (X = I, Br, Cl) between 1.58 and 2.29 eV using an in situ E_g tuning device and a CIGS bottom cell. The 2T PVK/CIGS TSC had 10.9% PCE. Han et al. [72] further investigated the importance of reliable contacts between the sub-cells in a 2T TSC, emphasizing that the smoothness of the ICL plays a critical role in improving performance. To overcome the challenges posed by the rough surface of the CIGS layer, they applied a boron-doped ZnO (BZO) layer on the CIGS, facilitating the deposition of $Cs_{0.09}FA_{0.77}MA_{0.14}Pb(I_{0.86}Br_{0.14})_3$ PVK with an E_g of 1.59 eV. This method allowed for effective work function adjustment between the poly[bis(4-phenyl)(2,4,6-trimethylphenyl)amine] (PTAA) and BZO; as a result, the 2T PVK/CIGS TSC achieved a validated PCE of 22.4%. Lang et al. [77] studied the stability of TSCs under high energy levels. Unlike PVK/Si tandem cells, which experienced a significant efficiency decline (down to just 1%) after exposure to 68 MeV proton irradiation, the PVK/CIGS tandem cells retained over 85% of their initial efficiency. These results highlighted the potential for PVK/CIGS TSCs in space environments. In another study, Al-Ashouri et al. [73] replaced the PTAA HTL with self-assembled monolayers in a single-junction PSC. As a result, they achieved a certified efficiency of 23.26% on an active area of 1.03 cm². These developments underline the significant progress in PVK/CIGS TSCs, particularly in enhancing their efficiency and radiation resistance, with innovative fabrication techniques improving both performance and stability.

7.3.5 Which TSC Is the Most Optimal?

PVK TSCs demonstrate the significant potential of PSCs as they begin to transition from laboratory research to commercial application in the solar industry. The growth of PVK TSCs includes several device configurations, raising the question of which type of TSC could be the most efficient. Among the various PVK TSC designs, PVK/Si, PVK/CIGS, and all-PVK TSCs are recognized as the most promising in terms of efficiency. High PCEs, efficient industrial processes, compatibility with flexible and lightweight substrates, cost-effectiveness, and large-scale industrial throughput make these designs prominent.

Currently, the most capable candidate is the PVK/Si TSC, primarily due to the dominance of Si cells in the solar market. The PVK/Si TSC offers a clear path to commercialization, as it can significantly progress the PCE of Si-based modules with minimal additional cost and scalable production for silicon modules. However, silicon cells still face specific constraints, such as difficulties in adjusting the E_g and challenges in fabricating flexible devices. However, PVK/CIGS TSCs have configurable E_g, easy production on rigid and flexible substrates, and space solar system applications. Additionally, large-scale CIGS manufacturing processes are well-established [77].

The ultimate goal is to develop all-PVK TSCs, which would offer elevated power conversion efficiencies, cost-effective production, and compatibility with flexible devices. Achieving this requires addressing key challenges related to long-term stability and NiO recombination contacts [77].

7.4 MEASUREMENTS

PVK TSC measurements are essential for device PCE evaluation and PCE loss analysis. These measures reveal sub-cell and TSC performance. EQE and I–V measurements in TSCs are more complex and challenging than in SJSCs due to the interconnection of sub-cells. This section will discuss PVK TSC measurement methods and how they can be used to analyze and improve TSC efficiency and performance.

7.4.1 Current Voltage Characteristics

While recording the related photocurrent, the bias voltage is systematically altered during this I–V experiment. In compliance with the global standards set by IEC 60904-3:2008 and ASTM G-173-03, the I–V measurement needs to be carried out at a temperature of 25°C utilizing a device holder and an illumination intensity of 100 mW cm^{-2}. The following equation is used to obtain important performance metrics like I_{sc}, V_{oc}, and FF from the I–V measurement:

$$PCE = \frac{I_{sc} \times V_{oc} \times FF}{A \times P_{in}} \tag{7.8}$$

The area of the device (A) and the amount of incident light (P_{in}) that the test cell received are both represented here.

It is important to recognize that the area of a solar cell is typically categorized into three types:

1. Overall area—The total area of the solar cell, including any regions not active in power generation
2. Aperture area—The area through which light can enter the solar cell and contribute to energy generation
3. Defined illumination area—The specific region of the solar cell that is illuminated and contributes to the photocurrent during testing

These distinctions help accurately assess the performance of the solar cell and ensure that the measured PCE reflects the actual active area and the light that is effectively used by the cell [78]. A xenon (Xe) bulb simulates in this measurement system's solar simulator. The system has a black temperature-regulated sample holder to reduce light reflection. Additionally, you will receive an I–V measurement program, a computer for data analysis, a source meter, and a four-point probe. The solar simulator's light intensity is calibrated using a reference cell whose photocurrent has been certified by a public testing facility. This calibration ensures accurate and precise PCE calculation from the I–V curve, delivering dependable performance data [79].

The spectral response of the reference cell used for I–V measurements must closely align with that of the measured solar cell to ensure accuracy. For c-Si solar cells, a KG-3 or KG-5 filter is often employed as the reference standard [80]. When compared to the I–V measurements used for SJSC, the I–V measurements for TSCs, particularly PVK TSCs, are more complex, as shown in Figure 7.9. This complexity arises due to several factors unique to PSCs, which exhibit anomalous I–V hysteresis.

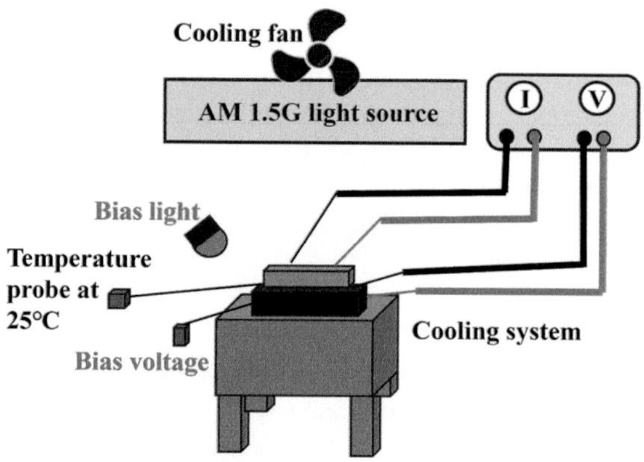

FIGURE 7.9 Setup for I–V measurement.

These characteristics contribute to the non-ideal behavior observed in the I–V of PSCs [6, 81] complicating the measurement and analysis of their performance.

To get accurate I–V for PVK TSCs, it is essential to carefully set and then verify several parameters, including the scan rate, scan direction, voltage step, step delay time, and light soaking conditions. The process should be repeated until the variations between backward and forward I–V scans become minimal, ensuring precision in the measurement [32, 82]. Currently, there are no IEC test standards available for the I–V of PSCs, making it important to develop standardized procedures for reliable characterization. In PVK TSCs, stabilized power output (SPO) measurements and MPP tracking have been improved to reduce hysteresis, which can affect the PCE evaluation of both PSCs and PVK TSCs. Calculating the SPO involves analyzing the I_{sc} and changing the voltage bias near the MPP. The gadget design significantly impacts PVK TSC I–V readings. A comparison of I–V values between 2T and 4T TSC, including sub-cells, shows performance variances due to structural and operational variables. I–V measurements for PVK TSCs are highly dependent on the specific device configurations. The I–V characteristics of 4T TSCs and their sub-cells differ from those of 2T TSCs, with the latter showing a more direct correlation between the sub-cells and the overall device performance [18].

The precise evaluation of I–V curves for 2T TSCs presents a significant challenge, mainly due to spectral mismatch as observed in simulations. To mitigate spectrum mismatch in an SJSC, it is recommended to calibrate the light intensity using a certified reference cell, such as a Si cell. While this method is effective for SJSCs, it poses difficulties when applied to series-connected TSCs. The reason for this is that many commercial solar simulators exhibit substantial spikes in the near-infrared (NIR) region, which can lead to erroneous readings of the I_{sc} [83]. TSCs cannot be calibrated for light intensity because only the bottom cell can absorb NIR photons where spikes occur. To coordinate the simulated sunlight with AM 1.5G irradiation, xenon and halogen lights are often used together. Proper documentation of I–V curves for 2T TSCs requires careful monitoring of parameters like light soaking, voltage step, and step delay time, similar to 4T TSCs [32]. Properly verifying and adjusting settings during I–V measurement is essential for generating trustworthy PCE data for TSCs. To assess SPO, the bias voltage is calculated by averaging the MPP voltages from the I–V curves. The background of semi-transparent cells is usually black to prevent reflection and ensure accurate evaluation. The measurement procedure follows these steps.

The initial step involves measuring the V_{oc} and analyzing its relationship with the I_{sc} for both sub-cells and the TSC. The V_{oc} of the top and bottom cells should be determined from their respective I–V curves under actual operating illumination conditions. To evaluate the combined voltage in comparison to the V_{oc} of the tandem cell, the bottom cell voltage can be inferred under operational conditions and vice versa for the top cell voltage. When measuring the I–V characteristics of the top cell, a bias light should be applied to limit photocurrent generation in the sub-cell, ensuring that the measurement reflects the top cell's current-limiting condition. To minimize the influence of the bottom cell during top cell measurements, a voltage bias close to the bottom cell's V_{oc} should be applied. The I–V curve of the top cell should then be measured under illumination conditions that mimic real-world operation. In the next step, the I–V characteristics of the bottom cell are measured by following a similar procedure. The TSC should remain in a current-limiting state while applying appropriate bias light and voltage to obtain accurate I–V curves for the sub-cells. To ensure precise readings, it is essential to minimize interferences such as electric field effects or shunt leakage, even when using bias light. Since optimizing the bias voltage is complex, a trial-and-error approach is often required to achieve the most accurate results.

To address spectral mismatch and other challenges faced during I–V in 2T TSCs, Park et al. [83] developed a design that allows sub-cell I–V characteristics to be examined without bias light, voltage, or filtering. TSC performance was accurately measured using this 3T construction. This design improves on the typical 2T design by resolving spectrum mismatch and current-limiting circumstances during I–V measurement.

7.4.2 External Quantum Efficiency

While I–V measurements are crucial for assessing device performance, they do not reveal the device's spectrum response. In TSCs, spectral response analysis is essential for identifying the limiting sub-cell and understanding how increased sub-cells or ICL changes can improve PCE.

The EQE measurement plays a key role in accurately assessing the J_{sc}, as it helps determine the spectral mismatch factor. This factor is used to correct deviations between the white light spectrum produced by a solar simulator and the standard AM 1.5G solar range [84]. The EQE setup plays a crucial role in analyzing electrical and optical losses in solar cells, including:

- Reflection losses
- Parasitic absorption losses
- Recombination losses

By evaluating the EQE spectra of individual sub-cells, it is possible to determine which sub-cell limits the overall performance of the TSC. This analysis helps in enhancing PCE through improvements in materials and device design. As depicted in Figure 7.10, EQE measurements provide valuable insights into efficiency trends, loss mechanisms, and effective optimization approaches.

The AL efficiently collects electron–hole pairs from photons hitting the solar cell. The electron–hole conversion ratio (EQE) shows how many incident photons contribute to photocurrent. The photon flux is typically evaluated by measuring the wavelength (λ) of a calibrated photodiode, which is exposed to the same light source as the solar cell. Once a bias voltage of 0 V is applied, the measured photocurrent density directly corresponds to the J_{sc}, as described by the following equations:

$$EQE = \frac{I_{ph}(\lambda)}{q\psi_{ph,\lambda}} \tag{7.9}$$

$$\psi_{ph,\lambda} = \frac{I_{ph}^{ref}(\lambda)}{qEQE^{ref}(\lambda)} \tag{7.10}$$

$$EQE(\lambda) = EQE^{ref}(\lambda)\frac{I_{ph}(\lambda)}{I_{ph}^{ref}(\lambda)} \tag{7.11}$$

$$J_{sc} = -q\int_{\lambda_1}^{\lambda_2} EQE(\lambda)\phi_{ph,\lambda}^{AM1.5}\,d\lambda \tag{7.12}$$

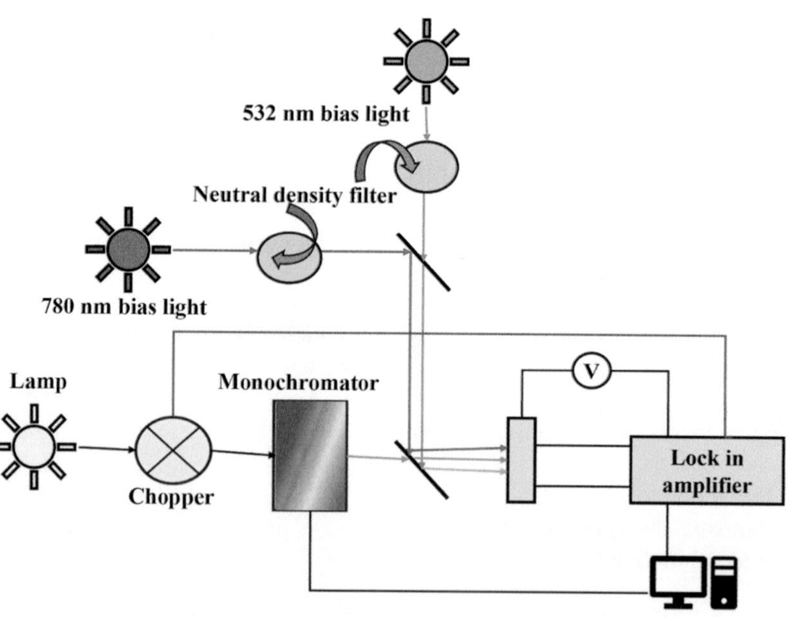

FIGURE 7.10 Configuration for the EQE.

Here, q represents the elementary charge and I_{ph} is photocurrent, $\psi_{ph,\lambda}$ indicates the spectrum photon flux affecting the solar cell, I_{ph}^{ref} is used to refer to the photocurrent of the calibrated photodiode, and $EQE^{ref}(\lambda)$ represents the EQE. A standard EQE typically consists of the following components:

- Computer—For data acquisition and system control
- Lock-in amplifier—Enhances signal detection by reducing noise
- Current meter—Measures the photocurrent generated by the solar cell
- Calibrated light detector—Ensures accurate light intensity measurements
- Chopper—Modulates the light source for improved signal-to-noise ratio
- Filter wheel—Selects specific wavelength bands for precise spectral analysis
- Monochromator—Provides narrowband light for EQE measurements
- Shutter—Controls light exposure to prevent overheating and degradation
- Wavelength-selective light source—Illuminates the solar cell with specific photon energies [85]

A shutter is crucial for proper system operation, allowing precise control over light exposure during measurement cycles. The most commonly used light source is a xenon gas discharge lamp, which produces a broad spectral range of light emission. By incorporating a filter wheel and monochromators, the system can selectively generate precise photon energies, enabling accurate EQE characterization across different wavelength ranges. The EQE of an SJSC should be measured under short-circuit conditions using a modest quantity of monochromatic light and a suitable bias light to replicate AM 1.5G solar spectrum operation conditions. IEC 60904-8-1 test techniques are used for measurement. EQE measurement is more difficult for TSCs since the intermediate contact is inaccessible and sub-cells cannot be individually addressed, especially in 2T TSCs [86].

7.4.2.1 EQE Measurement for Different TSC Architectures

- 4T Tandem Solar Cells:
  - The semi-transparent top cell is illuminated from the glass side, and its EQE is measured separately.
  - The bottom cell Si or CIGS is analyzed without bias light or bias voltage, as semi-transparent top cell acts as a spectral filter during the EQE measurement.
- 2T Tandem Solar Cells:
  - EQE measurement is more challenging due to the electrical interdependence of the two sub-cells.
  - Advanced bias light techniques must be applied to isolate and analyze each sub-cell's spectral response.

For the bottom cell EQE, the semi-transparent top cell filters light, requiring specialized methods for spectrum characterization. These issues aside, TSC EQE measurement can be done using the same methods as SJSCs, with adjustments to account for optical filtering and electrical interdependence in TSC. Measuring the EQE of sub-cells in 2T TSCs is more difficult and challenging than in 4T TSCs, similar to

I–V measurements. This intricacy requires specialized measurement techniques due to the electrical dependency of the two sub-cells.

7.4.2.2 2T TSC EQE Measurement Procedure

- Biasing the Test Sub-Cell:
  - The sub-cell under evaluation must be biased at one sun intensity to replicate real operating conditions.
- Biasing the Non-Tested Sub-Cell:
  - The other sub-cell (not under test) must be exposed to a higher intensity of light and voltage bias to ensure it does not interfere with the measurement.
  - This technique ensures that the non-tested sub-cell generates negligible photocurrent, allowing the evaluated sub-cell to be current-limited for precise measurement.
- Top Cell EQE Measurement:
  - The top cell is illuminated using:
    - A solar simulator at one sun intensity as the primary light source.
    - A second bias light source to adjust the photocurrent balance in the TSC.
  - The purpose of this setup is to restrict the overall TSC current to match the top cell's current output.
- Maintaining Voltage-Free Conditions:
  - To ensure the test sub-cell remains voltage-free, an external voltage of approximately V_{oc} (bottom cell) is applied.
  - This adjustment prevents the non-tested sub-cell from influencing the measurement, ensuring accurate EQE characterization.

This approach enables accurate spectral response measurements in two-terminal (2T) TSCs, addressing challenges related to electrical coupling and current matching limitations. The next step in EQE measurement for 2T TSCs involves directly assessing the spectral response over the full operational range of the top cell.

7.4.2.3 Procedure for Measuring the Bottom Cell EQE

1. The bottom cell is exposed to simulated sunlight at one sun bias to replicate real-world operating conditions.
2. A secondary light source is used to filter specific wavelengths, ensuring that the tandem structure operates in a current-limited state due to the bottom cell.
3. To maintain an unbiased condition in the tested sub-cell, an external voltage close to V_{oc} (top cell) is applied.
4. After establishing these conditions, the spectral response is recorded across the bottom cell's active wavelength range. Example of Bias Light Sources: For 2T all-PVK TSC, such as $Cs_{0.2}FA_{0.8}PbI_{1.8}Br_{1.2}$ with an E_g of 1.77 eV and $MA_{0.3}FA_{0.7}Pb_{0.5}Sn_{0.5}I_3$ with an E_g of 1.22 eV, specific bias illumination sources are used.

For the EQE of PVK/Si TSCs, a 100 W tungsten excitation source is directed through a Princeton Instruments SpectraPro 150 monochromator to generate monochromatic light. To measure the EQE of the PVK and Si sub-cells, a 980 nm laser diode is used to illuminate the Si sub-cell, while a white light LED array is used to illuminate the PVK sub-cell [87]. During the EQE measurement, the bottom Si cell is saturated with infrared light to evaluate the top PVK cell, and the top PVK cell is saturated with blue light to evaluate the bottom Si cell [18, 87]. These carefully controlled conditions ensure precise measurement of the spectral response of each sub-cell, enabling accurate efficiency characterization in PVK/Si TSC.

A lock-in amplifier from Stanford Research Systems, model SR830, is employed to track the PVK signal at the chopping frequency usually about 73 Hz during the EQE of PVK/Si TSC [87]. Once the setup is ready, the reference photodetector is positioned correctly for the test and covered with a dark cloth or other opaque material to prevent any external light from contaminating the results. For reference measurements in PVK/CIGS TSCs, calibrated diodes, such as Si diodes with identified spectrum sensitivities, are used. Before starting the measurement, it is important to shut down the lights in the room and launch the EQE measurement program. After turning on the light source, the equipment should be allowed to warm up for at least 30 minutes before measurements begin. The accuracy of the EQE measurement is further ensured by setting a bandwidth of no more than 20 nm for all wavelength tests. The photodetector should also be covered with a black material, and the room lights should be completely turned off to maintain a controlled environment. Once the photocurrent is measured, the EQE can be recorded, and, using the previously provided equation, the J_{sc} can be calculated.

7.4.3 Light Beam–Induced Current

Conventional classification methods, such as I–V and EQE measurements, do not provide detailed insights into localized chemical and physical variations within PVK TSCs. Manufacturing defects and weak diodes can introduce photocurrent non-uniformities, which become even more detrimental in the presence of stability issues, significantly impacting cell performance [88]. To address these challenges, Song et al. [88] employed the light beam–induced current (LBIC) method to analyze photocurrent collection non-uniformities in individual PVK/Si TSC sub-cells. This approach enables the spatial resolution of photocurrent generation and collection, helping identify processing defects in both cells and modules [89, 90]. The LBIC technique utilizes laser light with wavelengths ranging from 332 to 1064 nm at a low power density (~800 mW cm^{-2}) to excite charge carriers in the sub-cells. This controlled excitation is designed to prevent photodamage to the PSC while allowing for precise local bias current mapping. The photogenerated current is then corrected using EQE measurements, enabling a detailed analysis of photocurrent uniformity and performance across the device.

7.5 FABRICATION TECHNIQUES AND SCALABILITY OF PVK TSC

The scalable fabrication of PVK TSCs is crucial for their transition from laboratory research to large-scale industrial production. The unique properties of PVK materials, such as their tunable E_g, high absorption coefficient, and compatibility with solution processing, create opportunities for the development of cost-effective and high-efficiency solar technologies. However, ensuring consistency, reliability, and high production rates in large-scale fabrication remains a significant challenge.

To address these issues, researchers have explored a variety of fabrication techniques, including:

1. Solution processing—A cost-effective and low-temperature method that enables scalable deposition but faces challenges related to film uniformity and large-area reproducibility
2. Vapor deposition—Produces high-quality, uniform films but is resource-intensive and increases production costs
3. Hybrid deposition techniques—Combines solution and vapor-based methods to optimize film quality, uniformity, and scalability
4. Large-area coating approaches—Includes methods such as roll-to-roll (R2R) printing, blade coating, slot-die coating, spray coating, and inkjet printing, which enable high-throughput industrial production

This section provides a comprehensive analysis of these fabrication techniques, discussing their advantages, limitations, and impact on the scalability of PVK TSC for commercial deployment.

7.5.1 Solution Processing

Solution processing is one of the most extensively studied fabrication techniques for PVK thin films, primarily due to its cost-effectiveness and ease of implementation. This method is particularly attractive because it operates at low temperatures, is easily scalable, and is compatible with flexible substrates. The solution processing approach involves dissolving PVK precursors in a suitable solvent, followed by deposition onto a substrate. The process is then completed with a crystallization phase, ensuring the formation of a uniform thin film. A key advantage of solution processing is its high-throughput manufacturing capability. Unlike conventional vacuum-based deposition methods, solution-based fabrication can be performed under ambient conditions, leading to lower energy consumption and reduced equipment costs. Additionally, solution processing enables precise control over film morphology, grain size, and composition, which are essential for enhancing PSC performance.

Despite these advantages, solution processing faces several challenges [91]. Achieving uniform films over large areas remains difficult due to issues such as uncontrolled crystallization, varying solvent evaporation rates, and defect formation. To achieve high-quality films, precise control over solution concentration, deposition speed, and drying conditions is necessary. Moreover, the intrinsic instability of PVK materials when exposed to moisture and oxygen underscores the importance of protective encapsulation strategies to ensure long-term device stability.

To address scalability challenges, various solution processing techniques have been explored, including:

- Spin coating
- Blade coating

- Slot-die coating
- Spray coating
- Inkjet printing

Each of these techniques offers distinct advantages and trade-offs in terms of film uniformity, material efficiency, and processing speed. Current research efforts focus on enhancing these methods and developing innovative strategies to improve scalability, reproducibility, and efficiency for large-scale production of PVK TSCs. Table 7.5 provides a comparative analysis of different solution processing techniques, while Figure 7.11 illustrates the spin-coating process, highlighting its role in thin-film fabrication.

7.5.2 Vapor Deposition

Vapor deposition techniques are extensively utilized for the fabrication of high-quality PVK films, providing enhanced uniformity, reproducibility, and greater device stability in comparison to solution processing. The methods generally entail transforming PVK precursors into a vapor phase, which is subsequently deposited onto a substrate with precision [92]. Vapor deposition processes remove the necessity for solvents, which minimizes contamination risks and allows for enhanced control over film morphology and crystallinity. Nonetheless, these techniques necessitate specific

TABLE 7.5
Different Techniques in Solution Processing

Technique	Process	Advantages	Challenges	Applications
Spin-coating	Applying a precursor solution onto a substrate and then rapidly spinning it to achieve uniform film coverage.	High-film uniformity in small areas, easy parameter control, and rapid processing	Poor scalability due to limited control over film thickness uniformity on large substrates	Primarily used for small-area research and device optimization
Blade coating	A solution is applied to a substrate using a blade, allowing for continuous thin-film deposition.	More scalable than spin-coating, reduces material waste, better thickness control	Requires optimization of coating speed and drying conditions for consistent quality	Suitable for mid-scale processing and laboratory-scale production
Slot-die coating	Uses a narrow nozzle to deposit a continuous film of perovskite precursor on the substrate	High-throughput, precise film thickness control, and good scalability	Sensitivity to solution viscosity and drying conditions	Feasible for roll-to-roll processing, making it a strong candidate for large-area TSC manufacturing

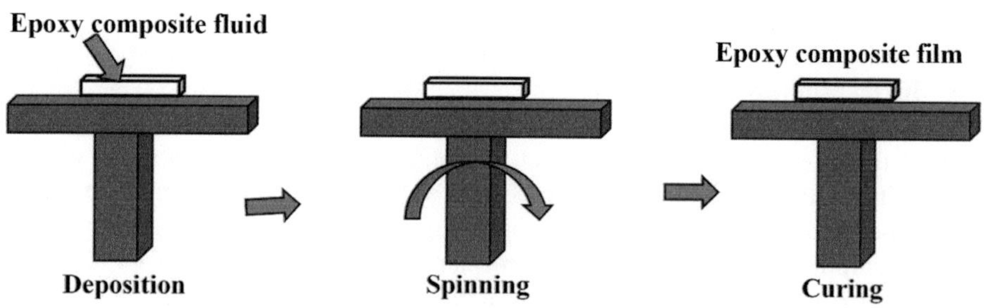

FIGURE 7.11 Setup for spin-coating process.

TABLE 7.6
Types of Vapor Deposition

Vapor Deposition Type	Process	Advantages	Challenges
CVD	PVK precursors are vaporized and deposited onto a heated substrate, enabling controlled crystallization.	Excellent film uniformity, high purity, and enhanced device stability	Requires high vacuum conditions and cost-intensive process
Thermal evaporation	PVK components are evaporated and deposited onto a substrate in a vacuum chamber.	Well-controlled deposition rate, uniform large-area films, high material purity	High energy consumption and expensive vacuum systems

equipment and are typically costlier than solution-based methods, and Table 7.6 shows different types of vapor deposition.

7.5.3 Hybrid Deposition Methods

Hybrid deposition methods integrate the benefits of both solution-based and vapor deposition techniques, aiming to optimize film quality, enhance reproducibility, and improve scalability. These techniques utilize the advantages of both methods to achieve high-performance PVK TSCs, characterized by improved film morphology and regulated crystallization. Hybrid deposition methods facilitate enhanced defect passivation, resulting in increased efficiency and greater device stability, and Table 7.7 shows different types of hybrid deposition methods.

7.5.4 Large-Area Coating Techniques

Techniques for large-area coating are crucial for advancing the fabrication of PVK TSCs, facilitating mass production while ensuring uniformity and efficiency of the films. The emphasis of these techniques lies in attaining high throughput, reducing material waste, and guaranteeing reproducibility over extensive substrates. With the rising demand for PVK-based PVs, it is essential to refine these techniques for successful commercialization, and Table 7.8 shows different types of large-area coating techniques.

The transition of PVK TSC from laboratory research to large-scale production requires a careful selection of fabrication techniques. While solution processing methods offer a cost-effective and simple approach, they face scalability challenges, particularly in achieving uniform and defect-free films over large areas. Vapor deposition techniques produce high-quality films with excellent uniformity but come with high resource demands and increased manufacturing costs. A promising alternative is hybrid deposition methods, which combine solution and vapor processes to improve film uniformity, reproducibility, and large-area scalability.

For high-throughput industrial production, techniques such as R2R printing and inkjet printing offer significant potential for large-area PVK coating, enabling cost-efficient mass production. Continued advancements in scalable fabrication techniques will be critical for the widespread adoption of PVK TSC in the PV industry, accelerating their commercial viability and integration into the global energy market.

TABLE 7.7
Types of Hybrid Deposition Methods

Hybrid Deposition Type	Process	Advantages	Challenges
Sequential deposition	Metal–halide and organic components are deposited separately, allowing in situ perovskite formation.	Improved film uniformity and better control over composition	Requirement of precise process optimization
Hybrid CVD-solution processing	A combination of vapor-based and solution-based deposition techniques for controlled crystallization.	Enhances crystallinity and film uniformity and reduces defects	More complex process integration

TABLE 7.8
Types of Large-Area Coating Techniques

Large-Area Coating Type	Process	Advantages	Challenges
Roll-to-roll (R2R) processing	It is a continuous process where a flexible substrate moves through different coating and drying stages.	High-throughput, minimal material waste, cost-effectiveness	Requires specialized equipment and precise process control
Inkjet printing	PVK precursor solution is ejected onto the substrate in a controlled manner.	Digital control over deposition, reduced material waste, flexible substrate compatibility	Requires optimization of ink formulation and printing parameters

7.6 CHALLENGES OF PVK TSC

7.6.1 The Structure of the Device

Two sub-cells in a 4T TSC function electrically independently. If one sub-cell dies, the device works but outputs less power. In contrast, a 2T TSC has direct electrical communication between sub-cells; hence, any sub-cell or ICL failure causes device failure. Developing high-efficiency semi-transparent PVK top cells for 4T TSCs is difficult. To reduce parasitic absorption and reflection losses, clear electrodes and antireflection layers are prioritized. Due to interfaces and transparent electrodes, 4T TSCs have parasitic absorption and reflection losses, lowering efficiency compared to 2T TSCs. Most TSC experiments use 2T designs, and they are predicted to remain the most popular due to their compact design, lesser parasitic losses, and increased integration possibilities. However, 2T configurations must reduce energy losses and match sub-cell currents.

7.6.1.1 Key Considerations for 2T TSC Optimization

1. E_g tuning of PVK materials and surface engineering to improve efficiency
2. ICL design to effectively block halide ion migration, preventing interface degradation
3. Contamination control—Si surfaces are highly sensitive to impurities, making them susceptible to halide ion penetration, which can degrade performance

For 2T all-PVK TSCs, solution processing of the top PVK layer must not compromise the underlying PVK sub-cell. The ICL must provide sufficient protection to prevent solvent infiltration, ensuring structural and electrical integrity. Beyond double-junction TSCs, triple-junction TSCs are anticipated to attain even higher PCEs. This represents a promising research direction, as multi-junction structures can further optimize solar spectrum utilization [93].

7.6.2 Efficiency

Enhancing PVK sub-cell efficiency is a direct and effective way to improve TSC device performance in addition to light management measures. PSCs with an E_g of 1.5 eV are the main focus of research. However, wide E_g PSCs (~1.7 eV) and narrow E_g PSCs (~1.2 eV) have got less attention. Efficient wide E_g PVK, including organic–inorganic hybrid and all-inorganic PVK, and narrow E_g lead-tin (Pb–Sn) binary PVK, are essential for PVK-based TSC development. Typical tandem topologies use wide E_g PVK as the top cell to maximize spectrum usage. However, wide E_g PSC faces considerable energy losses, primarily due to the following reasons:

1. The efficiency and lifespan of charge carriers are reduced due to non-radiative recombination at grain boundaries and interfaces [12].
2. Inefficient charge extraction and higher voltage losses caused by the PVK layer and CTL not aligning optimally in terms of energy levels [94].

Addressing these challenges through material engineering, interface passivation, and band alignment optimization is crucial for the continued advancement of high-efficiency PVK-Si TSC.

7.6.3 Stability

Degradation of PVK films in the presence of oxygen, moisture, heat, and light is the primary cause of the stability issues linked to PVK TSCs. Improving the durability of devices requires manufacturing PVK films of superior quality with minimal flaws and grain boundaries; this will lessen the impact of water and oxygen degradation.

Several studies emphasize the incorporation of cesium (Cs^+) into PVK films as an effective method to enhance thermal stability [30, 95, 96]. The most commonly used PVK formulations for top cells involve a mixture of halides, which raises concerns about halide segregation, potentially affecting the photostability of devices [61, 97]. With an E_g of 1.73 eV, which is ideal for tandem applications, CsPbI3 stands out among other PVKs as a potential top cell choice for TSCs. The photostability and thermal stability it provides are better than those of mixed-halide PVK. One major obstacle that needs to be addressed for $CsPbI_3$ to work effectively in TSCs is its phase stability [62, 98, 99].

7.6.3.1 CTL Stability and Electrode Interactions

Maintaining stable CTLs is equally important as preventing PVK breakdown. Dopants such as Li-TFSI in spiro-OMeTAD as HTL have the potential to damage PVK films, according to reports [100]. This has led to a surge in interest in studying dopant-free HTLs that are highly charged mobile.

Also, the device can degrade due to chemical interactions between halides in PVK and metal electrodes including gold, silver, and aluminum [101, 102]. To lessen the impact of these negative consequences, scientists are investigating:

- Incorporation of buffer layers to prevent metal–halide reactions
- Utilization of chemically stable electrodes to improve device longevity

7.6.3.2 Encapsulation for Long-Term Stability

Encapsulation is a highly effective approach to improve overall TSC stability. Proper encapsulation prevents water and oxygen permeation while also reducing the risk of toxic Pb compound leakage in water-soluble environments. By addressing these stability concerns, PVK TSCs can be further optimized for long-term reliability and commercial viability.

7.6.4 Costs

In a commercialized PV system, it is critical to consider the balance-of-system costs [103], which encompass production, encapsulation, installation, and other associated expenses. The overall cost of a PV system is inversely proportional to the module PCE and directly proportional to the device surface area. Therefore, strategic cost reduction can be achieved by enhancing the PCE of PV devices. According to Cai et al. [104], the cost of a PSC module accounts for approximately one-third of the total cost of Si cells, estimated at roughly 0.25 USD/W. Due to the relatively low-cost PVK materials, PVK/Si TSC modules have only slightly higher production costs. The most straightforward approach to commercializing PVK-based PV systems is through TSC improvements. Several strategies can help reduce costs while maintaining high efficiency, including:

1. Replacing gold (Au) and silver (Ag) electrodes with copper (Cu), which offers a cost-effective alternative with comparable conductivity
2. Developing more affordable HTL materials to replace expensive options like spiro-OMeTAD and PTAA, thereby lowering material costs
3. Optimizing the ICL in 2T TSCs, ensuring minimal losses by keeping high PCE.

7.6.5 Lead Toxicity

Lead toxicity poses a major challenge for the commercialization of Pb-based PSCs. Currently, numerous research efforts are dedicated to developing lead-free PSCs using alternative elements such as tin (Sn), antimony (Sb), bismuth (Bi), and others [105, 106]. Among these, Sn-based PSCs have demonstrated the highest efficiency. However, despite extensive research to enhance their PCE, Sn-based PSCs still significantly lag behind their Pb-based counterparts. The primary limiting factor is the instability of Sn^{2+}, which readily oxidizes to Sn^{4+}, leading to poor device performance.

To address the toxicity concerns of Pb-based PSCs, we propose three key recommendations:

1. Encapsulation of Pb-based PSCs is essential to prevent lead leakage, ensuring environmental safety and long-term stability.
2. Given the increased Pb content in large-area PV modules, it is crucial to develop effective recycling strategies for Pb-based devices to mitigate environmental impact.
3. Advancing high-performance Sn-based PSCs requires precise control over crystallization and effective inhibition of Sn^{2+} oxidation to improve stability and efficiency.

Additionally, integrating theoretical modeling with experimental approaches will be crucial for exploring novel, high-efficiency, and environmentally friendly PV devices, driving the transition toward sustainable solar technologies.

7.6.6 Large-Area Fabrication

PVK TSCs cannot achieve mass production until they can be manufactured over large areas. Currently, most high-performance PSCs and PVK TSCs remain at the laboratory scale, with active surface areas smaller than 1 cm^2. For TSC modules to achieve higher efficiency, transparent electrodes with high transmittance and optimized CTLs are essential. Additionally, the incorporation of antireflection layers is crucial to minimize optical losses and enhance overall performance. In PVK-based

TSCs, transparent electrodes typically consist of metal layers and transparent conductive oxides (TCOs). One effective approach to fabricating semi-transparent PSCs involves using thermal evaporation to deposit ultrathin layers of Ag, Au, or Cu, which serve as highly conductive and transparent electrodes, enabling efficient light absorption and charge transport [17, 71]. These metal layers serve a dual purpose as recombination barriers in most 2T PVK/Org TSCs [59, 61, 65]. For 2T PVK/Si and PVK/CIGS TSCs, TCOs such as ITO and IZO are commonly deposited using sputtering techniques to function as recombination layers [87, 68, 107]. These layers facilitate efficient charge recombination while maintaining high transparency and conductivity, contributing to the overall performance of TSC.

Transparent electrodes in PVK sub-cells often require buffer layers. Sputtering can harm CTL and PVK material, but these layers protect them. Buffer layers typically consist of thermally formed MoO_3 or ALD-deposited AZO and SnO_2, enhancing device stability and performance [87, 108]. Conventional TCO ICLs in large-area 2T PVK/Si TSCs exhibit high lateral conductivity, which can lead to the formation of shunt paths through defects and pinholes in the PVK sub-cell [109, 110] In 2T all-PVK and PVK/Org TSCs, solution-processed organic materials have been utilized as ICLs in previous studies. The superior charge mobility and enhanced stability of SnO_2 have led to an increasing adoption of ALD-deposited SnO_2 layers as ICLs in 2T all-PVK TSCs. Since ALD is a low-temperature technique, it is particularly well-suited for scalable and widespread fabrication of TSC [52, 53]. High-efficiency, large-area TSCs require high-quality PVK films over vast surfaces. Inkjet, blade, slot-die, and R2R printing are used to make PSC modules. Further study is needed to fully realize the promise of these technologies for large-area PVK-based TSC applications. An alternative approach for depositing high-quality PVK films is thermal evaporation, which enables precise control over film thickness and uniformity. Additionally, since flexible substrates are compatible with the fabrication of PVKs, CIGS, and OSCs, the development of high-performance flexible TSCs presents an exciting avenue for future exploration.

7.7 SUMMARY AND PERSPECTIVE

As the efficiency of PVK and Si SJSCs approaches their S–Q limits, there is an increasing focus on the development of PVK/Si TSCs in the current PV landscape. After years of advancements, the certified PCE of PVK/Si TSCs has surpassed that of standalone PVK and Si SJSCs, reaching 29.5% for 2T TSCs and 28.2% for 4T TSCs.

To lay the foundation for the design of high-performance PVK/Si TSCs, this work begins with a comprehensive analysis of the structural configurations and operational principles of 2T and 4T TSC architectures. A concise overview of recent technological advancements in PVK/Si TSCs is provided, with a primary focus on the role of ICL, PVK material composition, and optical losses in determining overall device performance.

The discussion highlights how TSC performance can be further optimized by implementing various proposed strategies that enhance individual device components while effectively minimizing optical losses, thereby paving the way for next-generation high-efficiency TSC.

7.8 CONCLUSION

The advancements in PVK TSCs represent a significant milestone in the PV sector, providing an effective pathway to surpass the S–Q limit of SJSC. The rapid evolution of PVK-based PV, progressing from early single-junction efficiencies to well-optimized 2T and 4T TSC, underscores the potential of this technology. Notably, 2T PVK/Si TSCs have achieved efficiencies exceeding 32.9%, while 4T configurations have surpassed 36%, demonstrating their high-performance capabilities.

The scalability of PVK/Si TSCs is crucial for commercialization, leveraging the well-established Si solar industry while incorporating PVK as a high-efficiency top cell. The timeline of efficiency improvements, presented in this work, highlights ongoing breakthroughs in device engineering, E_g optimization, interfacial passivation, and large-area fabrication techniques. While 4T TSC architectures enable independent power optimization of sub-cells, 2T monolithic designs offer a more compact and cost-effective solution, making them particularly attractive for large-scale deployment.

Despite significant gains in efficiency, several challenges must be addressed before widespread commercialization is feasible. Key obstacles include stability concerns, large-area scalability, effective encapsulation strategies, and cost-efficient production methods. The intrinsic instability of PVK materials under environmental factors such as moisture, oxygen exposure, and temperature fluctuations remains a crucial issue, necessitating advanced encapsulation techniques and compositional engineering.

Additionally, various TSC technologies, including all-PVK TSCs, PVK-org TSCs, and PVK-CIGS TSCs, are gaining attention. These configurations offer distinct advantages such as flexibility, lightweight design, and tunable E_g, expanding the potential of PVK TSCs beyond conventional Si-based structures. These advancements pave the way for innovative applications, including space technology, building-integrated photovoltaics (BIPV), and next-generation flexible electronics.

Further research into long-term operational stability, efficient CTL, interfacial engineering, and scalable deposition techniques will be critical in unlocking the full potential of PVK TSCs. Combining high efficiency, E_g tunability, and cost-effectiveness, PVK/Si TSCs are poised to revolutionize the future of solar energy technology. Successfully addressing stability and scalability challenges could position PVK TSCs as the leading PV technology of the next decade, reshaping the global landscape of renewable energy adoption.

REFERENCES

[1] D. M. Chapin, C. S. Fuller, and G. L. Pearson, "A New Silicon p-n Junction Photocell for Converting Solar Radiation into Electrical Power," *J. Appl. Phys.*, vol. 25, no. 5, pp. 676–677, May 1954, https://doi.org/10.1063/1.1721711.

[2] Y. Jiang *et al.*, "High Efficiency Multi-Crystalline Silicon Solar Cell With Inverted Pyramid Nanostructure," *Sol. Energy*, vol. 142, pp. 91–96, Jan. 2017, https://doi.org/10.1016/j.solener.2016.12.007.

[3] W. Wang *et al.*, "Solution-Processed Electron-Selective Contacts Enabling 21.8% Efficiency Crystalline Silicon Solar Cells," *Sol. RRL*, vol. 4, no. 12, p. 2000569, Dec. 2020, https://doi.org/10.1002/solr.202000569.

[4] www.nrel.gov/pv/cell-efficiency.html.
[5] B. Ehrler, E. Alarcón-Lladó, S. W. Tabernig, T. Veeken, E. C. Garnett, and A. Polman, "Photovoltaics Reaching for the Shockley–Queisser Limit," *ACS Energy Lett.*, vol. 5, no. 9, pp. 3029–3033, Sep. 2020, https://doi.org/10.1021/acsenergylett.0c01790.
[6] M. A. Green, A. Ho-Baillie, and H. J. Snaith, "The Emergence of Perovskite Solar Cells," *Nat. Photonics*, vol. 8, no. 7, pp. 506–514, Jul. 2014, https://doi.org/10.1038/nphoton.2014.134.
[7] L. Chouhan, S. Ghimire, C. Subrahmanyam, T. Miyasaka, and V. Biju, "Synthesis, Optoelectronic Properties and Applications of Halide Perovskites," *Chem. Soc. Rev.*, vol. 49, no. 10, pp. 2869–2885, 2020, https://doi.org/10.1039/C9CS00848A.
[8] A. Kojima, K. Teshima, Y. Shirai, and T. Miyasaka, "Organometal Halide Perovskites as Visible-Light Sensitizers for Photovoltaic Cells," *J. Am. Chem. Soc.*, vol. 131, no. 17, pp. 6050–6051, May 2009, https://doi.org/10.1021/ja809598r.
[9] L. Meng, J. You, T.-F. Guo, and Y. Yang, "Recent Advances in the Inverted Planar Structure of Perovskite Solar Cells," *Acc. Chem. Res.*, vol. 49, no. 1, pp. 155–165, Jan. 2016, https://doi.org/10.1021/acs.accounts.5b00404.
[10] J. Albero, A. M. Asiri, and H. García, "Influence of the Composition of Hybrid Perovskites on Their Performance in Solar Cells," *J. Mater. Chem. A*, vol. 4, no. 12, pp. 4353–4364, 2016, https://doi.org/10.1039/C6TA00334F.
[11] C. Liu, C. Bing, and Z. Ge, "Understanding of Perovskite Crystal Growth and Film Formation in Scalable Deposition Processes," *Chem. Soc. Rev.*, vol. 49, Mar. 2020, https://doi.org/10.1039/C9CS00711C.
[12] Q. Jiang *et al.*, "Surface Passivation of Perovskite Film for Efficient Solar Cells," *Nat. Photonics*, vol. 13, no. 7, pp. 460–466, Jul. 2019, https://doi.org/10.1038/s41566-019-0398-2.
[13] J. Zhang and H. Yu, "Multifunctional Dopamine-Assisted Preparation of Efficient and Stable Perovskite Solar Cells," *J. Energy Chem.*, vol. 54, pp. 291–300, Mar. 2021, https://doi.org/10.1016/j.jechem.2020.05.061.
[14] L. M. Pazos-Outón, T. P. Xiao, and E. Yablonovitch, "Fundamental Efficiency Limit of Lead Iodide Perovskite Solar Cells," *J. Phys. Chem. Lett.*, vol. 9, no. 7, pp. 1703–1711, Apr. 2018, https://doi.org/10.1021/acs.jpclett.7b03054.
[15] S.-W. Lee, S. Bae, D. Kim, and H.-S. Lee, "Historical Analysis of High-Efficiency, Large-Area Solar Cells: Toward Upscaling of Perovskite Solar Cells," *Adv. Mater.*, vol. 32, no. 51, p. 2002202, Dec. 2020, https://doi.org/10.1002/adma.202002202.
[16] R. Cariou *et al.*, "III-V-On-Silicon Solar Cells Reaching 33% Photoconversion Efficiency in Two-Terminal Configuration," *Nat. Energy*, vol. 3, no. 4, pp. 326–333, Apr. 2018, https://doi.org/10.1038/s41560-018-0125-0.
[17] P. Löper *et al.*, "Organic–Inorganic Halide Perovskite/Crystalline Silicon Four-Terminal Tandem Solar Cells," *Phys. Chem. Chem. Phys.*, vol. 17, no. 3, pp. 1619–1629, 2015, https://doi.org/10.1039/C4CP03788J.
[18] J. P. Mailoa *et al.*, "A 2-Terminal Perovskite/Silicon Multijunction Solar Cell Enabled by a Silicon Tunnel Junction," *Appl. Phys. Lett.*, vol. 106, no. 12, p. 121105, Mar. 2015, https://doi.org/10.1063/1.4914179.
[19] M. H. Futscher and B. Ehrler, "Efficiency Limit of Perovskite/Si Tandem Solar Cells," *ACS Energy Lett.*, vol. 1, no. 4, pp. 863–868, Oct. 2016, https://doi.org/10.1021/acsenergylett.6b00405.
[20] C. H. Henry, "Limiting Efficiencies of Ideal Single and Multiple Energy Gap Terrestrial Solar Cells," *J. Appl. Phys.*, vol. 51, no. 8, pp. 4494–4500, Aug. 1980, https://doi.org/10.1063/1.328272.
[21] A. Martí and G. L. Araújo, "Limiting Efficiencies for Photovoltaic Energy Conversion in Multigap Systems," *Sol. Energy Mater. Sol. Cells*, vol. 43, no. 2, pp. 203–222, Sep. 1996, https://doi.org/10.1016/0927-0248(96)00015-3.

[22] G. E. Eperon, M. T. Hörantner, and H. J. Snaith, "Metal Halide Perovskite Tandem and Multiple-Junction Photovoltaics," *Nat. Rev. Chem.*, vol. 1, no. 12, p. 0095, Nov. 2017, https://doi.org/10.1038/s41570-017-0095.
[23] J. Werner *et al.*, "Efficient Near-Infrared-Transparent Perovskite Solar Cells Enabling Direct Comparison of 4-Terminal and Monolithic Perovskite/Silicon Tandem Cells," *ACS Energy Lett.*, vol. 1, no. 2, pp. 474–480, Aug. 2016, https://doi.org/10.1021/acsenergylett.6b00254.
[24] J. Werner *et al.*, "Parasitic Absorption Reduction in Metal Oxide-Based Transparent Electrodes: Application in Perovskite Solar Cells," *ACS Appl. Mater. Interfaces*, vol. 8, no. 27, pp. 17260–17267, Jul. 2016, https://doi.org/10.1021/acsami.6b04425.
[25] Y. Zhou *et al.*, "Composition-Tuned Wide Bandgap Perovskites: From Grain Engineering to Stability and Performance Improvement," *Adv. Funct. Mater.*, vol. 28, no. 35, p. 1803130, Aug. 2018, https://doi.org/10.1002/adfm.201803130.
[26] M. Filipic *et al.*, "CH3NH3PbI3 Perovskite/Silicon Tandem Solar Cells: Characterization Based Optical Simulations," *Opt. Express*, vol. 23, pp. A263–A278, Feb. 2015, https://doi.org/10.1364/OE.23.00A263.
[27] M. H. Futscher and B. Ehrler, "Modeling the Performance Limitations and Prospects of Perovskite/Si Tandem Solar Cells under Realistic Operating Conditions," *ACS Energy Lett.*, vol. 2, no. 9, pp. 2089–2095, Sep. 2017, https://doi.org/10.1021/acsenergylett.7b00596.
[28] P. Löper *et al.*, "Organic–Inorganic Halide Perovskites: Perspectives for Silicon-Based Tandem Solar Cells," *IEEE J. Photovolt.*, vol. 4, no. 6, pp. 1545–1551, Nov. 2014, https://doi.org/10.1109/JPHOTOV.2014.2355421.
[29] C. D. Bailie *et al.*, "Semi-Transparent Perovskite Solar Cells for Tandems With Silicon and CIGS," *Energy Env. Sci.*, vol. 8, no. 3, pp. 956–963, 2015, https://doi.org/10.1039/C4EE03322A.
[30] D. P. McMeekin *et al.*, "A Mixed-Cation Lead Mixed-Halide Perovskite Absorber for Tandem Solar Cells," *Science*, vol. 351, no. 6269, pp. 151–155, Jan. 2016, https://doi.org/10.1126/science.aad5845.
[31] B. Chen *et al.*, "Efficient Semitransparent Perovskite Solar Cells for 23.0%-Efficiency Perovskite/Silicon Four-Terminal Tandem Cells," *Adv. Energy Mater.*, vol. 6, 2016, [Online]. https://api.semanticscholar.org/CorpusID:99518045
[32] Z. Qiu *et al.*, "Monolithic Perovskite/Si Tandem Solar Cells Exceeding 22% Efficiency Via Optimizing Top Cell Absorber," *Nano Energy*, vol. 53, pp. 798–807, Nov. 2018, https://doi.org/10.1016/j.nanoen.2018.09.052.
[33] P. S. C. Schulze *et al.*, "25.1% High-Efficiency Monolithic Perovskite Silicon Tandem Solar Cell with a High Bandgap Perovskite Absorber," *Sol. RRL*, vol. 4, no. 7, p. 2000152, Jul. 2020, https://doi.org/10.1002/solr.202000152.
[34] B. Chen *et al.*, "Enhanced Optical Path and Electron Diffusion Length Enable High-Efficiency Perovskite Tandems," *Nat. Commun.*, vol. 11, no. 1, p. 1257, Mar. 2020, https://doi.org/10.1038/s41467-020-15077-3.
[35] F. Hao, C. C. Stoumpos, R. P. H. Chang, and M. G. Kanatzidis, "Anomalous Band Gap Behavior in Mixed Sn and Pb Perovskites Enables Broadening of Absorption Spectrum in Solar Cells," *J. Am. Chem. Soc.*, vol. 136, no. 22, pp. 8094–8099, Jun. 2014, https://doi.org/10.1021/ja5033259.
[36] T. Wu and J. Wang, "Global Discovery of Stable and Non-Toxic Hybrid Organic-Inorganic Perovskites for Photovoltaic Systems by Combining Machine Learning Method With First Principle Calculations," *Nano Energy*, vol. 66, p. 104070, Dec. 2019, https://doi.org/10.1016/j.nanoen.2019.104070.
[37] N. K. Noel *et al.*, "Lead-Free Organic–Inorganic Tin Halide Perovskites for Photovoltaic Applications," *Energy Env. Sci.*, vol. 7, no. 9, pp. 3061–3068, 2014, https://doi.org/10.1039/C4EE01076K.

[38] G. E. Eperon *et al.*, "Perovskite-Perovskite Tandem Photovoltaics With Optimized Band Gaps," *Science*, vol. 354, no. 6314, pp. 861–865, Nov. 2016, https://doi.org/10.1126/science.aaf9717.
[39] W. Liao *et al.*, "Fabrication of Efficient Low-Bandgap Perovskite Solar Cells by Combining Formamidinium Tin Iodide with Methylammonium Lead Iodide," *J. Am. Chem. Soc.*, vol. 138, no. 38, pp. 12360–12363, Sep. 2016, https://doi.org/10.1021/jacs.6b08337.
[40] D. Zhao *et al.*, "Low-Bandgap Mixed Tin–Lead Iodide Perovskite Absorbers With Long Carrier Lifetimes for All-Perovskite Tandem Solar Cells," *Nat. Energy*, vol. 2, no. 4, p. 17018, Mar. 2017, https://doi.org/10.1038/nenergy.2017.18.
[41] D. Zhao *et al.*, "Four-Terminal All-Perovskite Tandem Solar Cells Achieving Power Conversion Efficiencies Exceeding 23%," *ACS Energy Lett.*, vol. 3, no. 2, pp. 305–306, Feb. 2018, https://doi.org/10.1021/acsenergylett.7b01287.
[42] J. Tong *et al.*, "Carrier Lifetimes of >1 μs in Sn-Pb Perovskites Enable Efficient All-Perovskite Tandem Solar Cells," *Science*, vol. 364, no. 6439, pp. 475–479, May 2019, https://doi.org/10.1126/science.aav7911.
[43] B. Abdollahi *et al.*, "Tandem Solar Cells: Vacuum-Assisted Growth of Low-Bandgap Thin Films (FA 0.8 MA 0.2 Sn 0.5 Pb 0.5 I 3) for All-Perovskite Tandem Solar Cells," *Adv. Energy Mater.*, vol. 10, May 2020. https://doi.org/10.1002/aenm.202070021.
[44] F. Jiang *et al.*, "A Two-Terminal Perovskite/Perovskite Tandem Solar Cell," *J. Mater. Chem. A*, vol. 4, no. 4, pp. 1208–1213, 2016, https://doi.org/10.1039/C5TA08744A.
[45] J. H. Heo and S. H. Im, "CH3NH3PbBr3-CH3NH3PbI3 Perovskite–Perovskite Tandem Solar Cells with Exceeding 2.2 V Open Circuit Voltage," *Adv. Mater.*, vol. 28, no. 25, pp. 5121–5125, Jul. 2016, https://doi.org/10.1002/adma.201501629.
[46] J. Ávila *et al.*, "High Voltage Vacuum-Deposited CH3NH3PbI3-CH3NH3PbI3 Tandem Solar Cells," *Energy Env. Sci.*, vol. 11, no. 11, pp. 3292–3297, 2018, https://doi.org/10.1039/C8EE01936C.
[47] A. Rajagopal *et al.*, "Highly Efficient Perovskite–Perovskite Tandem Solar Cells Reaching 80% of the Theoretical Limit in Photovoltage," *Adv. Mater.*, vol. 29, no. 34, p. 1702140, Sep. 2017, https://doi.org/10.1002/adma.201702140.
[48] C. Li *et al.*, "Thermionic Emission–Based Interconnecting Layer Featuring Solvent Resistance for Monolithic Tandem Solar Cells with Solution-Processed Perovskites," *Adv. Energy Mater.*, vol. 8, no. 36, p. 1801954, Dec. 2018, https://doi.org/10.1002/aenm.201801954.
[49] D. Zhao *et al.*, "Efficient Two-Terminal All-Perovskite Tandem Solar Cells Enabled by High-Quality Low-Bandgap Absorber Layers," *Nat. Energy*, vol. 3, no. 12, pp. 1093–1100, Dec. 2018, https://doi.org/10.1038/s41560-018-0278-x.
[50] A. F. Palmstrom *et al.*, "Enabling Flexible All-Perovskite Tandem Solar Cells," *Joule*, vol. 3, no. 9, pp. 2193–2204, Sep. 2019, https://doi.org/10.1016/j.joule.2019.05.009.
[51] Z. Yu *et al.*, "Simplified Interconnection Structure Based on C60/SnO2-x For All-Perovskite Tandem Solar Cells," *Nat. Energy*, vol. 5, no. 9, pp. 657–665, Sep. 2020, https://doi.org/10.1038/s41560-020-0657-y.
[52] Z. Yang *et al.*, "Enhancing Electron Diffusion Length in Narrow-Bandgap Perovskites for Efficient Monolithic Perovskite Tandem Solar Cells," *Nat. Commun.*, vol. 10, Oct. 2019, https://doi.org/10.1038/s41467-019-12513-x.
[53] R. Lin *et al.*, "Monolithic All-Perovskite Tandem Solar Cells With 24.8% Efficiency Exploiting Comproportionation to Suppress Sn(ii) Oxidation in Precursor Ink," *Nat. Energy*, vol. 4, no. 10, pp. 864–873, Oct. 2019, https://doi.org/10.1038/s41560-019-0466-3.
[54] K. Xiao *et al.*, "All-Perovskite Tandem Solar Cells With 24.2% Certified Efficiency and Area Over 1 cm2 Using Surface-Anchoring Zwitterionic Antioxidant," *Nat. Energy*, vol. 5, no. 11, pp. 870–880, Nov. 2020, https://doi.org/10.1038/s41560-020-00705-5.

[55] T. Leijtens *et al.*, "Tin–Lead Halide Perovskites With Improved Thermal and Air Stability for Efficient All-perovskite Tandem Solar Cells," *Sustain. Energy Fuels*, vol. 2, no. 11, pp. 2450–2459, 2018, https://doi.org/10.1039/C8SE00314A.

[56] Z. Xiao, X. Jia, and L. Ding, "Ternary Organic Solar Cells Offer 14% Power Conversion Efficiency," *Sci. Bull.*, vol. 62, no. 23, pp. 1562–1564, Dec. 2017, https://doi.org/10.1016/j.scib.2017.11.003.

[57] L. Meng *et al.*, "Organic and Solution-Processed Tandem Solar Cells With 17.3% Efficiency," *Science*, vol. 361, no. 6407, pp. 1094–1098, Sep. 2018, https://doi.org/10.1126/science.aat2612.

[58] C.-C. Chen *et al.*, "Perovskite/Polymer Monolithic Hybrid Tandem Solar Cells Utilizing a Low-Temperature, Full Solution Process," *Mater Horiz*, vol. 2, no. 2, pp. 203–211, 2015, https://doi.org/10.1039/C4MH00237G.

[59] Y. Liu *et al.*, "High Efficiency Tandem Thin-Perovskite/Polymer Solar Cells with a Graded Recombination Layer," *ACS Appl. Mater. Interfaces*, vol. 8, no. 11, pp. 7070–7076, Mar. 2016, https://doi.org/10.1021/acsami.5b12740.

[60] Z. Li *et al.*, "Hybrid Perovskite-Organic Flexible Tandem Solar Cell Enabling Highly Efficient Electrocatalysis Overall Water Splitting," *Adv. Energy Mater.*, vol. 10, no. 18, p. 2000361, May 2020, https://doi.org/10.1002/aenm.202000361.

[61] X. Chen *et al.*, "Efficient and Reproducible Monolithic Perovskite/Organic Tandem Solar Cells with Low-Loss Interconnecting Layers," *Joule*, vol. 4, no. 7, pp. 1594–1606, Jul. 2020, https://doi.org/10.1016/j.joule.2020.06.006.

[62] P. Wang *et al.*, "Solvent-Controlled Growth of Inorganic Perovskite Films in Dry Environment for Efficient and Stable Solar Cells," *Nat. Commun.*, vol. 9, no. 1, p. 2225, Jun. 2018, https://doi.org/10.1038/s41467-018-04636-4.

[63] Y. Wang, T. Zhang, M. Kan, and Y. Zhao, "Bifunctional Stabilization of All-Inorganic α-CsPbI3 Perovskite for 17% Efficiency Photovoltaics," *J. Am. Chem. Soc.*, vol. 140, no. 39, pp. 12345–12348, Oct. 2018, https://doi.org/10.1021/jacs.8b07927.

[64] W. Chen *et al.*, "A Semitransparent Inorganic Perovskite Film for Overcoming Ultraviolet Light Instability of Organic Solar Cells and Achieving 14.03% Efficiency," *Adv. Mater.*, vol. 30, no. 21, p. 1800855, May 2018, https://doi.org/10.1002/adma.201800855.

[65] S. Xie *et al.*, "Efficient Monolithic Perovskite/Organic Tandem Solar Cells and Their Efficiency Potential," *Nano Energy*, vol. 78, p. 105238, Dec. 2020, https://doi.org/10.1016/j.nanoen.2020.105238.

[66] Q. Zeng *et al.*, "A Two-Terminal All-Inorganic Perovskite/Organic Tandem Solar Cell," *Sci. Bull.*, vol. 64, no. 13, pp. 885–887, Jul. 2019, https://doi.org/10.1016/j.scib.2019.05.015.

[67] Y. (Michael) Yang *et al.*, "Multilayer Transparent Top Electrode for Solution Processed Perovskite/Cu(In,Ga)(Se,S)2 Four Terminal Tandem Solar Cells," *ACS Nano*, vol. 9, no. 7, pp. 7714–7721, Jul. 2015, https://doi.org/10.1021/acsnano.5b03189.

[68] F. Fu *et al.*, "Low-Temperature-Processed Efficient Semi-Transparent Planar Perovskite Solar Cells for Bifacial and Tandem Applications," *Nat. Commun.*, vol. 6, no. 1, p. 8932, Nov. 2015, https://doi.org/10.1038/ncomms9932.

[69] F. Fu *et al.*, "High-Efficiency Inverted Semi-Transparent Planar Perovskite Solar Cells in Substrate Configuration," *Nat. Energy*, vol. 2, no. 1, p. 16190, Dec. 2016, https://doi.org/10.1038/nenergy.2016.190.

[70] H. Shen *et al.*, "Mechanically-Stacked Perovskite/CIGS Tandem Solar Cells With Efficiency of 23.9% and Reduced Oxygen Sensitivity," *Energy Environ. Sci.*, vol. 11, pp. 394–406, 2018.

[71] D. H. Kim *et al.*, "Bimolecular Additives Improve Wide-Band-Gap Perovskites for Efficient Tandem Solar Cells with CIGS," *Joule*, vol. 3, no. 7, pp. 1734–1745, Jul. 2019, https://doi.org/10.1016/j.joule.2019.04.012.

[72] Q. Han *et al.*, "High-Performance Perovskite/Cu(In,Ga)Se2 Monolithic Tandem Solar Cells," *Science*, vol. 361, no. 6405, pp. 904–908, Aug. 2018, https://doi.org/10.1126/science.aat5055.

[73] A. Al-Ashouri *et al.*, "Conformal Monolayer Contacts With Lossless Interfaces for Perovskite Single Junction and Monolithic Tandem Solar Cells," *Energy Env. Sci.*, vol. 12, no. 11, pp. 3356–3369, 2019, https://doi.org/10.1039/C9EE02268F.

[74] T. Feurer *et al.*, "Efficiency Improvement of Near-Stoichiometric CuInSe2 Solar Cells for Application in Tandem Devices," *Adv. Energy Mater.*, vol. 9, 2019, [Online]. Available: https://api.semanticscholar.org/CorpusID:201215193

[75] S. Gharibzadeh *et al.*, "2D/3D Heterostructure for Semitransparent Perovskite Solar Cells with Engineered Bandgap Enables Efficiencies Exceeding 25% in Four-Terminal Tandems with Silicon and CIGS," *Adv. Funct. Mater.*, vol. 30, no. 19, p. 1909919, May 2020, https://doi.org/10.1002/adfm.201909919.

[76] T. Todorov *et al.*, "Monolithic Perovskite-CIGS Tandem Solar Cells Via In Situ Band Gap Engineering," *Adv. Energy Mater.*, vol. 5, no. 23, p. 1500799, Dec. 2015, https://doi.org/10.1002/aenm.201500799.

[77] F. Lang *et al.*, "Proton Radiation Hardness of Perovskite Tandem Photovoltaics," *Joule*, vol. 4, no. 5, pp. 1054–1069, May 2020, https://doi.org/10.1016/j.joule.2020.03.006.

[78] M. A. Green *et al.*, "Solar Cell Efficiency Tables (Version 53)," *Prog. Photovolt. Res. Appl.*, vol. 27, no. 1, pp. 3–12, Jan. 2019, https://doi.org/10.1002/pip.3102.

[79] H. J. Snaith, "The Perils of Solar Cell Efficiency Measurements," *Nat. Photonics*, vol. 6, no. 6, pp. 337–340, Jun. 2012, https://doi.org/10.1038/nphoton.2012.119.

[80] H. Kanda *et al.*, "Facile Fabrication Method of Small-Sized Crystal Silicon Solar Cells for Ubiquitous Applications and Tandem Device With Perovskite Solar Cells," *Mater. Today Energy*, vol. 7, pp. 190–198, Mar. 2018, https://doi.org/10.1016/j.mtener.2017.09.009.

[81] B. Chen, M. Yang, S. Priya, and K. Zhu, "Origin of J-V Hysteresis in Perovskite Solar Cells," *J. Phys. Chem. Lett.*, vol. 7, no. 5, pp. 905–917, Mar. 2016, https://doi.org/10.1021/acs.jpclett.6b00215.

[82] Y. Wang *et al.*, "Reliable Measurement of Perovskite Solar Cells," *Adv. Mater.*, vol. 31, no. 47, p. 1803231, Nov. 2019, https://doi.org/10.1002/adma.201803231.

[83] I. J. Park, J. H. Park, S. G. Ji, M.-A. Park, J. H. Jang, and J. Y. Kim, "A Three-Terminal Monolithic Perovskite/Si Tandem Solar Cell Characterization Platform," *Joule*, vol. 3, no. 3, pp. 807–818, Mar. 2019, https://doi.org/10.1016/j.joule.2018.11.017.

[84] V. Shrotriya, G. Li, Y. Yao, T. Moriarty, K. Emery, and Y. Yang, "Accurate Measurement and Characterization of Organic Solar Cells," *Adv. Funct. Mater.*, vol. 16, no. 15, pp. 2016–2023, Oct. 2006, https://doi.org/10.1002/adfm.200600489.

[85] Y. Yang *et al.*, "Enhancing the Stability of Li-Rich Mn-Based Oxide Cathodes Through Surface High-Entropy Strategy," *Energy Storage Mater.*, vol. 71, p. 103587, Aug. 2024, https://doi.org/10.1016/j.ensm.2024.103587.

[86] J. Gilot, M. M. Wienk, and R. A. J. Janssen, "Measuring the External Quantum Efficiency of Two-Terminal Polymer Tandem Solar Cells," *Adv. Funct. Mater.*, vol. 20, no. 22, pp. 3904–3911, Nov. 2010, https://doi.org/10.1002/adfm.201001167.

[87] K. A. Bush *et al.*, "23.6%-Efficient Monolithic Perovskite/Silicon Tandem Solar Cells with Improved Stability," *Nat. Energy*, vol. 2, no. 4, p. 17009, Feb. 2017, https://doi.org/10.1038/nenergy.2017.9.

[88] Z. Song *et al.*, "Probing Photocurrent Nonuniformities in the Subcells of Monolithic Perovskite/Silicon Tandem Solar Cells," *J. Phys. Chem. Lett.*, vol. 7, no. 24, pp. 5114–5120, Dec. 2016, https://doi.org/10.1021/acs.jpclett.6b02415.

[89] J. Reinhardt, P. Apilo, B. Zimmermann, S. Rousu, and U. Würfel, "Determining the Photocurrent of Individual Cells Within an Organic Solar Module by LBIC and the Filtering Approach: Experiments and Simulations," *Sol. Energy Mater. Sol. Cells*, vol. 134, pp. 157–164, 2015, https://doi.org/10.1016/j.solmat.2014.11.032.

[90] P. Vorasayan, T. R. Betts, A. N. Tiwari, and R. Gottschalg, "Multi-Laser LBIC System for Thin Film PV Module Characterisation," *17th Int. Photovolt. Sci. Eng. Conf.*, vol. 93, no. 6, pp. 917–921, Jun. 2009, https://doi.org/10.1016/j.solmat.2008.10.019.
[91] D. Chen *et al.*, "A Stepwise Solvent-Annealing Strategy for High-Efficiency Four-Terminal Perovskite/Cu(InGa)Se2 Tandem Solar Cells," *Mater. Today Energy*, vol. 49, p. 101816, Apr. 2025, https://doi.org/10.1016/j.mtener.2025.101816.
[92] "Chemical Vapor Deposition: Process, Advantages, and Disadvantages—Easy Explanation—Chemistry Notes." Accessed: Feb. 24, 2025. [Online]. Available: https://chemistnotes.com/nanochemistry/chemical-vapor-deposition-process-advantages-and-disadvantages-easy-explanation/
[93] M. T. Hörantner *et al.*, "The Potential of Multijunction Perovskite Solar Cells," *ACS Energy Lett.*, vol. 2, no. 10, pp. 2506–2513, Oct. 2017, https://doi.org/10.1021/acsenergylett.7b00647.
[94] C. Zuo and L. Ding, "Modified PEDOT Layer Makes a 1.52 V Voc for Perovskite/PCBM Solar Cells," *Adv. Energy Mater.*, vol. 7, no. 2, p. 1601193, Jan. 2017, https://doi.org/10.1002/aenm.201601193.
[95] M. Saliba *et al.*, "Cesium-Containing Triple Cation Perovskite Solar Cells: Improved Stability, Reproducibility and High Efficiency," *Energy Env. Sci.*, vol. 9, no. 6, pp. 1989–1997, 2016, https://doi.org/10.1039/C5EE03874J.
[96] C. Yi *et al.*, "Entropic Stabilization of Mixed A-Cation ABX3 Metal Halide Perovskites for High Performance Perovskite Solar Cells," *Energy Env. Sci.*, vol. 9, no. 2, pp. 656–662, 2016, https://doi.org/10.1039/C5EE03255E.
[97] Z. Song, C. Chen, C. Li, R. A. Awni, D. Zhao, and Y. Yan, "Wide-Bandgap, Low-Bandgap, and Tandem Perovskite Solar Cells," *Semicond. Sci. Technol.*, vol. 34, no. 9, p. 093001, Jul. 2019, https://doi.org/10.1088/1361-6641/ab27f7.
[98] Z. Fang *et al.*, "CsPbI2.25Br0.75 Solar Cells with 15.9% Efficiency," *Sci. Bull.*, vol. 64, no. 8, pp. 507–510, Apr. 2019, https://doi.org/10.1016/j.scib.2019.04.013.
[99] X. Jia, L. Liu, and Z. Fang, "TBAB Additive for Inorganic CsPbI2.4Br0.6 Perovskite Solar Cells With Efficiency Beyond 15%," *J. Mater. Chem. C*, vol. 7, no. 24, pp. 7207–7211, 2019, https://doi.org/10.1039/C9TC02362C.
[100] W. Zhou, Z. Wen, and P. Gao, "Less is More: Dopant-Free Hole Transporting Materials for High-Efficiency Perovskite Solar Cells," *Adv. Energy Mater.*, vol. 8, no. 9, p. 1702512, Mar. 2018, https://doi.org/10.1002/aenm.201702512.
[101] K. Domanski *et al.*, "Not All That Glitters Is Gold: Metal-Migration-Induced Degradation in Perovskite Solar Cells," *ACS Nano*, vol. 10, no. 6, pp. 6306–6314, Jun. 2016, https://doi.org/10.1021/acsnano.6b02613.
[102] L. Zhao *et al.*, "Redox Chemistry Dominates the Degradation and Decomposition of Metal Halide Perovskite Optoelectronic Devices," *ACS Energy Lett.*, vol. 1, no. 3, pp. 595–602, Sep. 2016, https://doi.org/10.1021/acsenergylett.6b00320.
[103] F. Sahli *et al.*, "Fully Textured Monolithic Perovskite/Silicon Tandem Solar Cells With 25.2% Power Conversion Efficiency," *Nat. Mater.*, vol. 17, no. 9, pp. 820–826, Sep. 2018, https://doi.org/10.1038/s41563-018-0115-4.
[104] M. Cai, Y. Wu, H. Chen, X. Yang, Y. Qiang, and L. Han, "Cost-Performance Analysis of Perovskite Solar Modules," *Adv. Sci.*, vol. 4, no. 1, p. 1600269, Jan. 2017, https://doi.org/10.1002/advs.201600269.
[105] K. Nishimura *et al.*, "Lead-Free Tin-Halide Perovskite Solar Cells with 13% Efficiency," *Nano Energy*, vol. 74, p. 104858, Aug. 2020, https://doi.org/10.1016/j.nanoen.2020.104858.
[106] C. Zuo and L. Ding, "Lead-Free Perovskite Materials (NH4)3Sb2IxBr9–," *Angew. Chem. Int. Ed.*, vol. 56, no. 23, pp. 6528–6532, Jun. 2017, https://doi.org/10.1002/anie.201702265.
[107] K. A. Bush *et al.*, "Minimizing Current and Voltage Losses to Reach 25% Efficient Monolithic Two-Terminal Perovskite-Silicon Tandem Solar Cells," *ACS Energy Lett.*, vol. 3, no. 9, pp. 2173–2180, Sep. 2018, https://doi.org/10.1021/acsenergylett.8b01201.

[108] K. A. Bush *et al.*, "Thermal and Environmental Stability of Semi-Transparent Perovskite Solar Cells for Tandems Enabled by a Solution-Processed Nanoparticle Buffer Layer and Sputtered ITO Electrode," *Adv. Mater.*, vol. 28, no. 20, pp. 3937–3943, May 2016, https://doi.org/10.1002/adma.201505279.

[109] F. Sahli *et al.*, "Improved Optics in Monolithic Perovskite/Silicon Tandem Solar Cells with a Nanocrystalline Silicon Recombination Junction," *Adv. Energy Mater.*, vol. 8, no. 6, p. 1701609, Feb. 2018, https://doi.org/10.1002/aenm.201701609.

[110] L. Mazzarella *et al.*, "Nanocrystalline Silicon Oxide Emitters for Silicon Hetero Junction Solar Cells," *5th Int. Conf. Silicon Photovolt. SiliconPV 2015*, vol. 77, pp. 304–310, Aug. 2015, https://doi.org/10.1016/j.egypro.2015.07.043.

8 Enhancing Solar Cell Defect Detection through a Generalized Hybrid Transfer Learning Model Synergized with Machine Learning Classifier

Lakshay Madan Gopal, S. Aditya, Anushiya Banu, S. Dhaneshver, M. Maghutesan, and S. Sridevi

8.1 INTRODUCTION

Solar panels are one of the most powerful tools in the transition process from fossil fuels. However, their efficiency is no longer constant in the case of small defects such as micro-cracks, discoloration, and snail trails that affect their lifespan. Such defects cannot be easily detected but become very influential in reducing energy output as realized from solar panels, hence far-reaching implications for the sustainability of solar power systems. Detection of such issues in the initial stages would ensure that, over time, the solar panels are still working at optimum performance. Advanced monitoring and defect-detection systems are hereby required, but the problem arises because the majority of these defects are visually obscure and require techniques beyond simple manual inspection methods that are both time-consuming and liable to errors. This emphasis on high precision in the identification of defects has increased the demand for automated detection systems.

The current scenario is that most inspection approaches are visual and low-end imaging techniques. Such techniques limit the nature of inspection to mere low-end precision and unreliability. Since such inspections are predominantly based on manual checks and infrared camera views, they have major drawbacks. Such inspection techniques are generally hard to apply since identifying small defects is very challenging. These also tend to be labor-intensive to a great extent, consuming vast

DOI: 10.1201/9781003515784-8

human effort and time that is required to inspect a high number of panels. This makes even manual inspections very challenging for large-scale solar farms, where efficiency and accuracy are critical to ensure that the solar panels are functioning correctly. As such, never has the need been so great for systems that can pick on even the tiniest defects, not necessarily through human intervention.

The integration of deep learning and machine learning (ML) tools in this field saw tremendous strides in the automation of industrial processes—mostly through image-based application types. Currently, convolutional neural networks (CNNs) are turning out to become the core tool in automated defect detection as great performance in virtually any image classification task has been reported for this area. CNNs are designed to automatically detect features within images through multiple layers and have made this a really strong candidate for detecting the patterns in a solar panel image, especially when defects are present. It makes these deep models of learning very effective in processing lots of data and making learnings from them to subsequently deliver results without necessitating explicit feature extraction through manual modes—a nice advantage over traditional ML-based models. As shown in prior research [1, 2, 3], CNNs have been highly beneficial in many applications where the detection of subtle patterns in visual data is needed.

Although the conventional methods of inspecting solar panels are highly dependent on judgment and relatively simple imaging procedures, these methods are slow, labor-intensive, and often inexact. Such problems have therefore rapidly accelerated interest in developing automated systems that can consistently detect defects with much higher accuracy. In this context, CNNs, as well as other deep learning techniques, have gained immense attention in terms of hierarchical feature extraction from images. The recent advances in artificial intelligence (AI), especially in deep learning [4, 5] and ML [6], have improved the capacities of automated defect detection systems, mainly in terms of accuracy, speed, and scalability. Studies, such as Abdelsattar et al. [7], have shown that decision tree (DT), K-nearest neighbors (KNN), random forest (RF), and extra tree classifiers can significantly enhance the accuracy and efficiency of fault detection in solar panel systems. Such strong, complex patterns and visual features make faulty panels differ from functional ones, as based on studies [8, 9, 10]. CNNs specifically succeed in image classification tasks due to those characteristics.

This paper presents a new approach to identifying defects in a solar panel through the integration of CNNs with the traditional ML classifier [11]. Our model harnesses the strengths of CNNs to automatically extract features from images of solar panels, which may include subtle defects that traditional methods would fail to identify. These features are then fed into various ML classifiers like support vector machine (SVM), KNN, RF, and gradient boosting (GB) to determine the classification of the defects. This hybrid model combines the best of both worlds—CNNs for feature extraction and classical ML classifiers for accurate decision-making—aiming to improve overall detection accuracy while reducing computational overhead.

An important innovation of our approach lies in the incorporation of linear discriminant analysis (LDA) to enhance classification performance [12]. As LDA is a strong dimensionality reduction technique, it significantly contributes to the transformation of the high-dimensional feature space into one where different classes of defects are more distinctly separated. This simplifies data complexity and improves class separability, leading to enhanced classification results. Our work thoroughly examines the impact of LDA by comparing outcomes from CNN models with and without LDA. This analysis provides a deeper understanding of how dimension reduction affects fault detection, particularly for defects with subtle visual features that might be hard to differentiate otherwise.

We examine several CNN architectures, including MobileNetV3 [13], ResNet34 [14], VGG16 [15], InceptionV3 [16], InceptionResNetV2 [17], and EfficientNetB4 [18]. Each architecture presents unique benefits depending on available computational resources and the complexity of feature extraction needed. For instance, MobileNetV3 is especially suited for low-resource environments like drone-based inspections, where lightweight models are essential. In contrast, deeper architectures such as ResNet34 and InceptionV3 capture rich image features but require higher computational power. We aim to identify which model best balances high accuracy with low computational cost, a crucial consideration for scaling models to real-world applications.

The efficiency and safety of photovoltaic modules are compromised by microcracks, discoloration, and hotspots. Traditional electroluminescence (EL) imaging is indeed accurate in defect detection, but it is human-dependent, slow, and prone to errors. Deep learning offers an opportunity for the development of fully automated, highly accurate solutions; however, it requires massive amounts of data and often faces challenges with imbalanced representation. The current hybrid approach incorporates CNN-based feature extraction with classifiers such as SVM, RF, and KNNs, enhanced by LDA for dimensionality reduction and data augmentation. This approach supports reliable and scalable defect detection, guaranteeing optimum energy output while fostering sustainable energy systems [19].

The study "SPF-Net: Solar panel fault detection using U-Net based deep learning image classification" introduced an advanced method for solar panel fault detection using a U-Net-based deep learning model. This model combined the InceptionV3 architecture with image segmentation capabilities from the U-Net framework, achieving high precision and accuracy for detecting multiple types of faults. The SPF-Net method addressed issues such as dust accumulation, physical damage, and electrical faults, boasting a validation accuracy of 98.34% and a test accuracy of 94.35%. High precision was achieved using dense layers with LeakyReLU, batch normalization, and a softmax output layer. This study highlighted the necessity of accurate defect segmentation and classification for reliable and scalable solar fault detection, which is crucial for ensuring the efficiency of solar panels under diverse environmental conditions [20].

The paper "Machine learning for monitoring and classification in inverters from solar photovoltaic energy plants" explored the use of ML for real-time monitoring and fault classification in photovoltaic inverter systems. It underscored the importance of optimizing and maintaining inverters to enhance energy production while minimizing operational costs. Various ML algorithms, such as DTs, SVM, and neural networks, were designed to predict and classify faults including grid voltage issues, overcurrent faults, and device malfunctions. By leveraging historical inverter data and real-time performance metrics, this study established a system for predictive maintenance, enhancing the reliability of photovoltaic systems and ensuring consistent solar power output [21].

The dataset we used in our research includes common defects found in solar panels, such as micro-cracks, discoloration, and snail trails. These defects are particularly challenging to detect due to their subtle optical properties, especially when using standard electro-photometric imaging. Our model was tested on its ability to detect and classify visually similar defect types by incorporating a diverse set of defects. The varied dataset enhances the model's robustness, ensuring that it can handle different defect categories and improve its applicability in real-world settings.

Our model's effectiveness relies on its ability to detect subtle defects that can severely impact solar panel performance. Micro-cracks, though appearing minor, disrupt the electrical pathways within panels and lead to significant efficiency losses. Other defects, such as discoloration and snail trails, similarly reduce panel output. This paper evaluates how well our model identifies defects that are difficult to detect with the naked eye, making it a valuable tool for solar panel maintenance.

In conclusion, this paper proposes a hybrid model combining CNNs and classical ML classifiers for effective solar panel defect detection. The use of LDA to enhance classification accuracy is a key contribution of this research. We also evaluate various CNN architectures to identify the optimal model for balancing precision and computational cost. Our approach provides valuable insights into how AI and ML can be harnessed for automated solar panel inspection, reducing inspection time and enhancing detection accuracy. By leveraging lightweight models such as MobileNetV2, which has a relatively small size (14 MB), our method ensures scalability for deployment on low-resource devices like drones used in field inspections. This contributes to addressing the challenges of automated solar panel inspection and supports the shift toward more sustainable and efficient solar energy systems.

8.2 DATASET DESCRIPTION

In the current study, the main dataset used is known as the electroluminescence photovoltaic dataset (ELPV), and it has a good reputation in the domain of the search for defects in solar panels. The ELPV dataset is mostly publicly available and referenced many times in studies [22, 23, 24], since it provides a robust basis for studies focused on finding defects in solar cells. It comprises altogether 2,624 high-resolution EL images of individual solar cells, separately extracted from one of 44 photovoltaic modules. The images are standardized at 300 × 300 pixels with an 8-bit grayscale format, offering uniformity that enhances model training and testing reliability. The main objective of the dataset is to support a holistic analysis of defects in solar cells by attributing to each of the images binary labels indicating the status of cells: 0 as non-defective and 1 as defective. The given defect probability score is a floating point value ranging from 0 to 1, providing an estimation of the severity of every defect. This score in probability allows the scientists to make judgments regarding the nature of damage that has taken place in defective cells, which is much beyond the binary classification results. Besides, each cell image is assigned a type of module, namely either monocrystalline or polycrystalline, both variants of which are widely used in photovoltaic systems. The four major classes of the dataset are monocrystalline—defective, monocrystalline—non-defective, polycrystalline—defective and polycrystalline—non-defective. Table 8.1 summarizes the count for each class, and Figure 8.1 displays examples of images from these classes. To normalize the image size and perspective and also correct defects as a result of lens distortions, all images within the ELPV dataset received some kind of normalization process. This preprocessing ensures models trained on this dataset are less likely to encounter variability due to inconsistencies in imaging and will thus make defect identification more accurate. The dataset provides a useful tool for defect analysis and predictive modeling in solar systems [25] by capturing both intrinsic and extrinsic flaws that are known to negatively impact its power efficiency. The ELPV dataset further follows a structured architecture for ease of deployment with the ML workflows. It keeps images in a directory called images, and labels of the status of defects and probability scores are stored in a labels.csv file. A utility function, elpv_reader, is provided for hassle-free loading of data in Python. With its help, researchers can easily incorporate the dataset into the pipeline of their data analysis. The labeling structure with sample images in Figure 8.2 helps in a clear understanding of the organization of the dataset.

8.3 METHODOLOGY

This subsection illustrates the conceptual outline of the proposed method by explaining how integration with deep learning using architectures and ML classifiers based on CNNs ensures enhanced accuracy and reliability in the detection of defects for solar panels. The method leverages the strengths of the various steps involved to enable a comparative analysis of how the performance of the model differs without LDA-based dimensionality reduction, which effectively enhances the effectiveness of classification concerning defect recognition.

8.3.1 MobileNetV3

For this paper, the MobileNetV3 architecture was pre-trained on ImageNet weights as a basis for designing an image classification model. This particular architecture was strategically chosen since the MobileNetV3 architecture is very efficient in handling large-scale image data [13] and successfully maintains its performance by

TABLE 8.1
Defect Analysis of Monocrystalline and Polycrystalline Solar Modules

Solar Module	Defective	Non-Defective
Monocrystalline	430	644
Polycrystalline	580	970

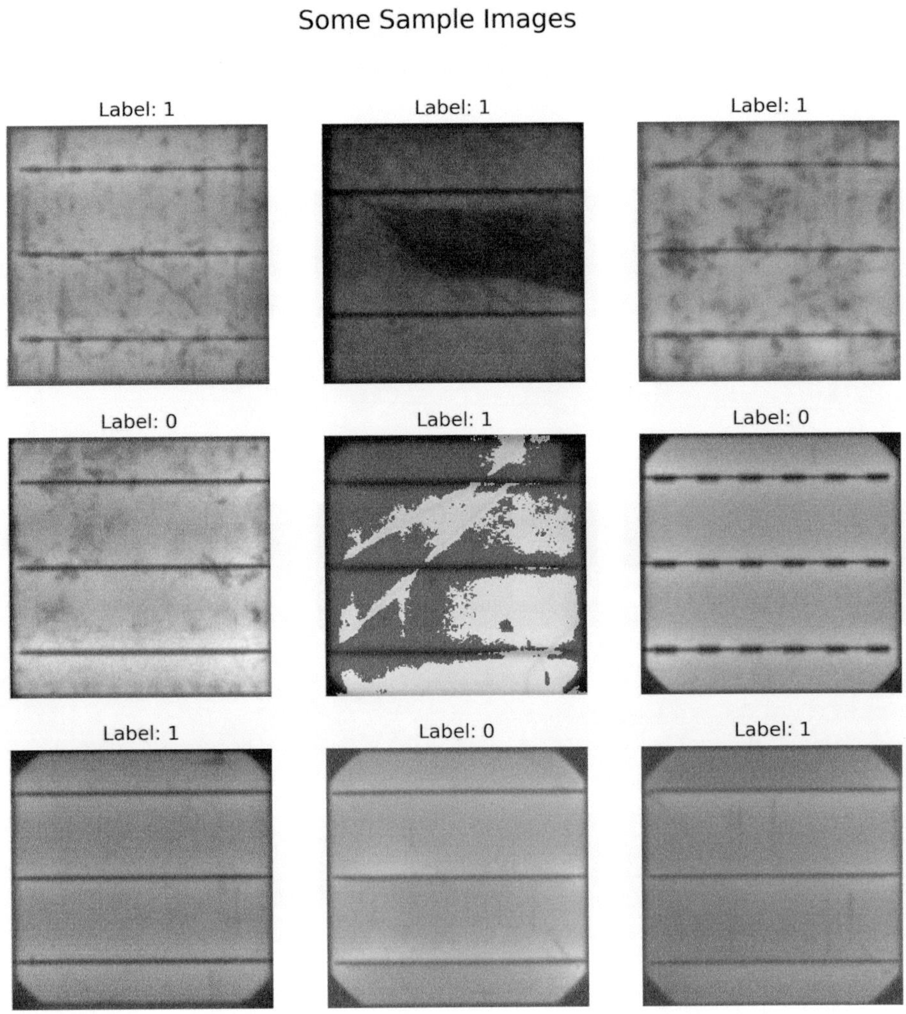

FIGURE 8.1 A collage of nine images of randomly selected silicon solar cells.

balancing computer resources consumed. With a lightweight design, MobileNetV3 accommodates various high-speed computation applications without reducing precision. For optimal performance on the MobileNetV3 architecture (Figure 8.3), we settled the size of the input image to 224 × 224 pixels; this fits within the configuration of the architecture and contributes to more efficient computation. Figure 8.3 depicts a detailed architecture of the MobileNetV3 model used in this work.

We have adjusted the MobileNetV3 model for use with binary classification tasks by setting include_top=False as it removes the classifier layer, which is enabled by default for the ImageNet weights. Instead of the intrinsic classifier, we designed a custom classification head. We started with a flattening layer that converts feature

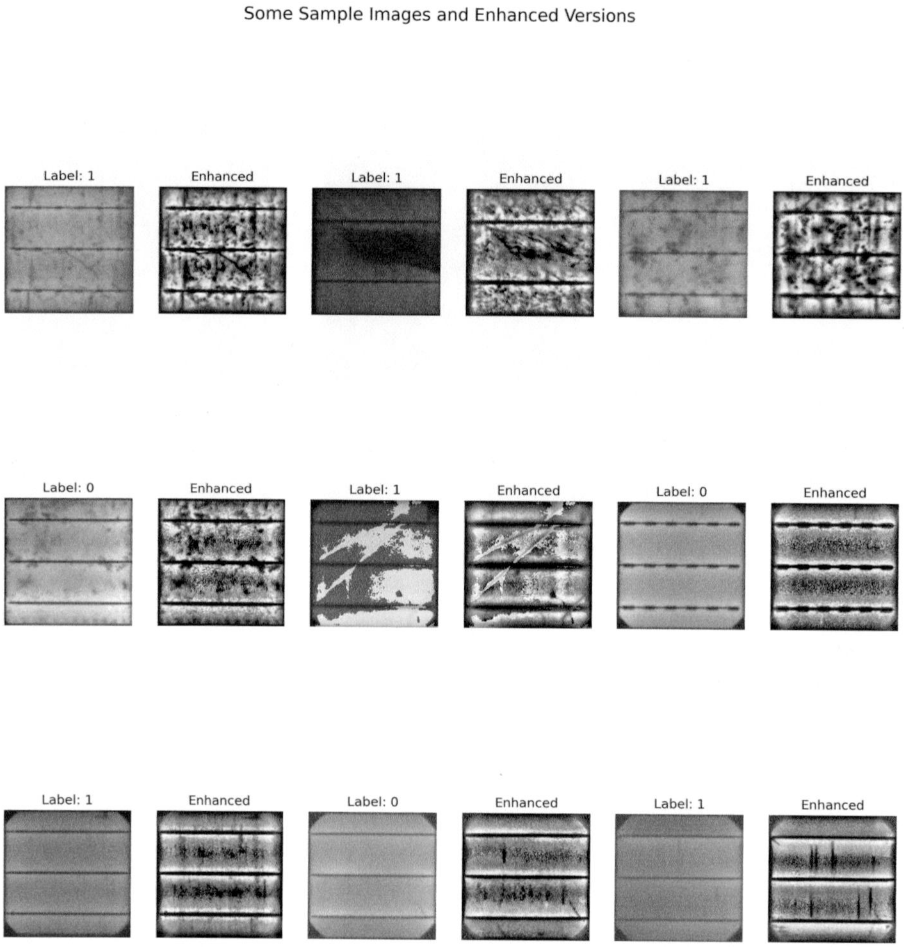

FIGURE 8.2 Surface photographs of materials, labeled as defective (1) or undefective (0), along with their enlarged versions. The enhanced images improve the detection of surface defects, making tiny flaws more visible. "Label: 1" highlights areas with visible defects, while "Label: 0" indicates clear, defect-free surfaces. The enhancement technique increases contrast and clarity, aiding in the identification of subtle flaws that are not apparent in the original images.

maps into a 1D vector. This is then followed by a fully connected dense layer that consisted of 512 units and had L2 regularization to improve generalization by penalizing large weights. Finally, a densely connected output layer with one neuron, and a sigmoid activation function was added; this is suitable for binary classification, whose outputs often fall into the range [0,1].

The model employs the Adam optimizer, which is very efficient in gradient-based optimization and most suitable when there are two or more layers involved in deep learning. We used a binary cross-entropy loss function, specifically suited for such problems in binary classification since the loss is estimated in relation to the probabilities predicted for two classes. The learning rate was 0.0001, taking into consideration both fast convergence and not overshooting the optimal solution. To further reduce the overfitting of the model on the training dataset, we implemented an early stopping callback. It monitors the validation accuracy, stopping the training in case of a lack of improvement of over 20 epochs. In such cases, it restores the model's weights from the epoch when the validation accuracy is highest.

$$\text{Sigmoid activation function } \hat{y} = \begin{cases} 1 \text{ } if \, \sigma(x) > 0.5 \\ 0 \text{ } if \, \sigma(x) \leq 0.5 \end{cases} \quad (8.1)$$

The sigmoid activation function in the output layer forms the heart of binary classification because it squashes the input values into between 0 and 1, which makes the predictions easier to interpret. Outputs nearing 1 have a higher probability of the positive class, for instance, a defect, while outputs close to 0 mean a high probability of the negative class, no defect. Using this threshold of 0.5, we classify these continuous output values into discrete class labels. We classify any predictions over 0.5 as the positive class, while below or equal to 0.5 would fall under the classification for being in the negative class. This setup lends interpretability of model confidence: greater sigmoid outputs mean greater certainty in predicting the positive class; lesser values mean confidence in predicting the negative class.

This MobileNetV3 configuration does that in the most effective way with striking efficiency compared to accuracy, leveraging transfer learning to process complex image data under minimal computational loads. Coupled with early stopping and L2 regularization, sigmoid-based classification ensures that the generalizability of the model is far maximized, thus offering it as a reliable solution to the detection defects of solar panels. This architecture is optimized to speed up and perform but also for reliability in binary classification tasks to ensure it accomplishes an overall goal of improving the accuracy in defect identification.

8.3.2 ResNet34

We proposed the ResNet34 [14, 16] inspired image classification architecture based on residual networks (ResNets) to overcome various training difficulties concerned with deep CNNs. The model starts from initial convolutional operations on the input RGB images of dimensions 224 × 224 × 3. This utilizes 64 filters of kernel sizes 7 × 7 and stride lengths of 2 and afterward batch normalization and rectified linear unit (ReLU) operations to add non-linearity. Then, a max-pooling layer with a kernel size of 3 × 3 and a stride of 2 reduces the spatial dimensionality of the input feature map. This phase, however, mainly aims at enhancing the low-level features of the images while, at the same time, causing a decline in the degree of resolution.

The central component of the model consists of residual blocks, which are included to solve the degradation problem of very deep networks. A pair of convolutional layers make up each residual block, which is again followed by batch normalization and

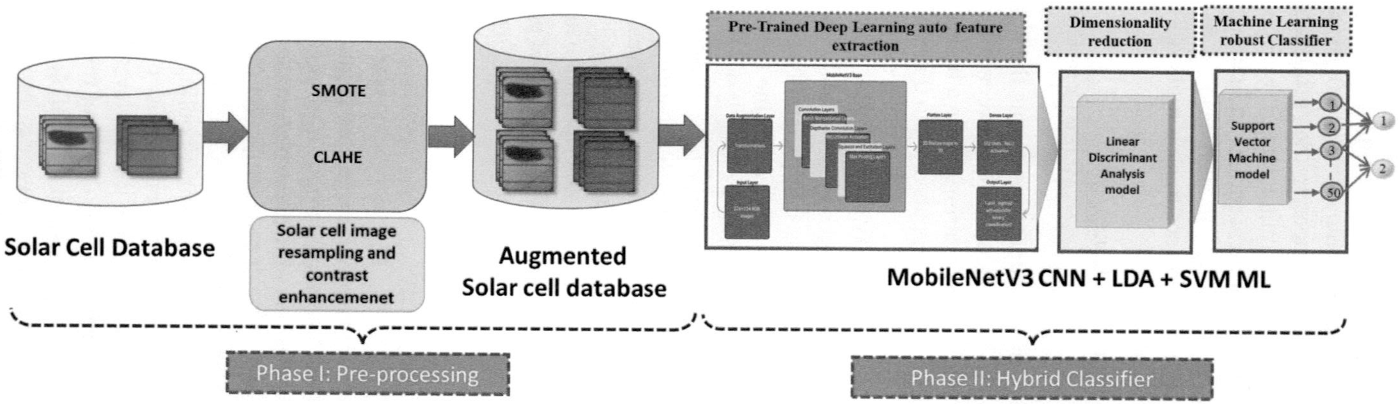

FIGURE 8.3 The preprocessed solar cell image database, enhanced with SMOTE, Contrast Limited Adaptive Histogram Equalization (CLAHE), and resampling for clearer features. MobileNetV3 extracts features, which are then reduced using LDA. These features are classified by an SVM, enabling efficient and accurate defect detection in solar cells.

ReLU activations. To realize the residual in connection to its name, this input to the block is treated as the output, even as these layers carry out their processing. Such a connection, termed the shortcut, is infused before the last ReLU activation is invoked by the convolutions. This enables the construction of the model to focus on learning residual mappings rather than learning direct mappings even for very deep networks, thus reducing the training difficulties associated with deeper networks. In the implementation, I used increasing numbers of filters for each of the next blocks 64, 128, 256, and 512 so that the model is capable of learning intricate features with each layer.

$$\text{softmax}(z_i) = \frac{e^{z_i}}{\sum_{j=1}^{n} e^{z_j}}, \text{ for } i = 1, 2, \ldots, n \tag{8.2}$$

The softmax function is computed by taking the exponent of every value and then normalizing by the sum of the exponentials of the values so that they form a probability distribution. This is further beneficial to the softmax function as it allows the softmax function to place the most emphasis on the highest-scoring category while still allowing all other categories to maintain positive probabilities, which is a great feature for multi-class problems.

After the residual blocks, a global average pooling layer was applied, which reduces the spatial dimensions of the feature map to a single vector, thus aggregating the learned features. This vector is then forwarded to the two fully connected layers, and the final output layer consists of a dense layer with a softmax activation function, which is used for multi-class classification. This design ensures that the model outputs class probabilities corresponding to the different categories.

For model training, an Adam optimizer was used and a sparse categorical cross-entropy loss function because the target labels are integers. An early stopping callback was implemented to avoid overfitting while generalizing my findings because this mechanism stops the training if the validation loss hasn't been improved during five consecutive epochs while restoring the best weights from the model. The training was done for a maximum of 200 epochs with a batch size of 32 on the training dataset and monitored the validation performance of a separate test set.

8.3.3 VGG16

At the outset, VGG16's [15, 16] weights are trained on ImageNet, where it has performed excellently on more than a few of the benchmarks. The depth of VGG16 is constituted by 16 trainable layers, which comprise convolutional layers, pooling layers, and fully connected layers. One of the characteristics which stand out about VGG16 is its design simplicity as all the convolutional layers have the same filter size of 3 × 3 and the pooling layers have a filter size of 2 × 2. Because of this simplicity and uniformity, it has also become easier to use in different image recognition tasks.

Choosing include_top=False, fully connected layers of VGG16 are removed; that is, only convolutional and pooling layers are left for feature extraction from input images. The features extracted in this manner are then forwarded to other layers to complete the classification task. For this purpose, we first introduce fully connected layers: we use a flattening layer, which transforms 2D feature maps into one-dimensional vectors; this is followed by inserting a fully connected layer consisting of 512 neurons and a ReLU activation function. Then a sigmoid activation function is assigned in the output layer since the task is a binary classification—determining whether the image depicts a defective or a non-defective solar panel.

Also, the model implemented employs early stopping, which is essentially useful in avoiding overfitting by controlling validation accuracy throughout training and stopping it when there is no upsurge in the validation accuracy for certain epochs. This serves to prevent the model from overfitting the training data by learning spurious details and helps the model to better adapt to data that it has not seen before.

8.3.4 InceptionV3

For image classification, the InceptionV3 [16] architecture is applied to extract the feature map. This model is also trained on the ImageNet weight, which means that the model is trained to recognize the features of the edges, textures, and objects in the images. A pre-trained model like InceptionV3 meaningfully contributes to the model's power by incorporating previously learnt information, making it less susceptible to overfitting problems in small datasets or where resources for computational power are limited. By setting include_top=False, the model is stripped of its original classification layer, and only the convolutional layers are used for feature extraction.

The base_model.trainable = False makes sure that the InceptionV3 pre-trained weights will be intact during training, which means they will not change. This is necessary because it prevents unneeded changes to the already trained model and restructures training only for the new layers on top of the model. This approach helps accelerate the training process and makes it less overfitting prone especially when the amount of data for training is very limited. The InceptionV3 layers that are held constant accept as input the image to be processed, and the output generated consists of feature maps depicting the important parts of the image.

In addition to the output provided by the InceptionV3 base once processing is done, the output is run through other layers, which are meant for the specific task. These are in addition to a flattening layer, which converts the feature maps into a single column of values, and a dense layer, which processes these features further. Finally, the output layer uses sigmoid activation to estimate the likelihood of a certain image being in a specific class, which is ideal for two-class classification problems.

8.3.5 InceptionResNetV2

InceptionResNetV2 [17] is a complex structure of several convolution layers that incorporates the best features of Inception and ResNet looks. It aims to perform efficiently by employing inception modules which perform numerous convolutions to the same input image using different filter sizes and pooling before fusing them to avoid the loss of details in the image. This gives InceptionResNetV2 the advantage of being able to interpret more intricate details and levels of structure than ordinary convolutional networks. The model also features residual connections like those found in ResNet, which help in solving the vanishing gradient effect, thus allowing the training of very deep networks as gradients can now be transported readily even in a constrained manner through backpropagation.

InceptionResNetV2, like VGG16, can also be applied for the purpose of feature extraction by excluding all its fully connected layers by setting include_top=False. This ensures that all the fully connected layers, which serve the purpose of classifying the input images, are removed and only the convolutional and pooling layers are retained. These features are thus subjected to further classification processes through other layers. In this regard, a flattening layer is implemented, which facilitates the transition of two-dimensional feature maps into a single-dimensional feature vector.

Then, there is a dense layer with 512 units and ReLU activation, and, for the output layer, the sigmoid activation is used for a two-class image classification task, where images are classified as either defective or non-defective solar panels. To reduce overfitting, we use an early stopping strategy, in which the training is discontinued when the validation accuracy does not improve within a certain number of epochs, thus enhancing the model's capability to generalize on new data other than the training data. To prevent overfitting, an early stopping strategy was adopted, and the InceptionResNetV2-based model was trained for 200 epochs.

8.3.6 EfficientNetB4

This implementation uses the EfficientNetB4 [18] architecture as a backbone for a binary classification task—specifically the identification of defects on solar panel surfaces. The model is based on the one pre-trained on ImageNet, yet the include_top argument is set to False to accommodate changes relevant to the task. This pre-trained model is then "frozen," meaning that all the weights of the base model during training will remain unchanged—the aim here is to modify and train exclusively the custom classifier that has been added on top.

Input images are converted to 224 × 224 × 3 to satisfy the expected input size requirement of the EfficientNetB4. The base model processes these supplied images as well without changing their weights. The modifications to the pre-trained base model incorporate the following additional layers: one fully connected layer of 512 neurons with ReLU activations connected to the input and one more fully connected layer, which provides outputs after sigmoid activation for the purpose of binary classification connected to the outmost layer of the model.

The model is built on an Adam optimizer with a learning rate set to 0.0001. The loss is binary cross-entropy, which is suitable considering that this is a binary classification task, and the metric observed while training is accuracy.

Binary Cross-Entropy Loss:

$$L(y, \hat{y}) = -(y \log(\hat{y}) + (1 - y) \log(1 - \hat{y})) \tag{8.3}$$

Binary cross-entropy loss measures the accuracy of binary classification by comparing actual labels with predicted probabilities. When the true label is positive, the loss increases if the predicted probability is low. Conversely, if the label is negative, the loss increases when the predicted probability is high. This method encourages predictions that are closer to the true labels, making it effective for training models in binary classification tasks.

8.4 HYBRID MODELS

The aim of this research is to propose and assess hybrid models which incorporate CNNs and standard models for classification. The hybrid models will be tested, with or without LDA, to optimize the features and the classification results. LDA is important because CNN-generated features are usually of high dimensionality, and LDA facilitates the representation of such features in a lower dimensional space with maximum class separation, which improves the classifying power and interpretability of ML algorithms. This way, we can avoid employing large and complicated feature vectors resulting from deep networks like MobileNetV3, ResNet34, VGG16, InceptionV3, EfficientNetB4, and InceptionResNetV2 and include more simple but effective and class separable features for the classic classifiers.

In contemporary society, sophisticated technologies like CNNs, ML, and LDA have permeated nearly all industries; these applications serve real-world problems in the most efficient and scalable manner. Such modern technological catchments can be found in all types of industries, including healthcare, e-commerce, agriculture, security, manufacturing, and transport; modern applications can involve defect detection, behavior analysis, resource optimization, and predictive analytics. Companies like GE Healthcare, Amazon, Tesla, and NASA used pure interpretation, absorbing the power of CNN in extraction from features with ML predictive modeling via LDA toward dimension reduction, thus making their systems faster, smarter, and more accurate. Table 8.2 summarizes several technologies industries use to innovate through CNN + ML + LDA in creating applications that affect technology advancement and efficiency.

In traditional classifiers, SVM [43] is often preferred because it can create a hyperplane within the n-dimensional feature space that separates different classes effectively. Also, with the radial basis function (RBF) kernel [44], the SVM classifier used here may be capable of classifying even the most non-linear data appropriate for SVM classifiers and CNN features such as MobileNetV3 (Figure 8.4a) and InceptionV3 (Figure 8.4d). In this case, the output probabilities hence can be used to compute different metrics such as AUC (Area Under the Curve—that refers to the Area Under the Receiver Operating Characteristic [ROC] Curve. It is a performance metric used to evaluate classification models, especially binary classifiers) and log loss when probability=True. The SVM is in a way designed to deal with high dimensionality, which is also the case for CNN architectures such as ResNet34, where rich and varied feature information is produced.

RF classification [45] provides another classification perspective as an ensemble classifier built of DTs, each applied on a different subset of the data and features. This aspect of randomness in the model enhances the ability of the model not to learn the features of the CNN too much. For instance, when n_estimators is set to 100, the forest classifier combines the results from all possible decision

TABLE 8.2
World-Class Industry Application of CNN with ML and LDA

Company	How They Leverage CNN, ML, and LDA
GE Healthcare	Utilizes CNN for analyzing medical images (e.g., X-rays, MRIs) [26] and ML for predictive diagnosis [27]
Siemens Healthineers	Applies CNNs in image segmentation for tumor detection [28], ML for disease prediction [29], and LDA for improving classification performance [30, 31]
Bosch	Uses CNNs for defect detection on manufacturing lines [32] and ML in predictive maintenance [33]
NASA	Employs CNNs for cloud detection algorithms for multi-modal satellite imagery [34], ML for disaster impact predictions [35], and LDA for reducing redundant features [36]
Amazon	Employs CNNs for event detection [37] and ML for personalized recommendations [38]
Wal-Mart	Uses CNNs for deep learning-based recommender systems [39] and ML for inventory management [40]
Alibaba	Leverages CNNs for product recommendation engines [41] and ML solves demand problems in retail [42]

paths up to the specified depth, which is advantageous for deep, structured data such as the ones produced by models VGG16 and EfficientNetB4. This strategy of combining multiple models is safe, particularly for problems where the interactions of features are many, as it allows the model to pay attention to important traits of the features obtained from the CNN; hence, it is a good classifier even with savage datasets.

The KNN [46] classifier works using a distance-based technique where it determines the class labels based on the nearest neighbors' majority class. In this configuration with n_neighbors=5 KNN model classifies the data based on how close the feature space is populated with the data. It is worth noting that KNN achieves better results when the data is not sparse. KNN's distance-dependent characteristic is suitable for less complex lower-dimensioned feature representation such as that obtained from the MobileNetV3; therefore, it is a relevant pillar comparison model within this hybrid framework. The benefits of KNN include understanding the extent of the effects of similarity in geographical space on classification, contributing to the comprehensiveness of the models built in an ensemble.

The GB framework used by LightGBM [47] addresses issues arising from high-dimensional data as it works by building one tree after the other in a sequential manner to correct the flaws of the previous tree. This is made more efficient through the use of histogram-based learning, which makes LightGBM both memory and computation efficient, which are essential benefits when it comes to training on large and complex training datasets of CNN's features. LightGBM also comes in handy when working with models such as InceptionResNetV2 and EfficientNetB4, which present deep feature vectors that are both complex and highly dimensional, as it is fast and able to model high-order feature interactions. Given its ability to model complex patterns and deal with a large number of features, LightGBM seems to fit well in such a hybrid framework, particularly when high accuracy comes alongside time constraints.

To conclude, the mixing of these ML classifiers with CNN models such as MobileNetV3, ResNet34, VGG16, InceptionV3, EfficientNetB4, and InceptionResNetV2, and then LDA transformation further extends the adaptability, scalability, and the performance of the hybrid model. Every classifier offers distinct advantages so assuredly the hybrid configuration maximizes accuracy and strength performance regardless of the complexity of the data and the features relations.

8.5 RESULT

The objective of the experiments described is to combine the properties of CNNs as feature extractors with conventional ML in the quest to increase prediction accuracy and efficiency. This combination adopts MobileNetV3 as the foremost CNN architecture for defect classification in solar panels, consistent with prior efforts where MobileNet helped improve the classification accuracy tremendously. MobileNetV3 was selected due to its optimally low parameters and computational expenses, especially for occasions that call for a small yet efficient model that can effectively process imageries as data. After feature extraction, traditional ML regressors like SVM, DT, and RF take over where each has its advantages and limitations.

8.5.1 Hybrid Model Design and Data Preprocessing

The exemplary hybrid model is designed into two phases where the first phase is for the extraction of features using CNNs, while the second involves training the model on different ML regressors. For this MobileNetV3 is utilized in image processing on the input images, which have a size of 224 × 224 and undergo CLAHE [45]. These preprocessing techniques are done to ensure the features that aid in detecting the defect are made clear enough for the model to produce clear and good feature maps. The CNN is given a stack with a dense layer of 512 neuron units; hence, the model generates a small yet dense feature vector for every input picture, minimizing the burden in subsequent stages without losing core visual information.

The second phase involves the fusion of multiple ML techniques to make use of the features that have been extracted for the purposes of defect classification. This study measured the performance and prediction power of five ML techniques: a SVM with RBF kernel, RF with 80 trees, KNNs, and light gradient boosting machine (LGBM). Every technique was identified as having certain advantages that have the ability to meet the needs of either regression or classification as required in the context of this study. This rich mix of the algorithms offers evaluation on several aspects across the algorithms: predictive accuracy, variance, and speed, giving a wide perspective of the suitability of each algorithm for the given task.

8.5.2 Comparative Performance Analysis of ML Regressors

Mean squared error (MSE), mean absolute error (MAE), *R*-squared (R^2) value, and explained variance are employed to evaluate and contrast different ML algorithms on the features derived from the CNN-based model. From the Results sections presented in Tables 8.3, 8.4, and 8.5 of the report, it can be said that the regressor used RF model yields a higher accuracy for regression than any other model versions evaluated including the cascaded CNNs. As a result, the CNN + RF model achieves an MSE and MAE of just 0.01 and 0.022, respectively, with an *R*-squared (R^2) value of 1, indicating that the predicted values line up almost precisely with the actual values.

SVM with an RBF kernel is also capable of achieving great results, especially when combined with LDA. In other research works focused on the classification of the defects in solar panels, the challenges of classification using SVM were found to be eased when LDA was incorporated at the beginning, leading to a reduction in error ratios further than MSE and MAE and an increase in separation of the classes. The SVM model with LDA preprocessing reached 97.41% accuracy, showing the importance of LDA in increasing the precision, recall, and F1 scores of the classifiers.

The RF regressor, with its ensemble approach leveraging multiple DTs, consistently provides robust predictions and benefits from LDA preprocessing, which emphasizes the value of dimensionality reduction in achieving stable predictions. The CNN + RF model's impressive performance underscores its ability to effectively capture feature interactions within the dataset, while the LDA-enhanced SVM model shines for its high classification accuracy, making both CNN + RF and LDA + SVM strong contenders for effective solar panel defect classification.

8.5.3 Deep Learning Model versus Hybrid Model Performance

To that end, we performed direct pairwise evaluations of both the standalone CNN architectures and hybrid ones and found that the hybrid architectures managed to raise the accuracy even with just a single dataset. The independent assessment of the standalone CNN model turned out to be ineffective in minimizing the MSLE loss (Mean Squared Logarithmic Error) is a regression metric that measures the ratio between predicted and actual values on a logarithmic scale because of the limited amount of data and its variability. Most importantly though, the hybrid CNN + RF model is able to achieve higher accuracy throughout the dataset, which indicates that features extracted from a CNN model can considerably boost the efficacy of conventional ML models, even in the presence of little training data.

All model variations clearly show the effect of LDA preprocessing, which tends to improve all these performance rates as it enhances the class separability, good, for instance, for SVM and KNN models. It was shown in the test results that LDA post-processing brings these benefits, along with other more significant ones, especially for those models where the input high-dimensional data is a problem. To this end, when employing the MobileNetV3 model, along with LDA and an SVM classifier, accuracy improves from 90.44% to 97.41%, indicating the effectiveness of the combined approaches of lowering the dimensionality and enhancing the features of the model. This highlights the effectiveness and flexibility of solar panel defect detection hybrid models with LDA preprocessing.

8.5.4 Evaluating the Robustness and Scalability of Hybrid Models

A pivotal takeaway from this evaluation is the ability of ML regressors to tolerate abuse relative to the amount of data available, which is unlike the sensitivity of deep learning methods. Even though models that include CNN architectures are considered data-hungry models that rely on large amounts of data to perform well, the ML parts of such hybrid structures are more resistant to such drawbacks as long as the extracted features are of good quality. This quality allows for the use of hybrid models in situations where the amount of data available is restricted in that, other than using CNNs, ML models can fill that gap.

The strength of the CNN + RF model integration proposed in this study is evidenced by a better performance in terms of MSE, MAE, and R^2 scores, indicating its applicability to various datasets. In addition, the performance of the models using CNN + SVM and CNN + RF was stable, suggesting the benefit of using deep feature extraction with other ML methods, which is especially useful in cases where the results need to be interpretable and the time efficient.

8.5.5 Concluding Remarks and Practical Implications

The use of CNNs and ML techniques for solar panel defect classification is a strong advantageous combination that employs both deep learning and ML techniques. Experimental findings suggest that whereas CNN models are notoriously data-hungry, ML regressors are quite elastic and robust, making the combination approach ideal for situations where data is scant.

Evaluation in a controlled setting has shown that always the RF regressor with CNN extracted features has been scored the highest, especially, for various data types. In addition, the increase in performance due to LDA preprocessing explains the benefits of reducing dimensionality in terms of improved performance and ease of use, particularly for high-dimensional data models.

To summarize, the hybrid method makes good use of the feature extraction power of CNNs and goes further to implement effective ML models for the difficult task of regression for the purposes of defect classification. Such results are particularly useful to practitioners wishing to deploy state-of-the-art models in applications where resources are limited and provides motivation for further advances in hybrid modeling techniques.

From Figure 8.4a, loss curves of MobileNetV3 with training and validation, which had a more pronounced drop within the first epoch, show rapid learning and adaptation. At later epochs, both losses stabilize with minor fluctuations; hence, the model is said to be reaching some optimum solution but at some minor variances within a steady state. These continuous fluctuations show that MobileNetV3 learns effectively while avoiding local minima, achieving a balance between convergence and adaptability.

The peaks in the loss curves of ResNet34 are witnessed in Figure 8.4b, but the model adapts within the initial epochs, especially with an almost dramatic drop in the training loss. Validation loss fluctuates, meaning that the model is fine-tuning the weights for a balance between train accuracy and validation performance. Toward the end, the validation loss stabilizes, which reflects that ResNet34 is indeed learning features properly with no overfitting as evidenced by the train and validation losses being relatively close to each other that should promise good generalization on unseen data.

In Figure 8.4c, the training and validation loss curves for VGG16, which fall sharply at the start, reflect efficient initial learning. However, the validation loss varies and includes a sharp spike toward overfitting. It means that the model VGG16 might be capturing some specifics in training data that are not very generalizable to the validation data, lowering its potential on new data. A tendency in the model: increasing complexity versus increasing generalization.

In Figure 8.4d, train and validation losses of InceptionV3 are more or less stable in both. Every now and then, validation loss spikes, though that is small, inferring some volatility yet not serious enough to disrupt the performance in general. InceptionV3 has achieved an excellent trade between learning to be able to train data and retaining generalization to do fairly all right on unseen data.

Figure 8.4e indicates the erratic behavior of InceptionResNetV2, where both training and validation losses have many ups and downs. The instability more noticeably in the validation loss indicates a failure of convergence of the model, and this indicates that this model is unstable with this architecture configuration. High levels of oscillation suggest that InceptionResNetV2 probably doesn't generalize so well, possibly due to its very complex architecture, and therefore will not be reliable for consistent performance.

Finally, in Figure 8.4f it is evident that EfficientNetB4 undergoes a steep valley in training and validation loss in the early epochs, which means that initial learning is highly fast. Following the steep fall, low and stable values of loss indicate that convergence is strongly achieved. The high proximity between curves of training and validation loss indicates that EfficientNetB4 really strikes an excellent balance between fitting the data used for training it and generalizing it to new data. This stability of the two curves shows that EfficientNetB4 is a resilient model in this scenario and maintains generalization capability with less overfitting, which is good for reliable performance in different samples.

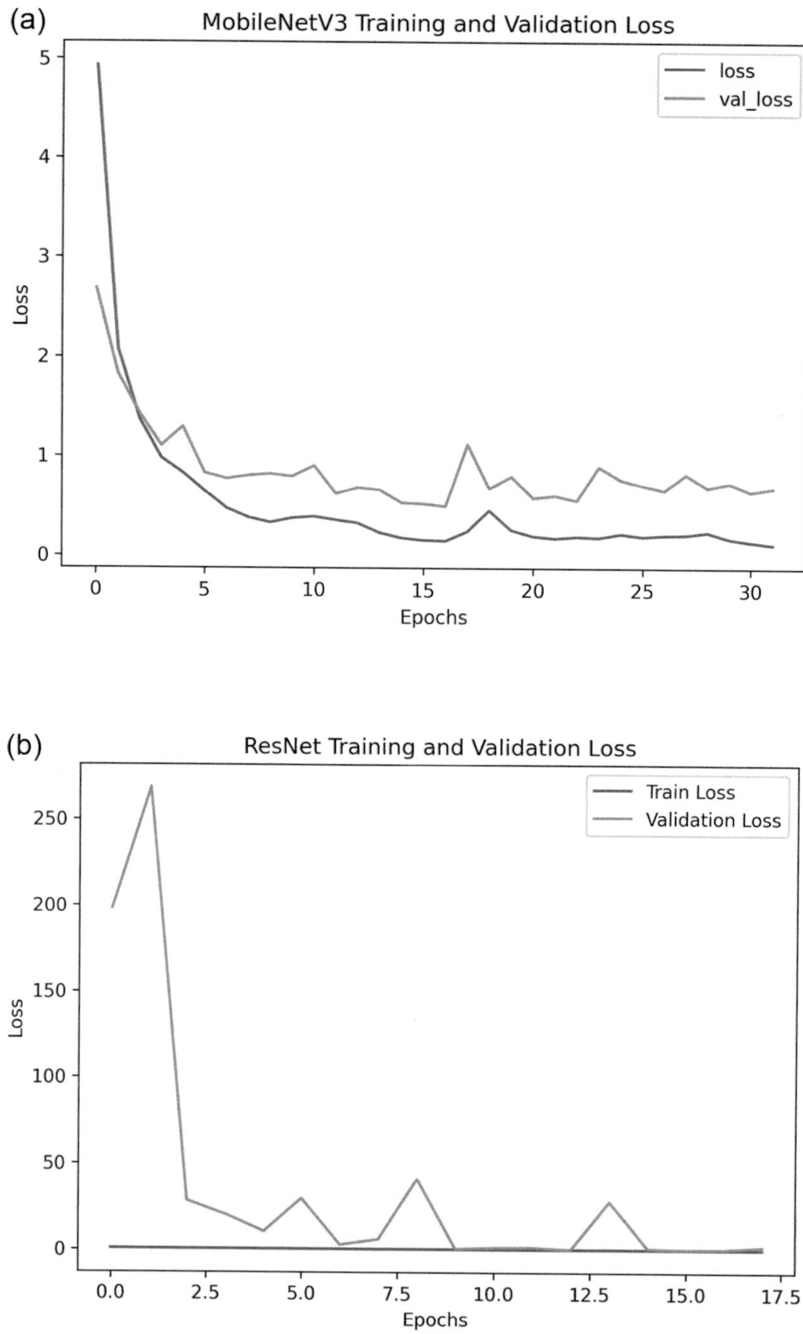

FIGURE 8.4 Loss and validation loss of various models.

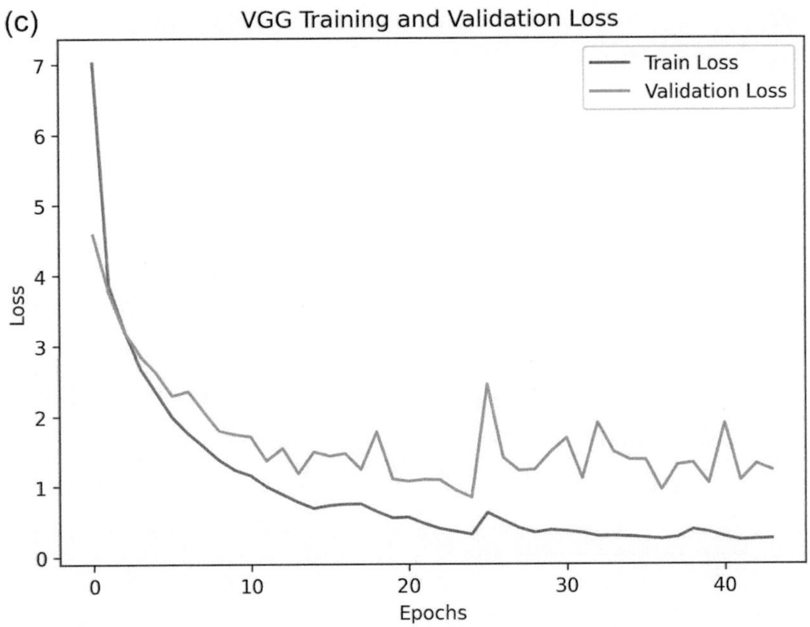

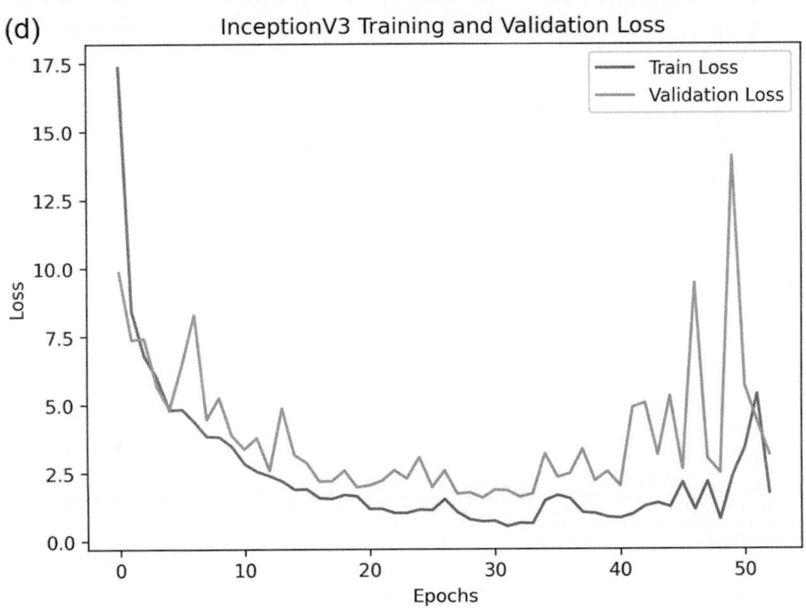

FIGURE 8.4 (Continued)

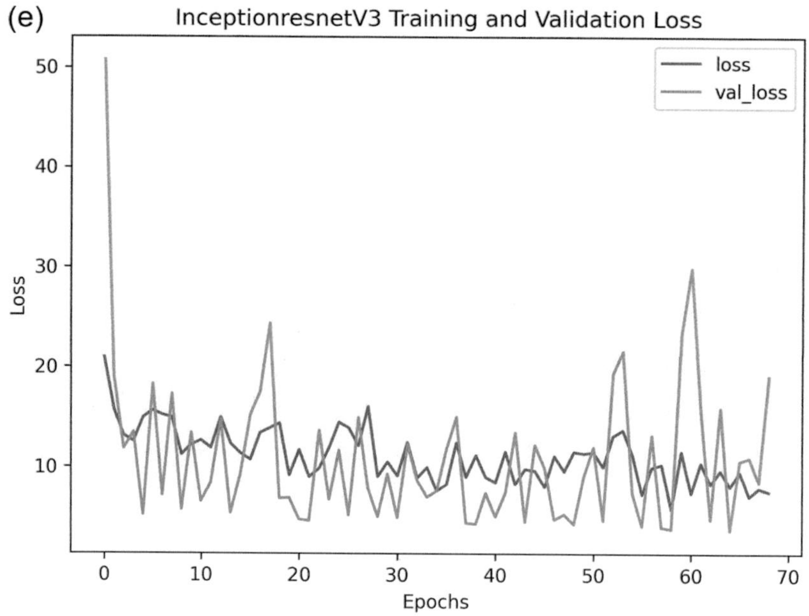

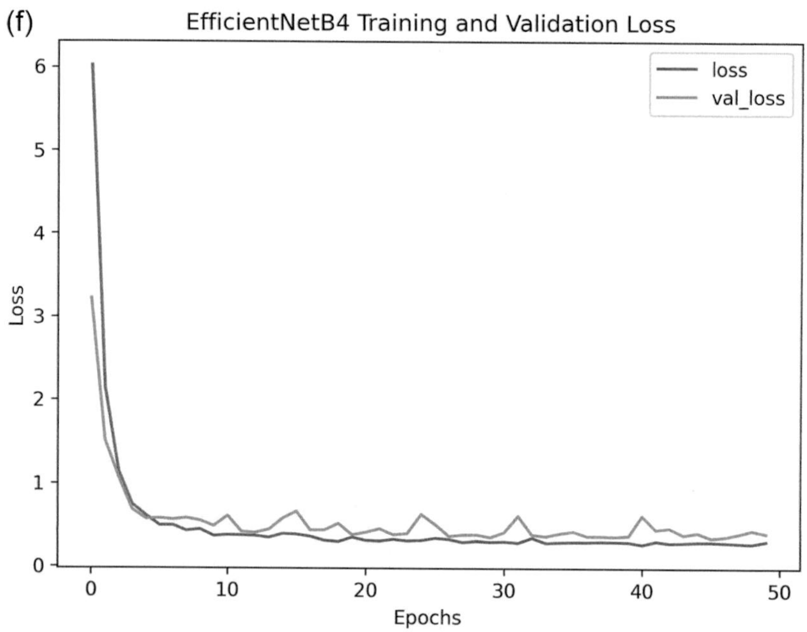

FIGURE 8.4 (Continued)

From Figure 8.5a, we notice that MobileNetV3 learns very fast and gives a steep decrease in validation loss during the initial epochs along with an increase in validation accuracy. Thus, the deep fall of validation loss in the initial phase implies that the model learns the data extremely quickly. Nevertheless, afterward, the validation accuracy keeps on moving above and below a mean level, and validation loss flattens. This trend suggests that while MobileNetV3 learns general features within the initial periods, it fails to significantly improve further accuracy by the latter stages of training. These fluctuations indicate that MobileNetV3 rapidly learns the basic structure of the data but sometimes faces difficulties in refining its learning to boost accuracy.

Figure 8.5b illustrates that the decrease in validation loss for ResNet34 is extremely significant in early epochs with a continued increase in accuracy for validation. After reaching this point, accuracy is constant and shows fluctuations only to a minute extent but is predominantly a constant one. This behavior of the network implies that ResNet34 processes learning very efficiently; hence, it reaches a point where high accuracy can be achieved very early during training. As can be observed from Figure 8.10, the stability of both validation loss and accuracy is an indicator of its effective generalization to the validation data with just a small risk of overfitting. ResNet34 appears to be particularly suitable for extracting meaningful features, and the convergence is also quite rapid, and therefore this is a sound model for this task.

In Figure 8.5c, we plot the learning dynamics of VGG16; in this instance, validation loss initially drops smoothly but eventually begins increasing after a while. This behavior, coupled with validation accuracy that was erratically fluctuating, might suggest that the network VGG16 is overfitting. It's quite clear that, in the initial phases, the model learns well because validation loss decreased, and accuracy increased over the start. Further training, though, causes validation loss to begin increasing while accuracy shows unstable spikes. This may indicate that VGG16 is overfitting to the details in the training data and, therefore, can't generalize well to new data.

In the plots in Figure 8.5d, InceptionV3 also exhibits a steadying downturn in validation loss coupled with increasing validation accuracy. Both metrics oscillate with their peaks and troughs but seem to be slowly improving and converging overall. These oscillations are nearly minimal, meaning that InceptionV3 is fairly trading off between learning and generalization. The slowly but steadily increasing validation accuracy in combination with a decline in validation loss demonstrates that InceptionV3 is absorbing useful patterns in the data and persists in these learned features over time, thus promoting generalization across different samples.

Figure 8.5e demonstrates the learning behavior of InceptionResNetV2. Validation loss and validation accuracy fluctuate very much during the course of the learning process. None of the metrics converge to a plateau during the training process, which verifies the instability of the training regarding the learning behavior of the model.

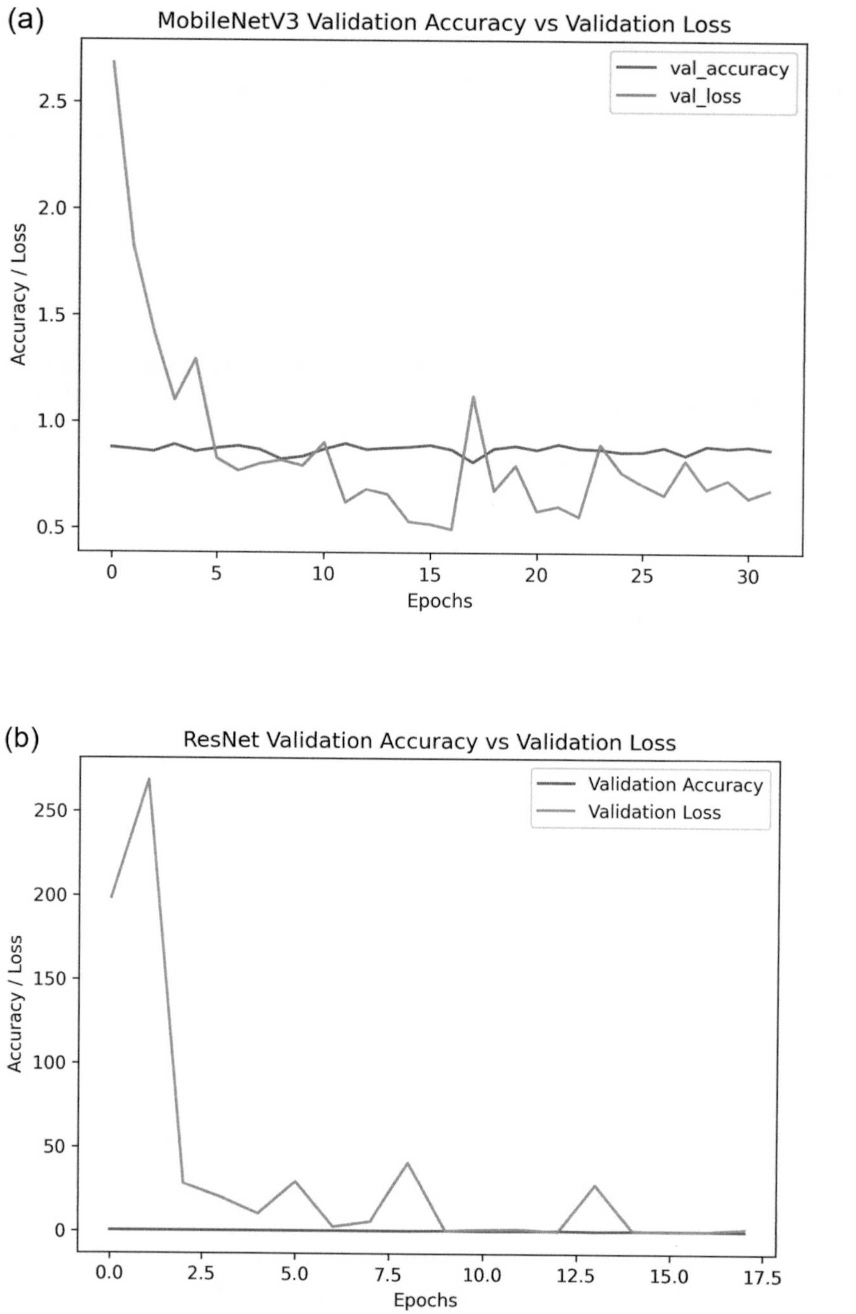

FIGURE 8.5 Validation accuracy versus validation loss of various models.

(c)

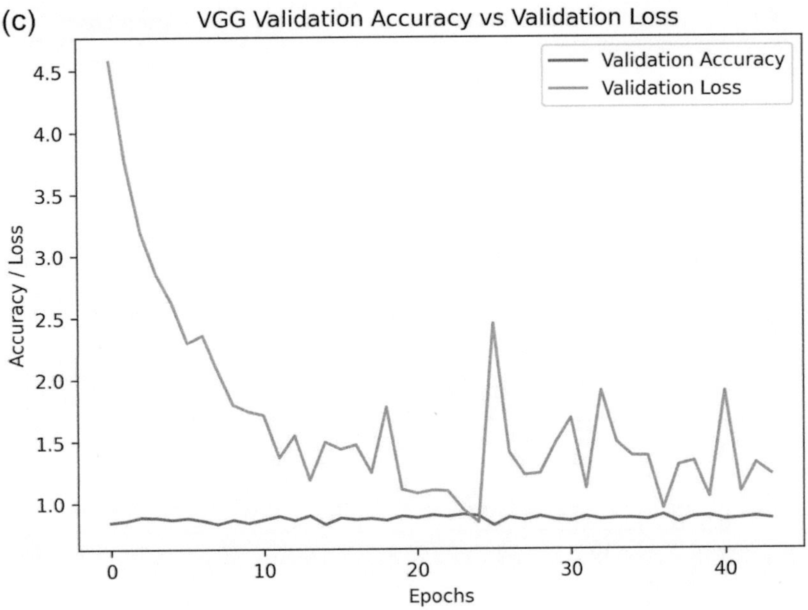

(d)

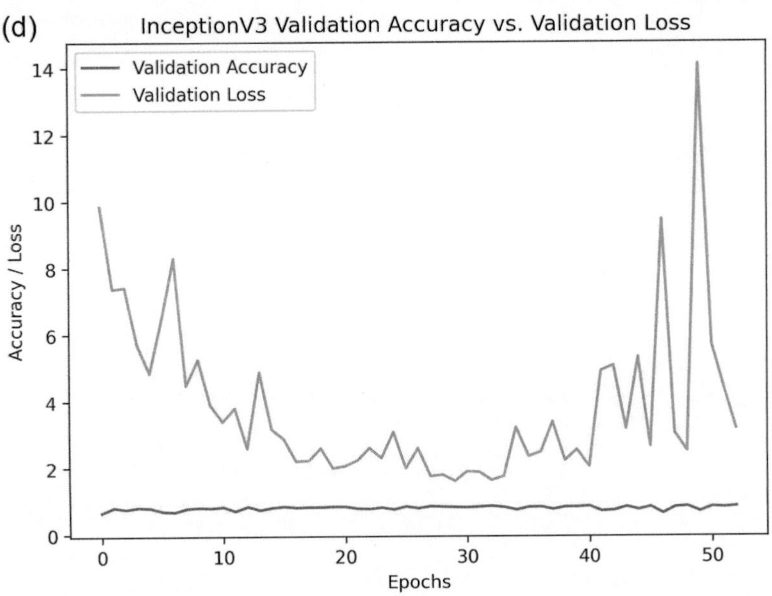

FIGURE 8.5 (Continued)

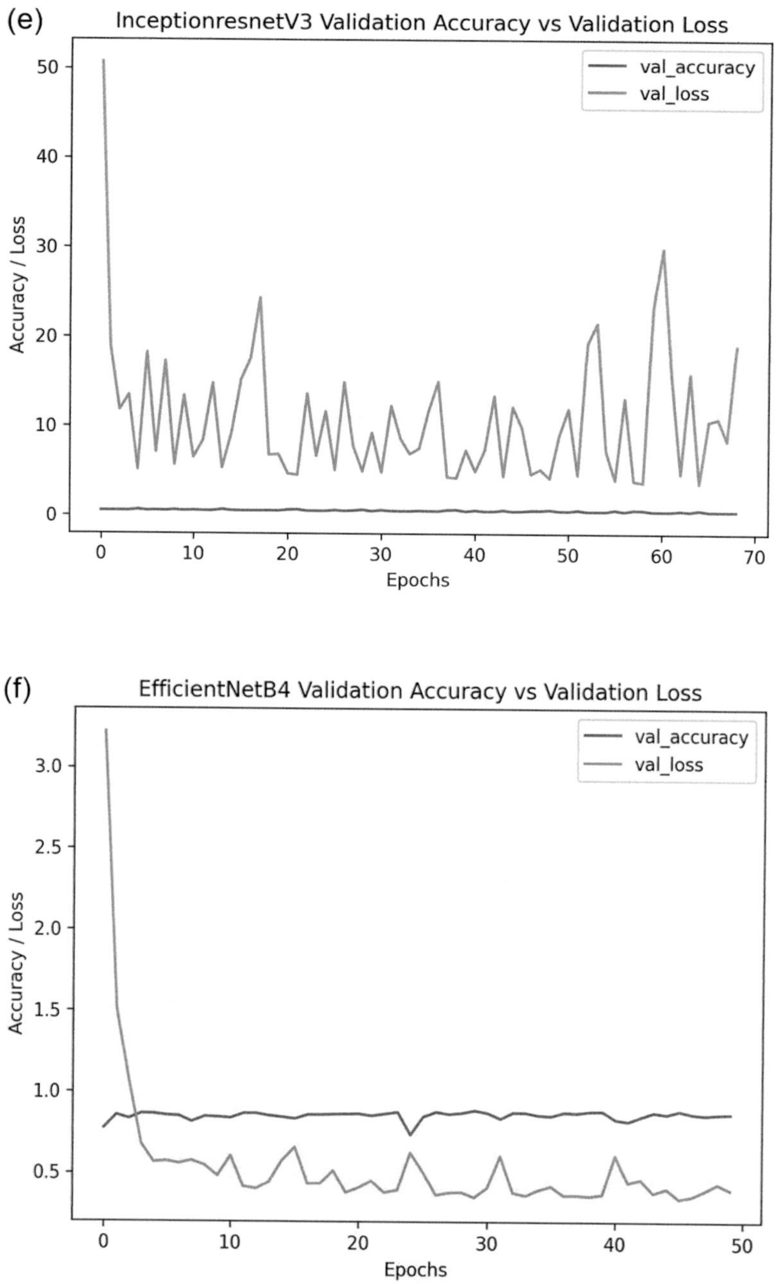

FIGURE 8.5 (Continued)

TABLE 8.3
Numerical Comparative Performance Analysis of MobileNetV3 Hybrid

Metrics/ Model	SVM	LDA + SVM	RF	LDA + RF	KNN	LDA + KNN	LGBM	LDA + LGBM
Accuracy	0.9044	0.9741	0.9058	0.9566	0.9058	0.9723	0.9058	0.9732
Precision	0.9056	0.9744	0.9087	0.9566	0.908	0.9725	0.908	0.9732
Recall	0.9044	0.9741	0.9058	0.9566	0.9058	0.9723	0.9058	0.9732
F1 score	0.9044	0.9741	0.9057	0.9566	0.9057	0.9723	0.9057	0.9732
Log loss	0.433	0.0974	0.432	0.4585	1.9236	0.4128	0.5734	0.0969

TABLE 8.4
Numerical Comparative Performance Analysis of ResNet34 Hybrid

Metrics/ Model	SVM	LDA + SVM	RF	LDA + RF	KNN	LDA + KNN	LGBM	LDA + LGBM
Accuracy	0.8684	0.9658	0.8753	0.9501	0.8726	0.9612	0.8767	0.964
Precision	0.8716	0.9668	0.8773	0.9502	0.8752	0.9621	0.8783	0.9645
Recall	0.8684	0.9658	0.8753	0.9501	0.8726	0.9612	0.8767	0.964
F1 score	0.8682	0.9658	0.8752	0.9501	0.8724	0.9612	0.8766	0.9639
Log loss	0.4724	0.1349	0.962	0.6454	2.5031	0.708	0.7607	0.1421

TABLE 8.5
Numerical Comparative Performance Analysis of VGG16 Hybrid

Metrics/ Model	SVM	LDA + SVM	RF	LDA + RF	KNN	LDA + KNN	LGBM	LDA + LGBM
Accuracy	0.8324	0.842	0.8629	0.8133	0.8476	0.8327	0.8601	0.8521
Precision	0.8423	0.8473	0.8672	0.8136	0.852	0.8346	0.8616	0.8547
Recall	0.8324	0.842	0.8629	0.8133	0.8476	0.8327	0.8601	0.8521
F1 score	0.8312	0.8413	0.8625	0.8133	0.8472	0.8325	0.86	0.8518
Log loss	0.3249	0.3774	0.3412	1.7067	1.0392	1.7572	0.3275	0.3645

TABLE 8.6
Numerical Comparative Performance Analysis of InceptionV3 Hybrid

Metrics/ Model	SVM	LDA + SVM	RF	LDA + RF	KNN	LDA + KNN	LGBM	LDA + LGBM
Accuracy	0.7645	0.7699	0.7701	0.695	0.7618	0.7384	0.7742	0.7652
Precision	0.7679	0.7777	0.7726	0.695	0.7645	0.739	0.7769	0.7695

(Continued)

TABLE 8.6 ***(Continued)***
Chemical Structure of Diverse Epigenetic Inhibitors

Metrics/ Model	SVM	LDA + SVM	RF	LDA + RF	KNN	LDA + KNN	LGBM	LDA + LGBM
Recall	0.7645	0.7699	0.7701	0.695	0.7618	0.7384	0.7742	0.7652
F1 score	0.7639	0.7682	0.7696	0.695	0.7612	0.7383	0.7737	0.7643
Log loss	0.4865	0.506	0.4731	2.6205	1.9687	2.3391	0.5012	0.4882

TABLE 8.7
Numerical Comparative Performance Analysis of InceptionResNetV2 Hybrid

	SVM	LDA + SVM	RF	LDA + RF	KNN	LDA + KNN	LGBM	LDA + LGBM
Accuracy	0.59	0.7717	0.7659	0.7043	0.7064	0.7569	0.7909	0.7828
Precision	0.7424	0.7782	0.7688	0.7046	0.7342	0.7592	0.7937	0.7851
Recall	0.59	0.7717	0.7659	0.7043	0.7064	0.7569	0.7909	0.7828
F1 score	0.5145	0.7703	0.7653	0.7041	0.6976	0.7564	0.7904	0.7823
Log loss	0.6552	0.5023	0.4778	2.1322	2.1066	1.935	0.4677	0.4787

TABLE 8.8
Numerical Comparative Performance Analysis of EfficientNetB4 Hybrid

Metrics/ Model	SVM	LDA + SVM	RF	LDA + RF	KNN	LDA + KNN	LGBM	LDA + LGBM
Accuracy	0.856	0.8447	0.8518	0.7985	0.8476	0.8355	0.8601	0.8429
Precision	0.8744	0.8574	0.856	0.7988	0.852	0.8399	0.8658	0.8484
Recall	0.856	0.8447	0.8518	0.7985	0.8476	0.8355	0.8601	0.8429
F1 score	0.8542	0.8433	0.8514	0.7985	0.8472	0.8349	0.8596	0.8422
Log loss	0.3069	0.3773	0.3553	1.9025	1.2792	1.8377	0.3162	0.3507

These oscillations mean that InceptionResNetV2 may suffer from problems with generalization and that it fails to find the right balance between overfitting and learning. This instability might imply that the model requires either some parameters or training for the convergence to be made better. The bias or jitter in metrics might reflect a tough task in the problem of learning robust features and hence does not give the model reliability.

Finally, as observed in Figure 8.5f, the curve of EfficientNet B4 shows a sharp decline in the validation loss right from the beginning with a consistently high validation accuracy during all periods of training. After some precipitous fall in loss, later stages of loss stabilize at very low values and accuracy remains at a very high

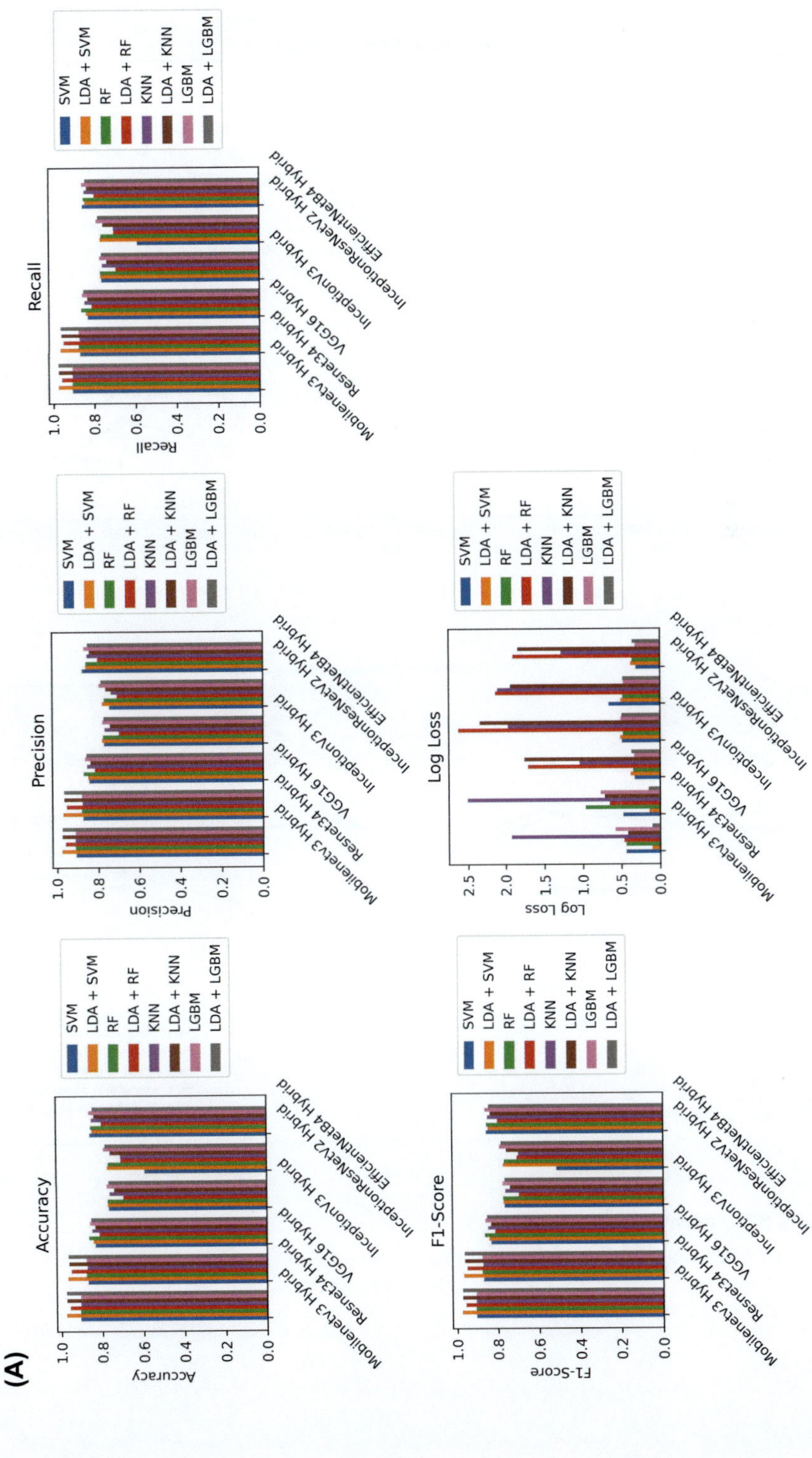

FIGURE 8.6 The plots demonstrate the comparison between machine learning models across various metrics.

(B)

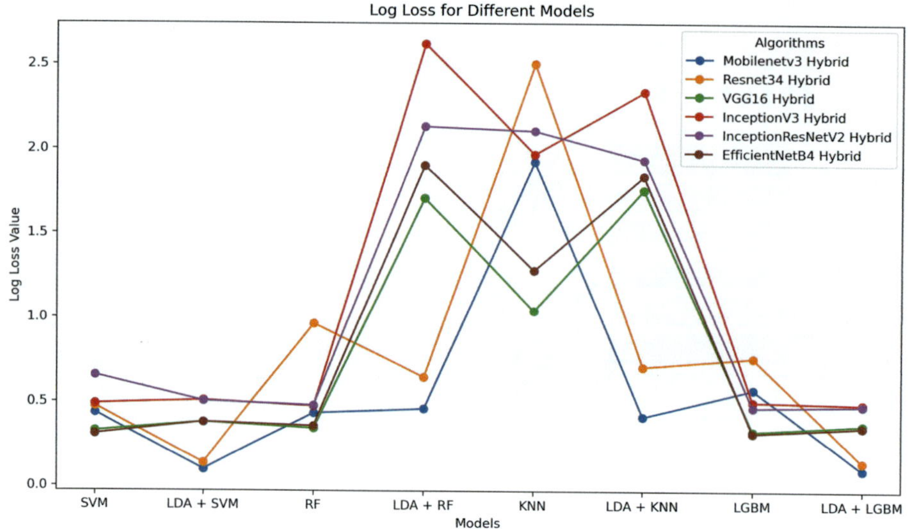

(C)

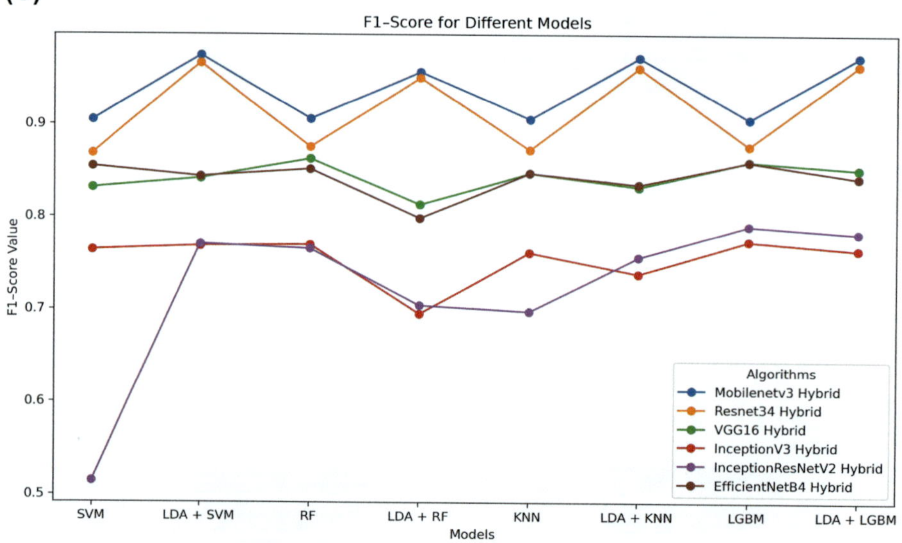

FIGURE 8.6 (Continued)

(D)

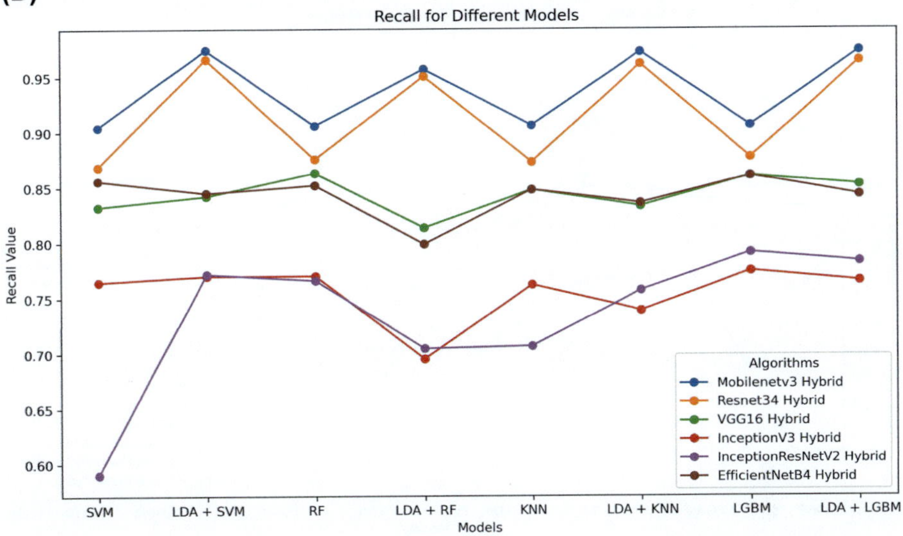

(E)

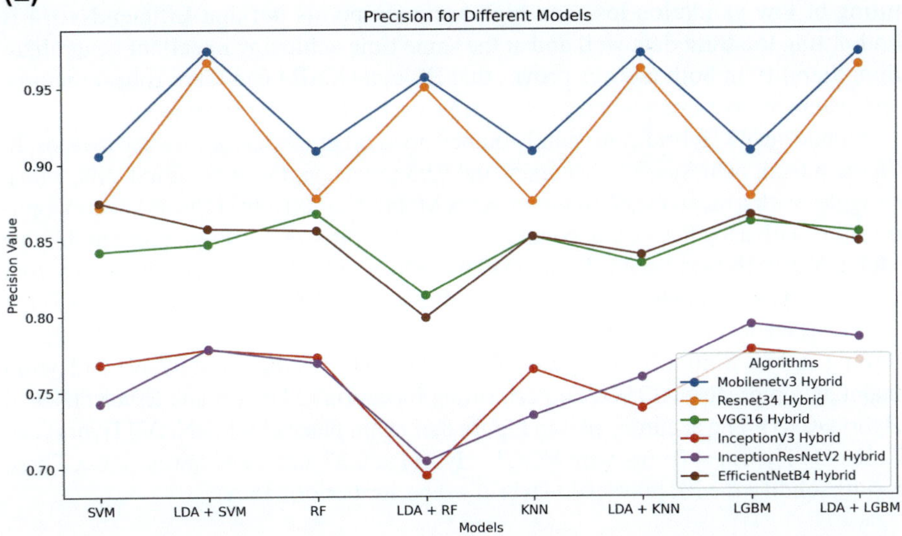

FIGURE 8.6 (Continued)

(F)

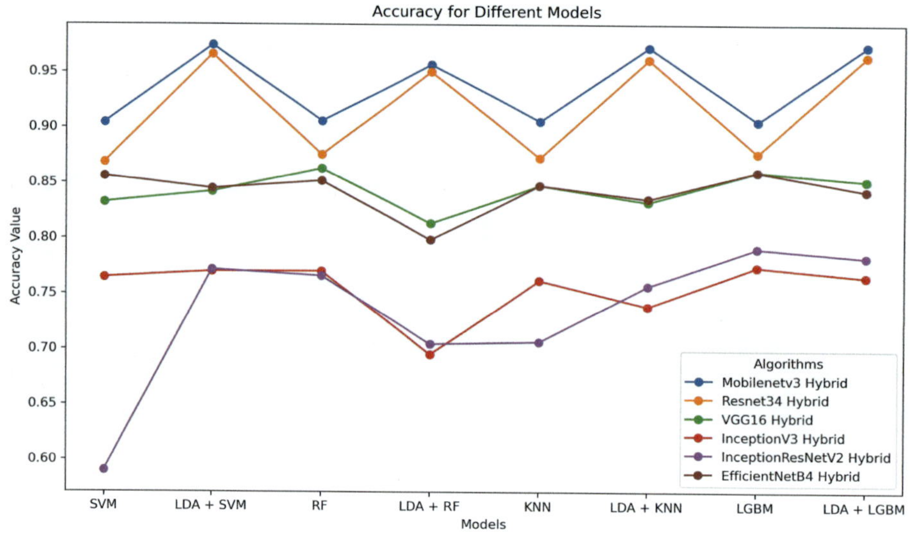

FIGURE 8.6 (Continued)

level, indicating a good convergence and effective learning process. The very good fitting of low validation loss and high accuracy points out that EfficientNetB4 is both fitting the train data well and at the same time achieving excellent generalization. Stability in both metrics proves that EfficientNetB4 can learn robust features applicable almost everywhere; hence, it is a reliable choice for this task.

MobileNetV3 Hybrid consistently outperforms other models across key metrics. In Figure 8.6a, it achieves 0.92 accuracy and 0.95 precision, far exceeding KNN, which struggles with a recall of 0.7. In Figure 8.6b, MobileNetV3 Hybrid shows superior optimization with a log loss of 0.5, compared to KNN's peak of 2.5. Figure 8.6c highlights MobileNetV3 Hybrid's stable F1 score (0.9–0.95), while LDA + LGBM improves from 0.5 to 0.8. The recall plot in Figure 8.6d confirms MobileNetV3 Hybrid and ResNet34 Hybrid as leaders (0.96 recall), with EfficientNetB4 Hybrid at 0.85 and LDA + KNN trailing at 0.75. Similarly, in Figure 8.6e, MobileNetV3 Hybrid and ResNet34 Hybrid maintain top precision (0.96), outperforming InceptionV3 Hybrid and EfficientNetB4 Hybrid (0.85). The accuracy plot in Figure 8.6f again places MobileNetV3 Hybrid and ResNet34 Hybrid at 0.96+, with VGG16 Hybrid at 0.85 and SVM lowest at 0.7. These results solidify the MobileNetV3 Hybrid's superior performance.

8.6 CONCLUSION

The results in Table 8.3 thus validate that MobileNetV3 is a better model compared to the others for defect detection in solar cells. With a high accuracy of 97.41%, it has surpassed the classifier LDA + SVM. MobileNetV3 also strikes the balance between high accuracy and computation efficiency, giving it a justified use, especially in real time. Earlier models such as [1] and [2] had efficiency issues which MobileNetV3 has overcome by design, through its lightweight approach. As illustrated in Table 8.4, the ResNet34 with LDA + SVM stood at an accuracy of 96.58% but was not as efficient. Accuracies of VGG16 and InceptionResNetV2, as shown in Table 8.5 and Table 8.6, were relatively low due to their larger sizes. In Table 8.7, InceptionV3 was even the least efficient in its computation with an accuracy standing at 77.42%. EfficientNetB4, as shown in Table 8.8, was competitive at 86.01% but required more computational resources than the model MobileNetV3. In conclusion, MobileNetV3 achieves an optimal balance between the two poles of accuracy and efficiency and eliminates the numerous challenges that had besieged previous models. Very compact and efficient, it may represent the best answer to solar panel real-time, scalable defect detection, supporting sustainability in solar power systems.

REFERENCES

[1] Mittal, S., Srivastava, S., & Jayanth, J. P. (2022). A survey of deep learning techniques for underwater image classification. *IEEE Transactions on Neural Networks and Learning Systems*, *34*(10), 6968–6982.

[2] Li, Q., Cai, W., Wang, X., Zhou, Y., Feng, D. D., & Chen, M. (2014, December). Medical image classification with convolutional neural network. In 2014 13th International Conference on Control Automation Robotics & Vision (ICARCV) (pp. 844–848). IEEE.

[3] Satish, S., & Sridevi, S. (2023). CNN and XGBoost based hybrid model in classification of fetal ultrasound scan planes images in detection of congenital heart defects. In: Joby, P. P., Balas, V. E., & Palanisamy, R. (eds.), IoT Based Control Networks and Intelligent Systems. Lecture Notes in Networks and Systems, vol 528. Springer, Singapore.

[4] Prabhakaran, S., Uthra, R. A., & Preetharoselyn, J. (2023). Deep learning-based model for defect detection and localization on photovoltaic panels. *Computer Systems Science & Engineering*, *44*(3).

[5] Li, Z., Zhang, S., Qu, C., Zhang, Z., & Sun, F. (2024). Research on multi-defects classification detection method for solar cells based on deep learning. *PLoS One*, *19*(6), e0304819.

[6] Mantel, C., Villebro, F., dos Reis Benatto, G. A., Parikh, H. R., Wendlandt, S., Hossain, K., . . . & Forchhammer, S. (2019, September). Machine learning prediction of defect types for electroluminescence images of photovoltaic panels. In *Applications of Machine Learning* (Vol. 11139, p. 1113904). SPIE.

[7] Abdelsattar Mohamed Saeed, M., Rasslan, A. A. A., & Emad-Eldeen, A. (2024). Comparative analysis of machine learning techniques for fault detection in solar panel systems. *SVU-International Journal of Engineering Sciences and Applications*, *5*(2), 140–152.

[8] Sinap, V., & Kumtepe, A. (2024). CNN-based automatic detection of photovoltaic solar module anomalies in infrared images: A comparative study. *Neural Computing and Applications*, *36*(28), 17715–17736.

[9] Joshua, S. R., Park, S., & Kwon, K. (2024, July). Solar panel fault detection: Applying convolutional neural network for advanced fault detection in solar-hydrogen system at university. In *2024 IEEE 24th International Conference on Software Quality, Reliability, and Security Companion (QRS-C)* (pp. 289–298). IEEE.

[10] Liu, Q., Liu, M., Wang, C., & Wu, Q. J. (2024). An efficient CNN-based detector for photovoltaic module cells defect detection in electroluminescence images. *Solar Energy*, *267*, 112245.

[11] Benghanem, M., Mellit, A., & Moussaoui, C. (2023). Embedded Hybrid Model (CNN–ML) for fault diagnosis of photovoltaic modules using thermographic images. *Sustainability*, *15*(10), 7811.

[12] Afjal, M. I., Mondal, M. N. I., & Mamun, M. A. (2023). Segmentation-based linear discriminant analysis with information theoretic feature selection for hyperspectral image classification. *International Journal of Remote Sensing*, *44*(11), 3412–3455.

[13] Yang, X., Liu, L., Song, X., Feng, J., Pei, Q., Yuan, X., & Li, J. (2024, May). An efficient lightweight satellite image classification model with improved mobileNetV3. In *IEEE INFOCOM 2024-IEEE Conference on Computer Communications Workshops (INFOCOM WKSHPS)* (pp. 1–6). IEEE.

[14] Subaar, C., Addai, F. T., Addison, E. C. K., Christos, O., Adom, J., Owusu-Mensah, M., . . . & Abbey, S. (2024). Investigating the detection of breast cancer with deep transfer learning using ResNet18 and ResNet34. *Biomedical Physics & Engineering Express*, *10*(3), 035029.

[15] Mayanja, A., Ozkan, I. A., & Taşdemir, Ş. (2023, September). Utilizing transfer learning on landscape image classification using the VGG16 model. In *Proceedings of the International Conference on Advanced Technologies* (Vol. 11, pp. 71–76).

[16] Kumar, S., Yadav, A., Varshney, A., Sharma, A., & Shivani, S. (2024, March). Comparative analysis of U-net models using ResNet34, inceptionV3, and VGG16 for the processing of satellite images. In *2024 IEEE International Conference on Interdisciplinary Approaches in Technology and Management for Social Innovation (IATMSI)* (Vol. 2, pp. 1–6). IEEE.

[17] Adegun, A. A., Viriri, S., & Tapamo, J. R. (2023). Review of deep learning methods for remote sensing satellite images classification: Experimental survey and comparative analysis. *Journal of Big Data*, *10*(1), 93.

[18] Zhalgasbayev, A., Aiteni, T., & Khaimuldin, N. (2024, April). Using the cnn architecture based on the efficientnetb4 model to efficiently detect deepfake images. In *2024 IEEE AITU: Digital Generation* (pp. 14–19). IEEE.

[19] Demirci, M. Y., Beşli, N., & Gümüşçü, A. (2024). An improved hybrid solar cell defect detection approach using Generative Adversarial Networks and weighted classification. *Expert Systems with Applications*, *252*, Part A, 124230.

[20] Rudro, R. A. M., Nur, K., Al Sohan, M. F. A., Mridha, M. F., Alfarhood, S., Safran, M., & Kanagarathinam, K. (2024). SPF-Net: Solar panel fault detection using U-Net based deep learning image classification. *Energy Reports*, *12*, 1580–1594.

[21] Pereira, F., & Silva, C. (2024). Machine learning for monitoring and classification in inverters from solar photovoltaic energy plants.

[22] Buerhop-Lutz, C., Deitsch, S., Maier, A., Gallwitz, F., Berger, S., Doll, B., Hauch, J., Camus, C., & Brabec, C. J. (2018). A benchmark for visual identification of defective solar cells in electroluminescence imagery. In *European PV Solar Energy Conference and Exhibition (EU PVSEC)*. https://doi.org/10.4229/35thEUPVSEC20182018-5CV.3.15

[23] Deitsch, S., Buerhop-Lutz, C., Sovetkin, E., Steland, A., Maier, A., Gallwitz, F., & Riess, C. (2021). Segmentation of photovoltaic module cells in uncalibrated electroluminescence images. *Machine Vision and Applications*, *32*(4). https://doi.org/10.1007/s00138-021-01191-9

[24] Deitsch, S., Christlein, V., Berger, S., Buerhop-Lutz, C., Maier, A., Gallwitz, F., & Riess, C. (2019). Automatic classification of defective photovoltaic module cells in electroluminescence images. *Solar Energy*, Elsevier BV, *185*, 455–468. https://doi.org/10.1016/j.solener.2019.02.067

[25] Demir, A. H. M. E. T., & Necati, A. T. A. B. E. R. K. (2024). *Defect Detection in Solar Panels Using a Customized 2D CNN: A Study on the ELPV Dataset*. 5th International Conference on Innovative Academic Studies. October 2024, 10–11, Konya, Turkey: All Sciences Academy.

[26] www.gehealthcare.in/insights/article/advancing-molecular-imaging-with-deeplearning-technology
[27] www.gehealthcare.com/insights/article/machine-learning-models-in-healthcare-enhancing-diagnostic-ai-models-with-smart-data-selection?srsltid=AfmBOooPc19lpjpmjOM3ytXn_jRr2sAcO_nX_r1lT1auZlcJzl3_IZie
[28] Pysch[1], D., Schlereth[1], M., Pomohaci, M., & Fischer, P. (2024, February). Katharina breininger[1] [1]department artificial intelligence in biomedical engineering, FAU Erlangen-Nürnberg, Erlangen, Germany 2 Siemens Healthcare GmbH, Forchheim, Germany. In *Bildverarbeitung für die Medizin 2024: Proceedings, German Conference on Medical Image Computing, Erlangen, March 10–12, 2024* (p. 279). Springer-Verlag.
[29] www.siemens-healthineers.com/digital-health-solutions/artificial-intelligence-in-healthcare/ai-to-help-predict-disease
[30] Arendt, C. T., Leithner, D., Mayerhoefer, M. E., Gibbs, P., Czerny, C., Arnoldner, C., . . . & Schernthaner, R. E. (2021). Radiomics of high-resolution computed tomography for the differentiation between cholesteatoma and middle ear inflammation: Effects of post-reconstruction methods in a dual-center study. *European Radiology, 31*, 4071–4078. Springer.
[31] Lan, X., Wang, X., Qi, J., Chen, H., Zeng, X., Shi, J., . . . & Zhang, J. (2022). Application of machine learning with multiparametric dual-energy computed tomography of the breast to differentiate between benign and malignant lesions. *Quantitative Imaging in Medicine and Surgery*, *12*(1), 810.
[32] Martin-Ramiro, P., de la Maza, U. S., Singh, S., Orus, R., & Mugel, S. (2023). Boosting defect detection in manufacturing using tensor convolutional neural networks. *arXiv preprint arXiv:2401.01373.*
[33] https://bosch-iot-suite.com/predictive-maintenance-4-0/
[34] https://airbornescience.nasa.gov/content/Cloud_detection_algorithm_for_multi-modal_satellite_imagery_using_T_convolutional_neural
[35] https://gpm.nasa.gov/sites/default/files/meeting_files/PMM%20Science%20Team%20Meeting%202024/Session%203_Applications/Galantowicz_2024%20PMM.pdf
[36] https://ntrs.nasa.gov/citations/20100019616
[37] Luo, S., Feng, Y., Liu, Z. (Joe), Ling, Y., Dong, S., & Ferry, B. (2023). High precision sound event detection based on transfer learning using transposed convolutions and feature pyramid network. In *Proceedings of the International Conference on Consumer Electronics (ICCE) 2023*. Retrieved from www.amazon.science/publications/high-precision-sound-event-detection-based-on-transfer-learning-using-transposed-convolutions-and-feature-pyramid-network
[38] https://aws.amazon.com/personalize/
[39] https://gsm.ucdavis.edu/blog/how-walmart-puts-ai-and-machine-learning-play
[40] https://tech.walmart.com/content/walmart-global-tech/en_us/blog/post/walmarts-ai-powered-inventory-system-brightens-the-holidays.html
[41] Zhang, Y., Pan, P., Zheng, Y., Zhao, K., Zhang, Y., Ren, X., & Jin, R. (2018, July). Visual search at alibaba. In *Proceedings of the 24th ACM SIGKDD International Conference on Knowledge Discovery & Data Mining* (pp. 993–1001).
[42] Deng, Yuming, Zhang, Xinhui, Wang, Tong, Wang, Lin, Zhang, Yidong, Wang, Xiaoqing, Zhao, Su, Qi, Yunwei, Yang, Guangyao, & Peng, Xuezheng. (2023). Alibaba realizes millions in cost savings through tntegrated demand forecasting, inventory management, price optimization, and product recommendations. *INFORMS Journal on Applied Analytics*, *53*, 32–46. https://doi.org/10.1287/inte.2022.1145
[43] Ahmad, A., Jin, Y., Zhu, C., Javed, I., Maqsood, A., & Akram, M. W. (2020). Photovoltaic cell defect classification using convolutional neural network and support vector machine. *IET Renewable Power Generation, 14*(14), 2693–2702.

[44] Chaganti, S. Y., Nanda, I., Pandi, K. R., Prudhvith, T. G., & Kumar, N. (2020, March). Image classification using SVM and CNN. In *2020 International Conference on Computer Science, Engineering and Applications (ICCSEA)* (pp. 1–5). IEEE.

[45] Amiri, A. F., Oudira, H., Chouder, A., & Kichou, S. (2024). Faults detection and diagnosis of PV systems based on machine learning approach using random forest classifier. *Energy Conversion and Management, 301*, 118076. Elsevier.

[46] Shijer, S. S., Jassim, A. H., Al-Haddad, L. A., & Abbas, T. T. (2024). Evaluating electrical power yield of photovoltaic solar cells with k-Nearest neighbors: A machine learning statistical analysis approach. *e-Prime-Advances in Electrical Engineering, Electronics and Energy*, *9*, 100674.

[47] Opoku, M., Weyori, B. A., Adekoya, A. F., & Adu, K. (2023). CLAHE-CapsNet: Efficient retina optical coherence tomography classification using capsule networks with contrast limited adaptive histogram equalization. *PLoS One*, *18*(11), e0288663.

9 Printable Solar Cells—Advancing Flexible and Sustainable Energy Solutions

Venkateswaran Pedinti Sankaran

9.1 INTRODUCTION

9.1.1 Solar Energy Harvesting for Modern Energy Solutions

The development of printable solar cells marks a significant advancement in renewable energy, particularly in solar energy harvesting. While traditional silicon-based solar cells have been instrumental in the shift to renewable energy, their high production costs, rigidity, and material limitations have presented challenges, especially in integrating them into diverse surfaces and products. Printable solar cells address these issues by using advanced printing techniques to deposit active materials onto flexible substrates, resulting in lightweight, thin, and bendable panels that can be seamlessly integrated into various surfaces, including those with complex geometries. This flexibility opens new possibilities for solar applications, especially in areas where conventional panels are impractical. Additionally, printable solar cells offer the potential to significantly reduce manufacturing costs, as they can be produced using low-cost materials and scalable methods like inkjet printing, screen printing, and roll-to-roll (R2R) processing. This cost efficiency makes them a viable option for widespread deployment, particularly in regions with limited access to clean energy. Furthermore, their lightweight and flexible nature allows for integration into a wide range of products, from building materials and vehicles to portable electronics and wearables, expanding the potential use cases for solar energy. Printable solar cells also align with sustainability goals, as they often use less resource-intensive materials and environmentally friendly processes compared to traditional silicon cells. As the demand for cleaner energy sources grows, the versatility, cost-effectiveness, and sustainability of printable solar cells position them as a crucial technology in the future of renewable energy [1–4].

This chapter aims to provide a comprehensive overview of the current state of printable solar cell technology, exploring the key advancements in materials and manufacturing techniques that have made this innovation possible. We will delve into the various types of materials used in printable solar cells, examining their

DOI: 10.1201/9781003515784-9

properties and performance characteristics. Additionally, the chapter will highlight the different printing techniques employed in the fabrication of these cells, discussing their respective advantages and challenges. The chapter will also address the critical issues of efficiency, durability, and scalability, which are essential factors in the commercialization of printable solar cells. As with any emerging technology, there are hurdles to overcome before printable solar cells can achieve widespread adoption. These include improving their long-term stability, optimizing their energy conversion efficiency, and developing reliable methods for large-scale production.

9.2 TECHNOLOGICAL BREAKTHROUGHS

9.2.1 Innovations in Printing Technologies

The field of printable solar cells has witnessed remarkable technological advancements that are driving the future of solar energy harvesting. Traditional solar cells, primarily based on crystalline silicon, have dominated the market for decades due to their high efficiency and relatively long lifespan. However, the rigid and expensive nature of these cells has prompted researchers and engineers to explore alternative materials and manufacturing processes that can offer more flexibility, lower costs, and greater versatility. The technological breakthroughs in printable solar cells revolve around innovations in materials science, printing technologies, and device architecture, each contributing to the development of more efficient, durable, and commercially viable solar cells [5].

One of the most significant technological breakthroughs in printable solar cells is the development of advanced materials that can be processed at low temperatures and applied to flexible substrates. Organic photovoltaics (OPVs), perovskite solar cells (PSCs), and quantum dot solar cells are among the leading materials that have been extensively researched for their potential in printable solar technology. OPVs, for example, are made from organic polymers and small molecules that can be dissolved in solvents and printed onto flexible substrates using various printing techniques. These materials offer the advantage of being lightweight, flexible, and capable of being produced in large areas, making them ideal for applications in portable electronics and wearable devices [6].

PSCs have also emerged as a game-changer in the field of printable solar cells due to their high efficiency and ease of fabrication. Perovskites are a class of materials with a unique crystal structure that allows for efficient light absorption and charge transport. The ability to process perovskite materials at low temperatures and their compatibility with solution-based printing techniques have made them a focal point of research in printable solar cells. The rapid advancements in PSCs have led to significant improvements in their efficiency, with some devices now rivalling or even surpassing the performance of traditional silicon-based solar cells [7, 8].

Another breakthrough in printable solar cells is the development of quantum dot solar cells, which utilize nanoscale semiconductor particles to absorb light and generate electricity. Quantum dots offer several advantages, including tunable bandgaps, which allow for the absorption of a broader spectrum of light and the potential for high-efficiency energy conversion. These properties make quantum dot solar cells particularly promising for applications where high efficiency and flexibility are required. The ability to print quantum dots onto various substrates using inkjet or other printing techniques has opened new possibilities for integrating solar cells into unconventional surfaces, such as windows, clothing, and even packaging materials [9–11].

9.2.2 Impact on Solar Cell Manufacturing

In addition to advancements in materials, significant progress has been made in the printing technologies used to fabricate printable solar cells. Techniques such as inkjet printing, screen printing, and R2R processing have been refined to improve the quality, consistency, and scalability of printed solar cells. Inkjet printing, for example, allows for the precise deposition of functional materials onto substrates, enabling the production of complex device architectures with minimal material waste. Screen printing, on the other hand, is a well-established technique that has been adapted for use in solar cell manufacturing, offering a cost-effective method for producing large-area solar cells. R2R processing is another critical advancement, as it allows for continuous production of solar cells on flexible substrates, significantly reducing manufacturing costs and increasing throughput [5, 12].

These technological breakthroughs have also led to innovations in device architecture, with researchers exploring new ways to enhance the performance and stability of printable solar cells. Multi-junction solar cells, tandem cells, and other advanced configurations have been developed to improve the efficiency of printable solar cells by combining different materials and optimizing their layer structures. Additionally, encapsulation techniques have been improved to protect the cells from environmental factors such as moisture and oxygen, which can degrade their performance over time [13, 14].

9.2.3 Enhancement of Printable Techniques Compared to Conventional Solar Technologies

Printable solar cells offer transformative advantages over conventional silicon-based solar technologies due to their flexibility, scalability, and cost-effectiveness. Traditional rigid crystalline silicon solar cells, while highly efficient, suffer from high production costs, energy-intensive manufacturing processes, and limited adaptability to various surfaces. In contrast, printable solar cells leverage innovative techniques such as inkjet printing, screen printing, and R2R processing to deposit functional materials on flexible substrates. These methods enable the production of lightweight, thin, and conformable solar panels, significantly reducing material wastage and manufacturing expenses [15].

Industries such as construction, automotive, and wearable technology are embracing printable solar cells due to their versatility. Building-integrated photovoltaics (BIPV) utilize printable solar cells for solar façades and energy-generating windows, seamlessly blending aesthetics and functionality. The automotive sector employs flexible panels for auxiliary power systems integrated into car roofs and windows. Similarly, textile manufacturers are embedding solar cells into fabrics for wearable electronics, offering on-the-go charging capabilities for portable devices. By addressing the limitations of conventional solar cells, printable solar technologies have opened avenues for diverse applications, making renewable energy more accessible and widespread [16].

Despite these advancements, several challenges remain in the commercialization of printable solar cells. Issues such as long-term durability, environmental stability, and the scalability of manufacturing processes must be addressed to ensure that printable solar cells can compete with traditional solar technologies in the market. Ongoing research and development efforts are focused on overcoming these challenges, to create printable solar cells that are not only efficient and cost-effective but also reliable and sustainable in the long term.

9.3 KEY MATERIALS

The advancement of printable solar cells is largely driven by the development and integration of innovative materials that enable the creation of flexible, efficient, and cost-effective solar devices. Unlike traditional silicon-based solar cells, which rely on a single material, printable solar cells utilize a variety of materials, each offering unique properties that contribute to the overall performance and versatility of the cells. This section explores the key materials used in printable solar cells, focusing on OPVs, perovskites, and quantum dots, which have been at the forefront of research and development in this field.

9.3.1 Organic Photovoltaics

OPVs have garnered significant attention due to their potential for low-cost production and mechanical flexibility. OPVs are composed of organic polymers or small molecules that can be processed in solution and printed onto flexible substrates. These organic materials absorb light and generate charge carriers (electrons and holes) that are collected at the electrodes to produce electricity. One of the primary advantages of OPVs is their ability to be processed at low temperatures, which not only reduces manufacturing costs but also allows them to be printed on a wide range of substrates, including plastic, fabric, and even paper [11, 17].

However, the efficiency of OPVs has traditionally lagged that of inorganic solar cells, primarily due to issues related to charge carrier mobility and stability. Recent research has focused on improving the molecular design of OPV materials to enhance their light absorption, charge transport, and overall efficiency. By carefully engineering the chemical structure of these materials, researchers have achieved significant improvements in OPV performance, making them a viable option for printable solar cells. Additionally, advances in multi-junction OPVs, where multiple layers of organic materials with different absorption spectra are stacked together, have further boosted their efficiency [18].

9.3.2 Perovskites

PSCs have emerged as one of the most promising materials for printable solar cells, thanks to their exceptional efficiency and ease of fabrication. Perovskites are a class of materials with a unique crystal structure that allows for efficient light absorption and charge transport. The most used perovskite material in solar cells is a hybrid organic–inorganic compound, typically composed of a lead halide framework. Such cells have more than 17% power efficiency due to stable memristive switching at low voltages, thereby retaining the efficiency over thousands of cycles [19]. PSCs have achieved rapid gains in efficiency over the past decade, with some devices now approaching the performance of traditional silicon-based cells. One of the key advantages of perovskites is their compatibility with solution-based processing techniques, which makes them well-suited for printing onto flexible substrates. Perovskite inks can be formulated and deposited using techniques such as inkjet printing, spin coating, and slot-die coating, enabling the production of large-area solar cells at low cost. Additionally, the tuneable bandgap of perovskite materials allows for the optimization of light absorption across different regions of the solar spectrum, further enhancing their efficiency. However, challenges related to the long-term stability of PSCs, toxicity due to the presence of lead, particularly their sensitivity to moisture and heat, remain a focus of ongoing research [20].

9.3.3 Quantum Dots

Quantum dots represent another exciting material for printable solar cells, offering unique optical and electronic properties that can be tailored by adjusting their size and composition. Quantum dots are nanoscale semiconductor particles that exhibit quantum confinement effects, which enable them to absorb and emit light at specific wavelengths. This tunability makes quantum dots particularly attractive for use in multi-junction solar cells, where different layers can be optimized to absorb different parts of the solar spectrum. Quantum dot solar cells can be fabricated using various printing techniques, such as inkjet printing and spray coating, allowing for the creation of flexible and lightweight devices. Additionally, quantum dots can be integrated with other materials, such as perovskites or organic semiconductors, to create hybrid solar cells that combine the advantages of each material. Despite their potential, quantum dot solar cells are still in the early stages of development, and further research is needed to improve their efficiency, stability, and scalability [21, 22].

The success of printable solar cells hinges on the continued development and optimization of key materials such as OPVs, perovskites, and quantum dots. These materials offer unique advantages in terms of flexibility, cost, and efficiency, making them ideal candidates for the next generation of solar technologies. As research progresses, we can expect to see further improvements in the performance and commercial viability of printable solar cells, paving the way for their widespread adoption in a variety of applications.

9.4 PRINTING TECHNIQUES

The development of printable solar cells is inextricably linked to the advancements in printing techniques, which enable the fabrication of solar cells on a wide range of substrates with unprecedented flexibility and capability. These techniques have revolutionized the manufacturing process, allowing for the production of lightweight, flexible, and cost-effective solar cells that can be integrated into various surfaces, from building materials to wearable electronics. This section delves into the key printing techniques that have been instrumental in the progress of printable solar cells, including inkjet printing, screen printing, R2R processing, and spray coating.

9.4.1 Inkjet Printing

Inkjet printing is a highly versatile technique used in the fabrication of printable solar cells, enabling precise deposition of functional materials like organic semiconductors, perovskite inks, and quantum dots onto substrates. This non-contact method is ideal for creating complex, multi-layered architectures and customizable designs, making it particularly suitable for small-scale, specialized applications such as portable electronics and sensors [23].

- **Types of Materials (Suitability to Different Types of Solar Cells)**: Inkjet printing is compatible with a broad range of materials, including OPV materials, perovskites, and quantum dots, making it suitable for various types of printable solar cells. The ability to print on flexible substrates like plastic and paper expands the potential applications of these cells in wearable electronics and portable power sources [24].
- **Methodology and Printing Parameters**: As shown in Figure 9.1, the process involves ejecting droplets of ink from a nozzle onto the substrate, with

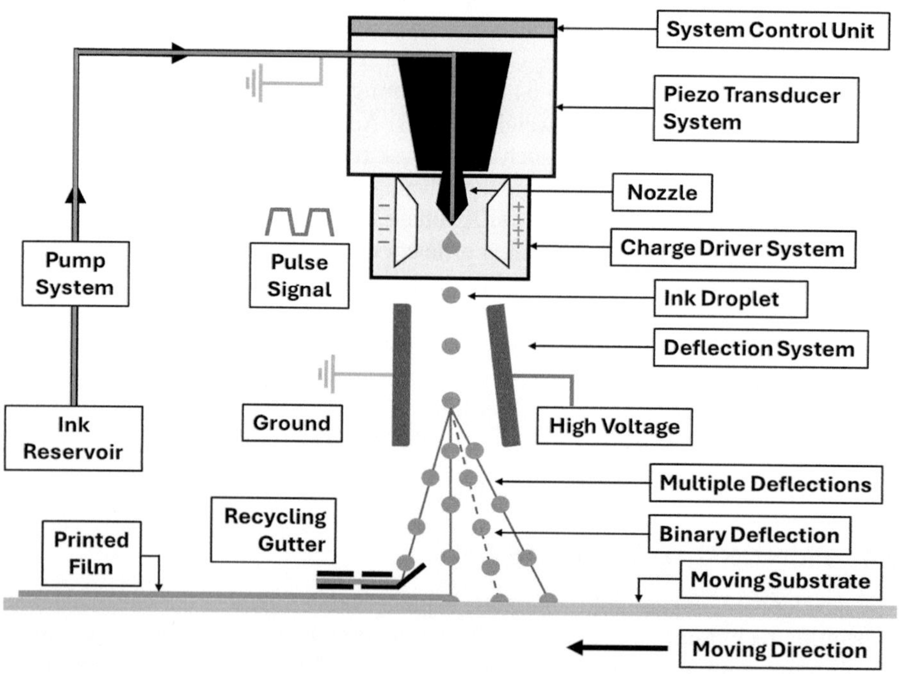

FIGURE 9.1 Inkjet printing of solar cells. Image created by the author.

control over parameters like droplet size, ink viscosity, and substrate temperature. Key factors such as the surface tension of the ink, nozzle diameter, and printing speed must be carefully optimized to achieve uniform layer deposition without defects like coffee-ring effects or nozzle clogging. The stepwise procedure for the inkjet printing process is as follows:

Step 1—Preparation of Ink

- **Formulation**: Develop an ink formulation with the appropriate concentration of functional materials, ensuring that the viscosity is suitable for the printing process.
- **Characterization**: Test the ink for properties such as surface tension and stability to ensure optimal printing performance.

Step 2—Printer Setup

- **Nozzle Calibration**: Adjust the nozzle settings, including diameter and ejection frequency, to control droplet size and deposition accuracy.
- **Substrate Preparation**: Clean and prepare the substrate to ensure good adhesion and uniform deposition of the ink. This may involve surface treatments or coating.

Step 3—Printing Process

- **Droplet Ejection**: Eject droplets of ink from the nozzle onto the substrate in a controlled manner. Monitor and adjust parameters like droplet size, printing speed, and substrate temperature.
- **Layer Formation**: Deposit multiple layers if necessary, allowing each layer to dry or cure before applying the next to avoid defects.

Step 4—Post-Printing Treatment

- **Drying/Curing**: Depending on the ink formulation, apply appropriate drying or curing methods, such as thermal or UV curing, to ensure proper film formation and performance.

Step 5—Quality Control

- **Inspection**: Examine the printed layers for uniformity, consistency, and defects. Use techniques such as microscopy or profilometry to assess layer quality.
- **Advantages**: The primary advantages of inkjet printing include its material efficiency, as the process minimizes waste by depositing material only where needed. It also offers high resolution and patterning flexibility, allowing for the integration of solar cells into unconventional designs and surfaces.
- **Challenges and Limitations**: Challenges in inkjet printing include the need for precise control over ink properties to avoid issues like nozzle clogging and to ensure consistent droplet formation. Additionally, achieving

high throughput for large-scale manufacturing remains a challenge, as the process can be slower compared to other techniques.

- **Applications**: Inkjet printing is particularly suited for applications requiring custom shapes or small, flexible solar cells, such as those used in portable chargers, wearable electronics, or integrated sensors in smart textiles.

9.4.2 Screen Printing

Screen printing is a widely adopted technique known for its scalability and cost-effectiveness in the production of printable solar cells. This method involves transferring functional materials through a mesh screen onto a substrate, which is especially useful for producing large-area solar cells with uniform thickness.

- **Types of Materials (Suitability to Different Types of Solar Cells)**: Screen printing is compatible with various materials, including metal pastes for electrodes, conductive inks, and organic or perovskite inks for active layers. It is particularly suited for the mass production of silicon-based and PSCs, where uniform layer deposition is essential for device performance [25].
- **Methodology and Printing Parameters**: As shown in Figure 9.2, the process requires the formulation of inks with appropriate viscosity and particle size to pass through the screen without clogging. Parameters like screen mesh size, squeegee speed, and pressure are critical for controlling the

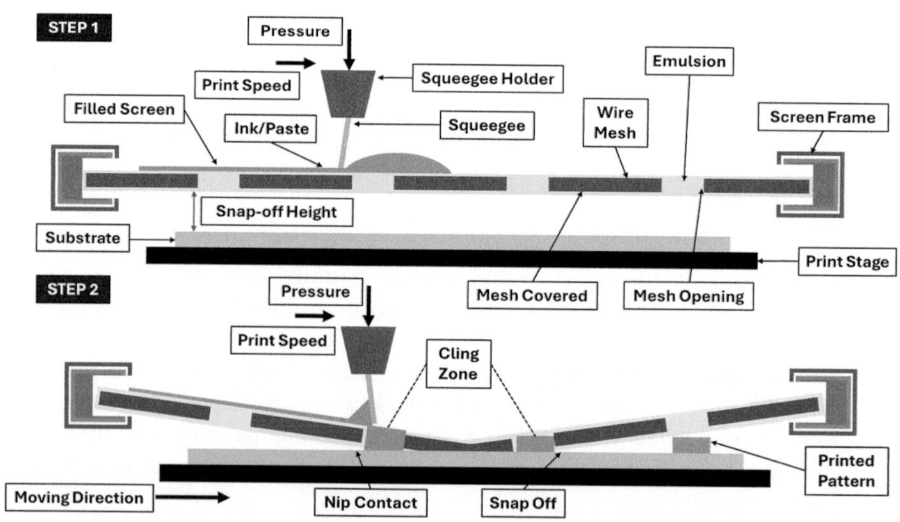

FIGURE 9.2 Screen printing of solar cells. Image created by the author.

thickness and uniformity of the deposited layers [26]. This method is often used for printing conductive lines or large-area active layers in solar cells. The stepwise procedure for the inkjet printing process is as follows:

Step 1—Ink Preparation

- **Formulation**: Develop inks with the right viscosity and particle size to ensure smooth passage through the screen mesh. Ensure the ink is well-mixed and free of air bubbles.
- **Testing**: Conduct tests to confirm that the ink's consistency meets the requirements for printing.

Step 2—Screen and Substrate Setup

- **Screen Mesh**: Choose an appropriate mesh size for the screen based on the desired resolution and layer thickness. Coarser meshes are used for thicker layers, while finer meshes are used for detailed patterns.
- **Substrate Preparation**: Clean and treat the substrate to enhance ink adhesion and prevent defects.

Step 3—Printing Process

- **Ink Application**: Spread the ink across the screen using a squeegee. Ensure even coverage to avoid inconsistencies.
- **Squeegee Operation**: Adjust the speed and pressure of the squeegee to control ink flow and layer thickness. Maintain consistent motion to ensure uniform deposition.
- **Layer Formation**: Print the ink onto the substrate in a controlled manner. Allow each layer to dry or cure before applying subsequent layers if necessary.

Step 4—Post-Printing Treatment

- **Drying/Curing**: Use appropriate methods to dry or cure the printed layers, such as heating or UV curing, to ensure proper film formation and functionality.

Step 5—Quality Control

- **Inspection**: Inspect the printed layers for uniformity and defects. Use tools like microscopes to check layer consistency and screen resolution.
- **Advantages**: The main advantage of screen printing is its high throughput and low cost, making it suitable for large-scale manufacturing. It also allows for the deposition of thick, robust layers, which contribute to the durability and longevity of solar cells.
- **Challenges and Limitations**: One limitation is the difficulty in achieving fine pattern resolution, which may restrict its use in applications requiring high-precision patterning. Additionally, material wastage can occur during the process, and the method may not be as effective for printing extremely thin or intricate layers.

- **Applications**: Screen printing is widely used in the production of traditional and emerging solar cells, including large-area solar panels for rooftops and BIPV. It is also employed in producing flexible solar cells for portable devices and integrated systems.

9.4.3 R2R Processing

R2R processing is a high-throughput technique that enables the continuous production of printable solar cells on flexible substrates, such as plastic or metal foils. This method is ideal for large-scale manufacturing, offering the potential for cost-effective mass production of solar cells [27, 28].

- **Types of Materials (Suitability to Different Types of Solar Cells)**: R2R processing is compatible with a wide range of materials, including OPVs, perovskites, and quantum dots, making it suitable for producing flexible and lightweight solar cells. The use of solution-processable materials is particularly advantageous in this technique [26].
- **Methodology and Printing Parameters**: As shown in Figure 9.3, in R2R processing, a flexible substrate is unrolled, and various layers of the solar cell are deposited sequentially using techniques like slot-die coating, gravure printing, or inkjet printing. Parameters such as substrate tension, deposition speed, and drying conditions are crucial for ensuring uniform layer formation and high device performance. The process concludes with the substrate being rolled up again, resulting in a continuous sheet of solar cells. The stepwise procedure for the inkjet printing process is as follows:

Step 1—Substrate Preparation

- **Substrate Selection**: Choose a flexible substrate (e.g., plastic, metal foil) that suits the specific type of solar cell being produced.
- **Surface Treatment**: Clean and treat the substrate surface to ensure proper adhesion and smooth deposition of the functional layers.

Step 2—Unrolling the Substrate

- **Tension Control**: Unroll the substrate under controlled tension to maintain a consistent, flat surface for layer deposition.
- **Alignment**: Ensure precise alignment of the substrate to avoid defects and ensure uniformity across the roll.

Step 3—Layer Deposition

Deposition Techniques: Sequentially deposit the various layers of the solar cell (e.g., active layer, electrode layers) using one or more of the following methods:

- **Slot-Die Coating**: Applies a uniform, thin layer of material across the substrate.
- **Gravure Printing**: Transfers ink from a patterned cylinder to the substrate for patterned layers.

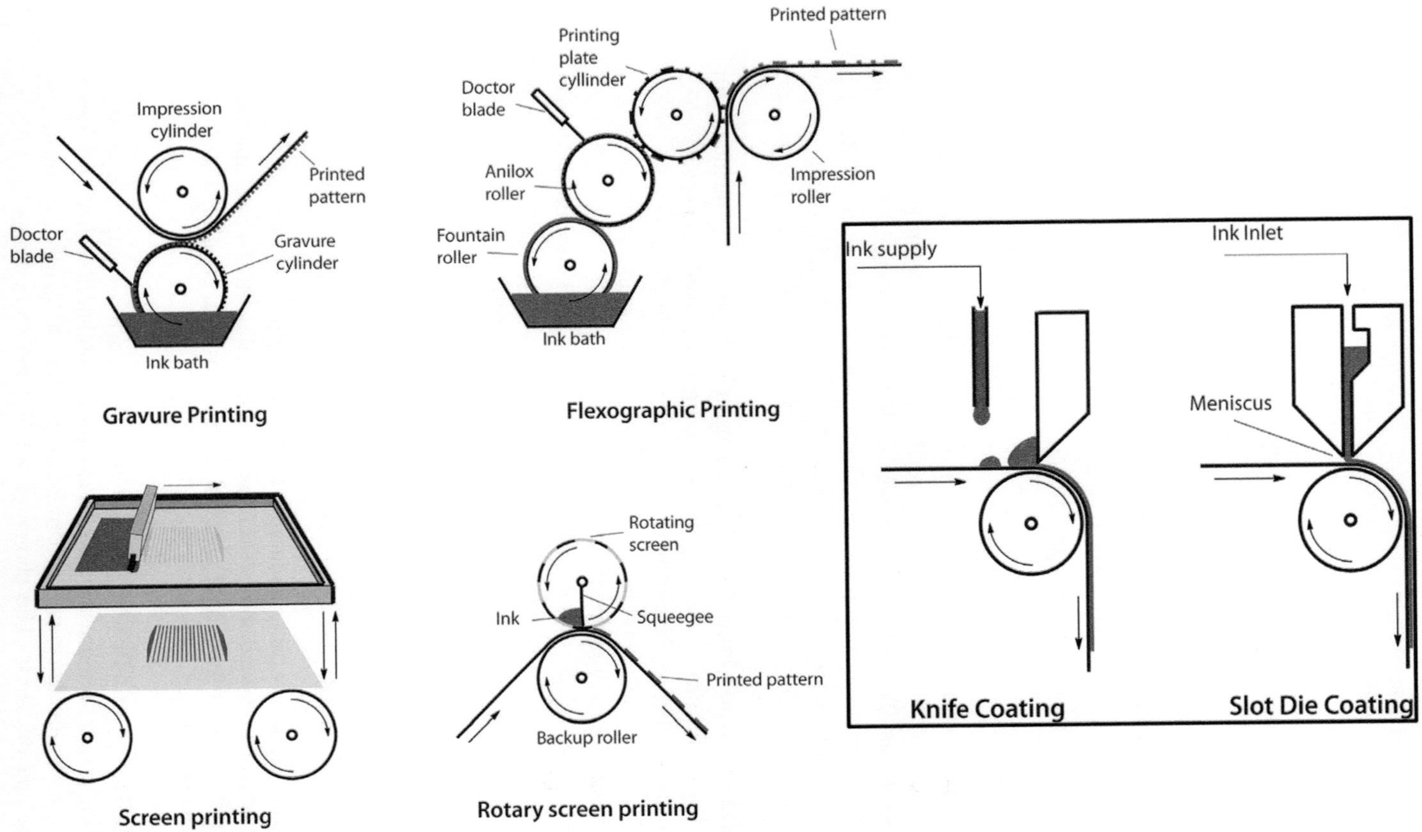

FIGURE 9.3 Roll-to-roll processing of solar cells covering gravure printing, flexographic printing, screen printing, and rotary screen printing with differences in knife and slot-die coating techniques.

Reproduced with permission from Elsevier. [29]

- Inkjet Printing: Precisely deposits materials in specific areas, useful for more intricate designs.
- Speed and Pressure Control: Adjusts the speed of the roll and the pressure applied during deposition to ensure consistent layer thickness and quality.

Drying and Curing: Implement controlled drying or curing processes (e.g., thermal, UV) after each deposition step to solidify the layers and prevent defects.

Step 4—Re-Rolling the Substrate

- **Tension Management**: Maintain appropriate tension while re-rolling the substrate to avoid wrinkling or damage.

Step 5—Quality Control

- **Final Inspection**: Conduct a final inspection of the rolled substrate to identify and correct any defects before further processing or cutting.
- **Advantages**: The primary advantage of R2R processing is its scalability, enabling the production of large-area solar cells at a lower cost. This technique is well-suited for producing flexible, lightweight solar panels that can be integrated into a variety of products, from building façades to portable solar chargers.
- **Challenges and Limitations**: Challenges in R2R processing include maintaining uniformity across large areas, ensuring consistent material deposition, and managing the curing or annealing processes. The need for precise control over these parameters can complicate the manufacturing process, especially for high-efficiency devices.
- **Applications**: R2R processing is primarily used for large-area solar panels, flexible photovoltaic (PV) modules, and integrated PV systems for buildings. It also has potential applications in creating lightweight, portable solar chargers and other energy-harvesting devices for mobile or off-grid use.

9.4.4 Spray Coating

Spray coating is an emerging technique in the fabrication of printable solar cells, known for its simplicity and ability to cover large areas quickly. This method involves spraying a fine mist of functional material onto a substrate, creating a uniform thin film that serves as an active layer in the solar cell [30].

- **Methodology and Printing Parameters**: As shown in Figure 9.4, the spray coating process requires careful control over parameters such as spray pressure, nozzle distance, and substrate temperature to achieve uniform coverage and thickness. The choice of solvent, ink formulation, and drying

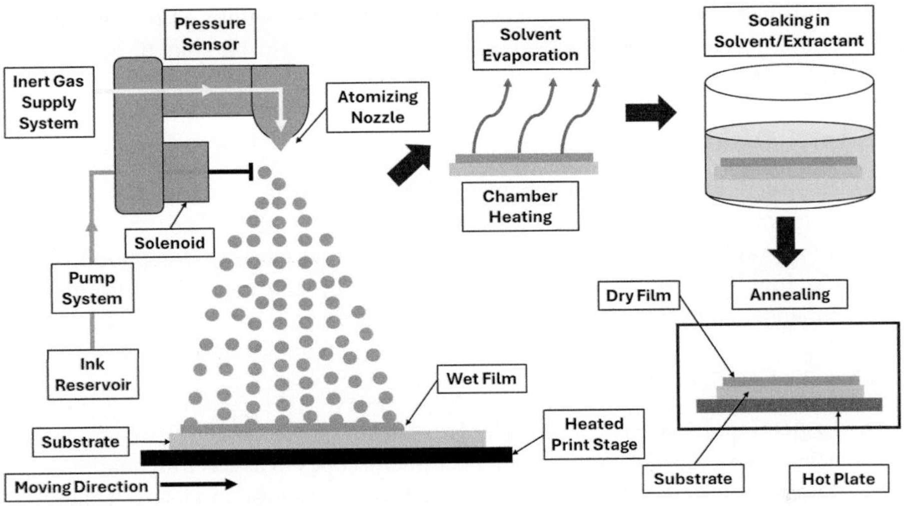

FIGURE 9.4 Spray coating of solar cells. Image created by the author.

conditions are also critical to ensure optimal film formation without defects. The stepwise procedure for the inkjet printing process is as follows:

Step 1—Ink Preparation

- **Formulation**: Develop an ink with the appropriate concentration of functional materials and solvent properties to ensure smooth spraying and film formation.
- **Solvent Selection**: Choose a solvent that balances drying time and film uniformity. The solvent should evaporate at a controlled rate to avoid defects.

Step 2—Substrate Setup

- **Surface Preparation**: Clean and treat the substrate surface to improve the adhesion and uniformity of the sprayed film. Ensure the substrate is compatible with the ink.
- **Temperature Control**: Set the substrate temperature to optimize ink adhesion and drying. This helps in achieving a uniform film without defects like cracks or pinholes.

Step 3—Spray Coating Process

- **Nozzle Setup**: Select an appropriate nozzle type and size to control the droplet size and spray pattern.
- **Spray Pressure and Distance**: Adjust the spray pressure and the distance between the nozzle and the substrate. Proper control is essential to ensure even coverage and avoid overspray or uneven thickness.

- **Motion Control**: Move the nozzle in a consistent pattern across the substrate, ensuring even deposition. The motion should be smooth and at a controlled speed to avoid overlapping or gaps.

Step 4—Drying and Curing

- **Drying Conditions**: Implement a drying process to allow the solvent to evaporate, forming a solid thin film. The drying method (e.g., thermal or air drying) should be chosen based on the ink and substrate compatibility.
- **Curing Process**: If necessary, apply a curing step (e.g., UV curing) to enhance the film's stability and performance.

Step 5—Quality Control

- **Inspection**: Inspect the coated film for uniformity, thickness consistency, and defects. Use tools like profilometers or microscopes to assess film quality.
- **Thickness Measurement**: Measure the film thickness to ensure it meets the desired specifications. Adjust the process parameters if necessary.
- **Types of Materials (Suitability to Different Types of Solar Cells)**: Spray coating is compatible with various materials, including perovskite inks, organic semiconductors, and other solution-processable materials. Its adaptability to different surface topographies makes it suitable for printing on curved or textured surfaces, expanding the range of possible applications [31].
- **Advantages**: The main advantage of spray coating is its ability to rapidly cover large areas with uniform thin films, making it a cost-effective option for large-scale production. It is also highly adaptable, allowing for the deposition of materials on non-flat surfaces, which is beneficial for unconventional solar cell designs.
- **Challenges and Limitations**: One limitation of spray coating is the potential for non-uniform thickness and the formation of defects if the process parameters are not well-controlled. Additionally, achieving high resolution and fine patterns can be challenging with this technique, which may limit its use in certain applications.
- **Applications**: Spray coating is particularly suited for applications where large-area coverage is required, such as in BIPV and automotive applications. It is also used for applying thin films in flexible and portable solar devices, where uniformity and surface adaptability are crucial. Printable solar cells benefit from various printing techniques, each with unique strengths. Inkjet printing offers precision, screen printing is cost-effective for large areas, and R2R processing supports high-volume production. As these technologies advance, they are expected to broaden the applications of printable solar cells, enhancing their role in renewable energy.

Various printing techniques play a crucial role in the fabrication of functional materials, each offering distinct advantages and limitations. Inkjet printing, screen printing, R2R printing, and spray coating are commonly used methods, differing in key parameters such as ink viscosity, critical linewidth, film thickness, roughness, printing speed, scalability, and design flexibility. As per the plot shown in Figure 9.5,

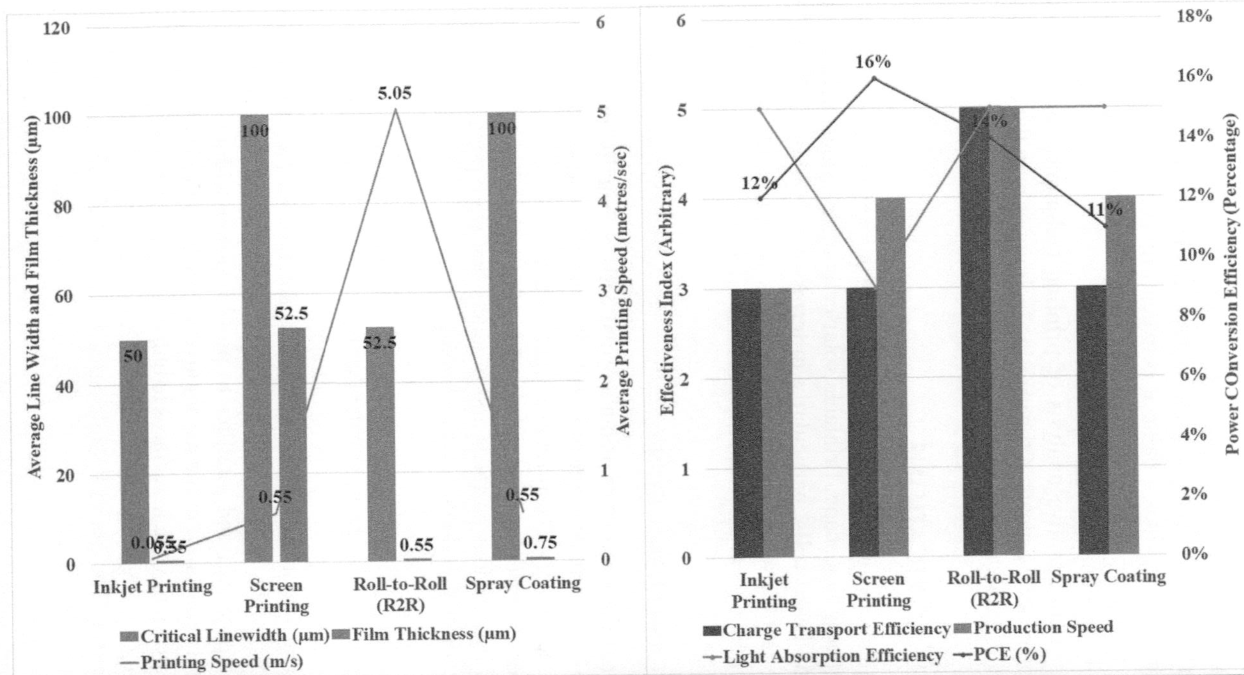

Printing Property	Inkjet Printing	Screen Printing	Roll-to-Roll Processing	Spray Coating
Ink Viscosity	5 – 50 cP	500 – 5000 cP	10 – 1000 cP	10 – 100 cP
Critical Linewidth	30 – 70 µm	50 – 150 µm	5 – 100 µm	50 – 150 µm
Film Thickness	0.1 – 1 µm	5 – 100 µm	0.1 – 1 µm	0.5 – 1 µm
Printing Speed	0.01 – 0.1 m/s	0.1 – 1 m/s	0.1 – 10 m/s	0.1 – 1 m/s

FIGURE 9.5 Technical performance comparison of printing techniques for printable solar cells. Image created by the author.

inkjet printing allows for high design flexibility and precise 2D patterning but has limitations in large-area scalability. Screen printing, with its higher ink viscosity and thicker films, supports large-area scalability but compromises design flexibility due to its coarser resolution. R2R (gravure) printing stands out for its high-speed continuous printing and scalability, making it suitable for large-scale production, although it offers lower design flexibility. Spray coating, while enabling moderate scalability and patterning capabilities, provides a balance between film roughness and speed. Understanding these characteristics is essential for selecting the appropriate technique based on specific application requirements [32].

Table 9.1 provides a comparative analysis of printing techniques for printable solar cells covering the general aspects of the printing techniques with the suitability of the techniques for the respective application areas.

9.5 DESIGN AND APPLICATIONS

The design and applications of printable solar cells are pivotal to their potential impact on the solar energy landscape. Printable solar cells offer unique advantages over traditional silicon-based cells, particularly in terms of flexibility, lightweight nature, and versatility in integration. This section explores how these attributes influence the design of printable solar cells and their diverse applications.

9.5.1 Lightweight and Flexible Panels

One of the most compelling features of printable solar cells is their lightweight and flexible nature, which opens a wide range of possibilities for innovative applications. Unlike traditional rigid solar panels, printable solar cells can be manufactured on thin, flexible substrates such as plastics or fabrics. This flexibility allows them to be seamlessly integrated into various surfaces and products that were previously challenging or impossible for conventional solar technologies.

- **Lightweight Design**: Printable solar cells are significantly lighter than their silicon counterparts due to the use of flexible materials and thin film technologies. This lightweight design is advantageous for applications where weight is a critical factor, such as in aerospace, automotive, and portable electronics. For example, in the aerospace industry, lightweight solar panels can be used on spacecraft and satellites to provide power without adding significant weight. Similarly, in the automotive sector, flexible solar panels can be incorporated into vehicle surfaces to enhance energy efficiency without compromising aerodynamic performance [33].
- **Flexible Substrates**: The use of flexible substrates in printable solar cells enables the creation of solar panels that can bend, stretch, and conform to different shapes. This flexibility is particularly useful for integrating solar cells into unconventional surfaces, such as curved building façades, wearable electronics, and even clothing. Flexible solar panels can be applied to the surface of a building to generate power while maintaining the building's aesthetic appearance. In wearable technology, flexible

TABLE 9.1
Comparative Analysis of Printing Techniques for Printable Solar Cells

Printing Technique	Key Features	Suitable Materials	Advantages	Challenges	PCE (Approx. %)	Light Absorption Efficiency	Charge Transport Efficiency	Resolution	Production Speed
Inkjet printing	—Precise, non-contact deposition	—Organic photovoltaics (OPVs), perovskites, quantum dots	—High resolution, material efficiency	—Nozzle clogging, slow for large-scale production	8–15	High	Moderate	High resolution (sub-micron)	Moderate
Screen printing	—Mesh-based, scalable technique	—Metal pastes for electrodes, conductive inks, organic and perovskite inks	—Cost-effective, high throughput	—Limited pattern resolution, material wastage	12–20	Moderate	Moderate	Moderate (100–200 microns)	High
Roll-to-roll (R2R)	—Continuous, large-scale production on flexible substrates	—Organic photovoltaics (OPVs), perovskites, quantum dots	—Scalability, lower cost per unit	—Uniformity control for large areas, complex curing processes	10–18	High	High	Moderate (50–150 microns)	Very high
Spray coating	—Simple, rapid coverage of large areas	—Perovskites, organic semiconductors, solution-processable materials	—Fast deposition, adaptable to non-flat surfaces	—Uniformity challenges, difficult to do fine patterns	8–14	High	Moderate	Low resolution (300–500 microns)	High

solar cells can be embedded in clothing or accessories to charge portable devices on the go [34].

- **Durability and Protection**: Despite their flexibility, printable solar cells must be designed to withstand environmental factors such as moisture, temperature fluctuations, and mechanical stress. Advances in encapsulation techniques have been developed to protect the delicate layers of printable solar cells from damage, ensuring long-term durability and reliable performance. Encapsulation materials such as flexible polymers and barrier layers help shield solar cells from external conditions while maintaining their flexibility and functionality [35].

9.5.2 Integration into Various Products

The versatility of printable solar cells extends beyond traditional solar panel applications, allowing for their integration into a wide range of products and technologies. This section highlights how printable solar cells are being incorporated into various products to harness solar energy in innovative and practical ways.

- **Transparent Solar Cells Using Printed Techniques**: Transparent solar cells represent a significant frontier in renewable energy innovation, particularly for urban and architectural applications. Printable techniques enable the fabrication of semi-transparent PV layers on glass and plastic substrates, making these solar cells suitable for energy-generating windows and transparent electronic displays. R2R and inkjet printing methods are particularly advantageous, allowing for precise material deposition over large areas with customizable transparency levels. Current research focuses on perovskite and OPV materials, which are highly compatible with printable techniques due to their tuneable optical properties. Transparent solar cells have already been implemented in pilot projects for smart windows and solar glazing systems in buildings, combining energy efficiency with natural lighting. However, challenges such as optimizing energy conversion efficiency and ensuring long-term stability under environmental conditions remain. As advancements in material science and printing processes continue, transparent solar cells are poised to revolutionize energy usage in architectural and consumer electronic applications [36, 37].
- **Building Integration**: One of the most promising applications of printable solar cells is their integration into building materials and infrastructure. Printable solar cells can be incorporated into BIPV, where they are integrated directly into the building's façade, roof, or windows. This approach allows buildings to generate their own electricity while maintaining architectural aesthetics when the design considers building orientation, location, cost, utility issues, appropriate safety codes, building codes, electrical loads, and so on. For instance, printable solar cells can be printed on transparent substrates to create solar windows that generate power without obstructing light or visibility. Solar glazing combines glass and PVs to control sunlight entering buildings through such windows. These BIPVs provide benefits

like construction-material cost offset, building envelops, sound- and heat-masking, insulation, and aesthetics [38, 39]

- **Consumer Electronics**: The lightweight and flexible nature of printable solar cells make them ideal for integration into consumer electronics. Portable devices such as smartphones, tablets, and wearables can benefit from the addition of flexible solar panels that provide a supplementary power source. By embedding printable solar cells into device cases or screens, manufacturers can extend the battery life of electronics and reduce the need for frequent recharging. Similarly, PSCs, along with OPVs and dye-sensitized solar cells, are solution-processable, customizable, and suitable for indoor lighting applications, with PSCs leading in efficiency and performance, especially in the growing Internet of things market [40, 41].
- **Textiles and Wearable Technology**: The integration of printable solar cells into textiles and wearable technology is another exciting application. Flexible solar panels can be incorporated into fabrics to create solar-charging clothing and accessories. One of the best examples would be organic solar cells (OSCs), which have high flexibility, portability, and low-cost solution-based processing with over 19% power efficiency, thus allowing users to charge electronic devices, such as smartphones or wearable sensors, while on the move. Additionally, solar-powered textiles can be used in outdoor gear and camping equipment, providing a convenient and sustainable power source in remote locations [42, 43].
- **Digital Healthcare Devices**: The integration of printable solar cells into medical devices offers promising advancements in healthcare technology. By incorporating flexible solar panels into medical implants or wearable health monitors, devices can maintain continuous power without the need for frequent battery replacements. This not only enhances patient comfort but also ensures the uninterrupted operation of critical health monitoring systems. Printable solar cells, with their lightweight and adaptable nature, are particularly suited for such applications, enabling the development of self-sustaining medical devices that can operate efficiently in various conditions [44, 45].
- **Packaging Materials**: Printable solar cells can also be integrated into packaging materials to create self-powered packaging solutions. This application is particularly relevant for packaging electronic products or food items that require real-time monitoring of conditions such as temperature and humidity. By incorporating printable solar cells into packaging, manufacturers can provide power for sensors and communication devices, enhancing the functionality and sustainability of the packaging [46, 47].
- **Automotive Applications**: In the automotive industry, printable solar cells can be integrated into vehicle surfaces to provide supplemental power for onboard systems. Flexible solar panels can be applied to the roofs or windows of vehicles, generating electricity to power auxiliary systems such as air conditioning, lighting, and infotainment systems. This integration can improve fuel efficiency and reduce the vehicle's reliance on external power sources [33, 42].

In summary, the design and applications of printable solar cells reflect their unique advantages in terms of flexibility, lightweight properties, and versatility. By leveraging these attributes, printable solar cells are transforming a wide range of industries, from architecture and consumer electronics to textiles and automotive technologies. As research and development continue to advance, the integration of printable solar cells into various products is expected to drive further innovation and adoption of solar energy solutions.

9.6 EFFICIENCY AND DURABILITY

9.6.1 Performance Metrics

The efficiency of printable solar cells is a critical parameter that influences their adoption and effectiveness in various applications. Efficiency refers to the ability of a solar cell to convert sunlight into electrical energy, and it is typically expressed as a percentage. The performance of printable solar cells is measured by several key metrics, including power conversion efficiency (PCE), light absorption, and charge transport efficiency [38].

- **PCE**: PCE is the most used metric to evaluate the effectiveness of a solar cell. It represents the ratio of electrical power output to the incident solar power input. Printable solar cells generally exhibit lower PCE compared to traditional silicon-based cells, primarily due to their different material properties and fabrication processes. However, advancements in materials and technology have led to significant improvements in the PCE of printable solar cells. Research is focused on optimizing active materials, improving device architectures, and enhancing light absorption to increase PCE [35].
- **Light Absorption**: Efficient light absorption is essential for maximizing the performance of solar cells. Printable solar cells often utilize materials that have broad absorption spectra to capture a wide range of sunlight. Techniques such as layer optimization, surface texturing, and light-trapping structures are employed to enhance light absorption and reduce reflection losses. The development of new materials, such as OPVs and perovskites, has also contributed to improved light absorption and overall efficiency [48].
- **Charge Transport Efficiency**: The efficiency with which charge carriers are transported through the solar cell is crucial for optimizing performance. Charge transport efficiency depends on the quality of the active layers, the interface between different materials, and the design of the cell structure. Enhancing charge mobility and reducing recombination losses are key areas of research aimed at improving charge transport efficiency. Techniques such as optimizing the thickness of active layers, using charge transport layers, and engineering interfaces contribute to better charge transport and higher efficiency [49].

9.6.2 Long-Term Stability Challenges

The long-term stability of printable solar cells is a significant challenge that impacts their commercial viability and practical applications. Unlike traditional silicon solar cells, printable solar cells are often fabricated using materials that are sensitive to environmental factors, which can affect their performance over time.

- **Degradation Mechanisms**: Printable solar cells face several degradation mechanisms that can reduce their longevity. Common degradation factors include moisture ingress, UV radiation, temperature fluctuations, and chemical instability. For instance, organic materials used in printable solar cells can degrade when exposed to moisture or UV light, leading to reduced performance and shorter lifespans. Researchers are actively working on developing better encapsulation techniques and protective coatings to mitigate these degradation effects and enhance the durability of printable solar cells [24].
- **Encapsulation and Protection**: Effective encapsulation is essential for protecting printable solar cells from environmental damage. Encapsulation materials such as polymers and barrier films are used to shield the active layers from moisture, oxygen, and UV radiation. Advances in encapsulation technologies are focused on improving the barrier properties, flexibility, and adhesion of encapsulation materials. Additionally, the development of self-healing materials and advanced coating techniques is being explored to further enhance the protection and longevity of printable solar cells [31].
- **Testing and Reliability**: Ensuring the reliability of printable solar cells under real-world conditions requires extensive testing and simulation. Accelerated aging tests, environmental exposure tests, and long-term stability assessments are conducted to evaluate the performance of solar cells over extended periods. These tests help identify potential failure modes and guide the development of more robust and durable solar cell designs. Ongoing research aims to improve the reliability of printable solar cells and address issues related to their long-term stability [3].

9.7 ARTIFICIAL INTELLIGENCE AND MACHINE LEARNING INTEGRATION

9.7.1 Enhancing Material and Process Optimization

Artificial intelligence (AI) and machine learning (ML) have increasingly become pivotal in advancing the development and optimization of printable solar cells. These technologies offer powerful tools for enhancing material performance, streamlining manufacturing processes, and improving overall efficiency.

- **Material Discovery and Optimization**: AI and ML algorithms are employed to accelerate the discovery and optimization of new materials for printable solar cells. By analysing vast datasets from experiments and simulations, AI can identify promising materials with desirable properties, such as high light absorption and efficient charge transport. Machine learning models can predict material behaviours under various conditions, enabling researchers to fine-tune compositions and structures to achieve optimal performance. This approach significantly reduces the time and cost associated with traditional material discovery methods [31, 50].

- **Process Optimization**: AI and ML also play a crucial role in optimizing the fabrication processes for printable solar cells. Techniques such as process control, quality monitoring, and defect detection are enhanced by AI-driven systems. Machine learning algorithms analyse data from production lines to identify patterns and correlations that can lead to improved process parameters and reduced variability. This optimization helps in achieving consistent quality and performance of solar cells while minimizing waste and production costs [51].
- **Predictive Maintenance**: Predictive maintenance, powered by AI, is used to anticipate and address potential issues in manufacturing equipment before they cause significant disruptions. By monitoring equipment performance and analysing historical data, AI models can predict failures or anomalies and suggest preventive measures. This proactive approach helps maintain smooth operations, reduce downtime, and ensure the reliability of the manufacturing process [52].
- **Future Directions**: The integration of AI and ML in printable solar cell technology is poised for continued growth and innovation. Future research may focus on developing more advanced ML models that can handle increasingly complex data and provide deeper insights into material and process optimization. Additionally, the application of AI in exploring novel materials and fabrication techniques could lead to breakthroughs in efficiency and stability. Collaboration between material scientists, engineers, and data scientists will be essential in driving these advancements and addressing the challenges of integrating AI and ML in solar cell technology [53].

9.8 CHALLENGES AND FUTURE OUTLOOK

9.8.1 Addressing Current Limitations

Despite significant advancements, printable solar cells face several challenges that need to be addressed to realize their full potential. These limitations impact their efficiency, durability, and overall commercial viability.

- **Efficiency Limitations**: One of the primary challenges is the relatively lower efficiency of printable solar cells compared to traditional silicon-based cells. While progress has been made in improving PCE, printable solar cells still generally lag in terms of performance. This limitation is partly due to the inherent properties of the materials used and the fabrication techniques employed. Research is ongoing to develop new materials and optimize device architectures to enhance efficiency [54].
- **Durability and Stability**: The long-term durability and stability of printable solar cells remain significant concerns. Unlike traditional solar cells, printable variants are often more susceptible to environmental factors such as moisture, UV radiation, and temperature fluctuations. Ensuring that these cells maintain performance over extended periods is crucial for their widespread adoption. Advances in encapsulation technologies and protective

coatings are addressing these issues, but achieving long-term reliability remains a challenge [27].

- **Scalability and Manufacturing**: Scaling up the production of printable solar cells from laboratory settings to commercial manufacturing poses another challenge. Ensuring consistent quality, minimizing defects, and reducing production costs are critical for commercial success. The transition from small-scale production to large-scale manufacturing requires huge technical strides in materials preparation, manufacturing platforms, installation platforms, as well as electronics, monitoring, and control systems followed by long-term monitoring and stability analysis to prove that printable solar cells are a potential for sustainable energy technology [55].
- **Regulatory and Market Acceptance**: The adoption of printable solar cells is shaped by both regulatory challenges and market acceptance. Compliance with international standards like ISO 9001 for quality management and ISO 14001 for environmental management is crucial to ensure high-quality manufacturing and minimize environmental impacts, essential for gaining market approval and consumer trust. Additionally, meeting performance and safety standards, such as those in IEC 61215 for PV modules, is necessary for obtaining certifications and ensuring reliable operation. Market acceptance also depends on factors like cost competitiveness, durability, and the availability of supporting infrastructure. Printable solar cells must compete with traditional silicon-based panels in terms of pricing, reliability, and efficiency. As the technology matures and becomes more cost-effective, overcoming regulatory hurdles and establishing robust infrastructure will be key to achieving broader market acceptance [56].

9.8.2 Prospects for Future Research

The future of printable solar cells holds promising prospects for overcoming current limitations and achieving broader adoption. Several areas of research are expected to drive advancements in this field.

- **Material Innovation**: Future research will likely focus on developing new materials with improved performance characteristics. Innovations in OPVs, perovskites, and quantum dots are expected to yield higher efficiency and better stability. Research into hybrid materials that combine the strengths of different types of PV materials could also lead to breakthroughs in performance [57, 58].
- **Advanced Fabrication Techniques**: Advancements in fabrication techniques will play a crucial role in enhancing the scalability and cost-effectiveness of printable solar cells. Techniques such as R2R processing, advanced inkjet printing, and scalable coating methods are expected to improve production efficiency and reduce costs. Research into automation and process control will also contribute to achieving consistent quality and performance [59].
- **Integration and Applications**: Exploring new applications and integration methods for printable solar cells will drive future research. Opportunities include integrating solar cells into innovative products, such as smart

textiles, flexible electronics, and building materials. Enhancing the versatility of printable solar cells and expanding their range of applications will contribute to their widespread adoption [60].

9.9 CONCLUSION

Printable solar cells represent a transformative approach to solar energy harvesting, leveraging innovative printing techniques and advanced materials for flexible, lightweight, and cost-effective solutions. By integrating OPVs, perovskites, and quantum dot technologies, these solar cells offer an alternative to conventional silicon-based counterparts. Despite existing challenges such as long-term stability, efficiency enhancement, and large-scale manufacturing, ongoing research continues to push the boundaries of their potential applications. The integration of AI and ML further optimizes material discovery and production processes. As the demand for sustainable energy grows, printable solar cells will play a crucial role in next-generation renewable energy solutions, enabling widespread adoption in consumer electronics, building-integrated PVs, and wearable technology.

REFERENCES

[1] Elhadad, A. A.; Rosa-Sainz, A.; Cañete, R.; Peralta, E.; Begines, B.; Balbuena, M.; Alcudia, A.; Torres, Y. Applications and Multidisciplinary Perspective on 3D Printing Techniques: Recent Developments and Future Trends. *Materials Science and Engineering: R: Reports*, **2023**, *156*, 100760. https://doi.org/10.1016/J.MSER.2023.100760.

[2] Htwe, Y. Z. N.; Mariatti, M. Printed Graphene and Hybrid Conductive Inks for Flexible, Stretchable, and Wearable Electronics: Progress, Opportunities, and Challenges. *Journal of Science: Advanced Materials and Devices*, **2022**, *7* (2), 100435. https://doi.org/10.1016/J.JSAMD.2022.100435.

[3] Mu, Y.; He, Z.; Wang, K.; Pi, X.; Zhou, S. Recent Progress and Future Prospects on Halide Perovskite Nanocrystals for Optoelectronics and Beyond. *iScience*, **2022**, *25* (11), 105371. https://doi.org/10.1016/J.ISCI.2022.105371.

[4] Li, L.; Han, L.; Hu, H.; Zhang, R. A Review on Polymers and Their Composites for Flexible Electronics. *Materials Advances*, **2023**, *4* (3), 726–746. https://doi.org/10.1039/D2MA00940D.

[5] Hunde, B. R.; Woldeyohannes, A. D. 3D Printing and Solar Cell Fabrication Methods: A Review of Challenges, Opportunities, and Future Prospects. *Results in Optics*, **2023**, *11*, 100385. https://doi.org/10.1016/J.RIO.2023.100385.

[6] Sharma, A.; Masoumi, S.; Gedefaw, D.; O'Shaughnessy, S.; Baran, D.; Pakdel, A. Flexible Solar and Thermal Energy Conversion Devices: Organic Photovoltaics (OPVs), Organic Thermoelectric Generators (OTEGs) and Hybrid PV-TEG Systems. *Applied Materials Today*, **2022**, *29*, 101614. https://doi.org/10.1016/J.APMT.2022.101614.

[7] Zhang, J.; Zhang, W.; Cheng, H. M.; Silva, S. R. P. Critical Review of Recent Progress of Flexible Perovskite Solar Cells. *Materials Today*, **2020**, *39*, 66–88. https://doi.org/10.1016/J.MATTOD.2020.05.002.

[8] Bhatnagar, P.; Zaferani, S. H.; Rafiefard, N.; Baraeinejad, B.; Vazifeh, A. R.; Mohammadpour, R.; Ghomashchi, R.; Dillersberger, H.; Tham, D.; Vashaee, D. Advancing Personalized Healthcare and Entertainment: Progress in Energy Harvesting Materials and Techniques of Self-Powered Wearable Devices. *Progress Materials Science*, **2023**, *139*, 101184. https://doi.org/10.1016/J.PMATSCI.2023.101184.

[9] Dutta, A.; Abdullah, A. M.; Biswas, M. A. S.; Das, S.; Zhu, J.; Lin, Y.; Cheng, H. Micro- and Nanostructured Semiconductor Materials for Flexible and Stretchable Electronics. *Reference Module in Materials Science and Materials Engineering*, **2024**. https://doi.org/10.1016/B978-0-323-96027-4.00014-0.

[10] Shahbaz, I.; Tahir, M.; Li, L.; Song, Y. Advancements in 2D Transition Metal Dichalcogenides (TMDs) Inks for Printed Optoelectronics: A Comprehensive Review. *Materials Today*, **2024**, *77*, 142–184. https://doi.org/10.1016/J.MATTOD.2024.06.008.

[11] Coll, M.; Fontcuberta, J.; Althammer, M.; Bibes, M.; Boschker, H.; Calleja, A.; Cheng, G.; Cuoco, M.; Dittmann, R.; Dkhil, B.; et al. Towards Oxide Electronics: A Roadmap. *Applied Surface Science*, **2019**, *482*, 1–93. https://doi.org/10.1016/J.APSUSC.2019.03.312.

[12] Pandey, S.; Karakoti, M.; Bhardwaj, D.; Tatrari, G.; Sharma, R.; Pandey, L.; Lee, M. J.; Sahoo, N. G. Recent Advances in Carbon-Based Materials for High-Performance Perovskite Solar Cells: Gaps, Challenges and Fulfillment. *Nanoscale Advances*, **2023**, *5* (6), 1492–1526. https://doi.org/10.1039/D3NA00005B.

[13] Kore, B. P.; Jamshidi, M.; Gardner, J. M. The Impact of Moisture on the Stability and Degradation of Perovskites in Solar Cells. *Materilas Advances*, **2024**, *5* (6), 2200–2217. https://doi.org/10.1039/D3MA00828B.

[14] Kim, D. H.; Whitaker, J. B.; Li, Z.; van Hest, M. F. A. M.; Zhu, K. Outlook and Challenges of Perovskite Solar Cells toward Terawatt-Scale Photovoltaic Module Technology. *Joule*, **2018**, *2* (8), 1437–1451. https://doi.org/10.1016/J.JOULE.2018.05.011.

[15] Du, T.; Qiu, S.; Zhou, X.; Le Corre, V. M.; Wu, M.; Dong, L.; Peng, Z.; Zhao, Y.; Jang, D.; Spiecker, E.; et al. Efficient, Stable, and Fully Printed Carbon-Electrode Perovskite Solar Cells Enabled by Hole-Transporting Bilayers. *Joule*, **2023**, *7* (8), 1920–1937. https://doi.org/10.1016/j.joule.2023.06.005.

[16] Wijewardane, S. Inventions, Innovations and New Technologies. *Solar Compass*, **2022**, *1*, 100012. https://doi.org/10.1016/j.solcom.2022.100012.

[17] Krishna, B. G.; Sundar Ghosh, D.; Tiwari, S. Progress in Ambient Air-Processed Perovskite Solar Cells: Insights into Processing Techniques and Stability Assessment. *Solar Energy*, **2021**, *224*, 1369–1395. https://doi.org/10.1016/J.SOLENER.2021.07.002.

[18] Abbasi, M. S.; Sultana, R.; Ahmed, I.; Adnan, M.; Shah, U. A.; Irshad, M. S.; Vu, H. N.; Do, L. T.; Thi Vu, H. H.; Pham, T. D.; et al. Contemporary Advances in Organic Thermoelectric Materials: Fundamentals, Properties, Optimization Strategies, and Applications. *Renewable and Sustainable Energy Reviews*, **2024**, *200*, 114579. https://doi.org/10.1016/J.RSER.2024.114579.

[19] Rogdakis, K.; Loizos, M.; Viskadouros, G.; Kymakis, E. Memristive Perovskite Solar Cells towards Parallel Solar Energy Harvesting and Processing-in-Memory Computing. *Materials Advances*, **2022**, *3* (18), 7002–7014. https://doi.org/10.1039/D2MA00402J.

[20] Zhang, Q.; Hao, F.; Li, J.; Zhou, Y.; Wei, Y.; Lin, H. Perovskite Solar Cells: Must Lead Be Replaced—and Can It Be Done? *Science and Technology of Advanced Materials*, **2018**, *19* (1), 425–442. https://doi.org/10.1080/14686996.2018.1460176.

[21] Almosni, S.; Delamarre, A.; Jehl, Z.; Suchet, D.; Cojocaru, L.; Giteau, M.; Behaghel, B.; Julian, A.; Ibrahim, C.; Tatry, L.; et al. Material Challenges for Solar Cells in the Twenty-First Century: Directions in Emerging Technologies. *Science and Technology of Advance Materials*, **2018**, *19* (1), 336–369. https://doi.org/10.1080/14686996.2018.1433439.

[22] Singh, A.; Dhau, J.; Kumar, R.; Badru, R.; Singh, P.; Kumar Mishra, Y.; Kaushik, A. Tailored Carbon Materials (TCM) for Enhancing Photocatalytic Degradation of Polyaromatic Hydrocarbons. *Progress in Materials Science*, **2024**, *144*, 101289. https://doi.org/10.1016/J.PMATSCI.2024.101289.

[23] Cramer, C. L.; Ionescu, E.; Graczyk-Zajac, M.; Nelson, A. T.; Katoh, Y.; Haslam, J. J.; Wondraczek, L.; Aguirre, T. G.; LeBlanc, S.; Wang, H.; et al. Additive Manufacturing of Ceramic Materials for Energy Applications: Road Map and Opportunities. *Journal of the European Ceramic Society*, **2022**, *42* (7), 3049–3088. https://doi.org/10.1016/J.JEURCERAMSOC.2022.01.058.

[24] Matias, M. L.; Pereira, C.; Almeida, H. V.; Jana, S.; Panigrahi, S.; Menda, U. D.; Nunes, D.; Fortunato, E.; Martins, R.; Nandy, S. 3D Printed MXene Architectures for a Plethora of Smart Applications. *Materials Today Advances*, **2024**, *23*, 100512. https://doi.org/10.1016/J.MTADV.2024.100512.

[25] Pereira, C.; Pereira, A. M.; Freire, C.; Pinto, T. V.; Costa, R. S.; Teixeira, J. S. Nanoengineered Textiles: From Advanced Functional Nanomaterials to Groundbreaking High-Performance Clothing. *Handbook of Functionalized Nanomaterials for Industrial Applications*, **2020**, 611–714. https://doi.org/10.1016/B978-0-12-816787-8.00021-1.

[26] Liu, J.; Zheng, D.; Wang, K.; Li, Z.; Liu, S.; Peng, L.; Yang, D. Evolutionary Manufacturing Approaches for Advancing Flexible Perovskite Solar Cells. *Joule*, **2024**, *8* (4), 944–969. https://doi.org/10.1016/J.JOULE.2024.02.025.

[27] Marques, M. J. M.; Lin, W.; Taima, T.; Umezu, S.; Shahiduzzaman, M. Unleashing the Potential of Industry Viable Roll-to-Roll Compatible Technologies for Perovskite Solar Cells: Challenges and Prospects. *Materials Today*, **2024**. https://doi.org/10.1016/J.MATTOD.2024.06.013.

[28] Ravikumar, K.; Dangate, M. S. Advancements in Stretchable Organic Optoelectronic Devices and Flexible Transparent Conducting Electrodes: Current Progress and Future Prospects. *Heliyon*, **2024**, *10* (13), e33002. https://doi.org/10.1016/J.HELIYON.2024.E33002.

[29] Søndergaard, R.; Hösel, M.; Angmo, D.; Larsen-Olsen, T. T.; Krebs, F. C. Roll-to-Roll Fabrication of Polymer Solar Cells. *Materials Today*, **2012**, *15* (1–2), 36–49. https://doi.org/10.1016/S1369-7021(12)70019-6.

[30] Chilvery, A.; Das, S.; Guggilla, P.; Brantley, C.; Sunda-Meya, A. A Perspective on the Recent Progress in Solution-Processed Methods for Highly Efficient Perovskite Solar Cells. *Science and Technology of Advanced Materials*, **2016**, *17* (1), 650–658. https://doi.org/10.1080/14686996.2016.1226120.

[31] Dananjaya, V.; Marimuthu, S.; Yang, R. (Chunhui); Grace, A. N.; Abeykoon, C. Synthesis, Properties, Applications, 3D Printing and Machine Learning of Graphene Quantum Dots in Polymer Nanocomposites. *Progress in Materials Science*, **2024**, *144*, 101282. https://doi.org/10.1016/J.PMATSCI.2024.101282.

[32] Khan, Y.; Thielens, A.; Muin, S.; Ting, J.; Baumbauer, C.; Arias, A. C. A New Frontier of Printed Electronics: Flexible Hybrid Electronics. *Advanced Materials*, **2020**, *32* (15). https://doi.org/10.1002/adma.201905279.

[33] Kanishka, K.; Acherjee, B. Revolutionizing Manufacturing: A Comprehensive Overview of Additive Manufacturing Processes, Materials, Developments, and Challenges. *Journal of Manufacturing Processes*, **2023**, *107*, 574–619. https://doi.org/10.1016/J.JMAPRO.2023.10.024.

[34] Siddiqui, S.; Surananai, S.; Sainath, K.; Zubair Khan, M.; Raja Pandiyan Kuppusamy, R.; Kempaiah Suneetha, Y. Emerging Trends in Development and Application of 3D Printed Nanocomposite Polymers for Sustainable Environmental Solutions. *European Polymer Journal*, **2023**, *196*, 112298. https://doi.org/10.1016/J.EURPOLYMJ.2023.112298.

[35] Shaker, L. M.; Mohammed, J. K.; Basem, A.; Halbos, R. J.; Mahdi, R. R.; Mohammed, S. A.; Fayad, M. A.; Al-Amiery, A.; Al Lami, M. H. Comparative Analysis of Solar Cells and Hydrogen Fuel: A Mini-Review. *Results in Engineering*, **2024**, *23*, 102507. https://doi.org/10.1016/J.RINENG.2024.102507.

[36] Ravikumar, K.; Dangate, M. S. Advancements in Stretchable Organic Optoelectronic Devices and Flexible Transparent Conducting Electrodes: Current Progress and Future Prospects. *Heliyon*, **2024**, *10* (13), e33002. https://doi.org/10.1016/j.heliyon.2024.e33002.

[37] Marques, M. J. M.; Lin, W.; Taima, T.; Umezu, S.; Shahiduzzaman, Md. Unleashing the Potential of Industry Viable Roll-to-Roll Compatible Technologies for Perovskite Solar Cells: Challenges and Prospects. *Materials Today*, **2024**, *78*, 112–141. https://doi.org/10.1016/j.mattod.2024.06.013.

[38] Faheem, M. B.; Khan, B.; Hashmi, J. Z.; Baniya, A.; Subhani, W. S.; Bobba, R. S.; Yildiz, A.; Qiao, Q. Insights from Scalable Fabrication to Operational Stability and Industrial Opportunities for Perovskite Solar Cells and Modules. *Cell Reports Physical Science*, **2022**, *3* (4), 100827. https://doi.org/10.1016/J.XCRP.2022.100827.

[39] Casini, M. Advanced Building Construction Methods. *Construction 4.0*, **2022**, 405–470. https://doi.org/10.1016/B978-0-12-821797-9.00006-4.

[40] Mishra, S.; Ghosh, S.; Boro, B.; Kumar, D.; Porwal, S.; Paul, M.; Dixit, H.; Singh, T. Solution-Processed Next Generation Thin Film Solar Cells for Indoor Light Applications. *Energy Advances*, **2022**, *1* (11), 761–792. https://doi.org/10.1039/D2YA00075J.

[41] Meena, J. S.; Choi, S. Bin; Jung, S. B.; Kim, J. W. Electronic Textiles: New Age of Wearable Technology for Healthcare and Fitness Solutions. *Materials Today Bio*, **2023**, *19*, 100565. https://doi.org/10.1016/J.MTBIO.2023.100565.

[42] Baskaran, P.; Rajasekar, M. Recent Trends and Future Perspectives of Thermoelectric Materials and Their Applications. *RSC Advances*, **2024**, *14* (30), 21706–21744. https://doi.org/10.1039/D4RA03625E.

[43] Wan, J.; Zhang, R.; Li, Y.; Li, Y. Applications of Organic Solar Cells in Wearable Electronics. *Wearable Electronics*, **2024**, *1*, 26–40. https://doi.org/10.1016/J.WEES.2024.03.001.

[44] Mishra, A.; Singh, P. K.; Chauhan, N.; Roy, S.; Tiwari, A.; Gupta, S.; Tiwari, A.; Patra, S.; Das, T. R.; Mishra, P.; et al. Emergence of Integrated Biosensing-Enabled Digital Healthcare Devices. *Sensors and Diagnostics*, **2024**, *3* (5), 718–744. https://doi.org/10.1039/D4SD00017J.

[45] Bhatnagar, P.; Zaferani, S. H.; Rafiefard, N.; Baraeinejad, B.; Vazifeh, A. R.; Mohammadpour, R.; Ghomashchi, R.; Dillersberger, H.; Tham, D.; Vashaee, D. Advancing Personalized Healthcare and Entertainment: Progress in Energy Harvesting Materials and Techniques of Self-Powered Wearable Devices. *Progress in Materials Science*, **2023**, *139*, 101184. https://doi.org/10.1016/J.PMATSCI.2023.101184.

[46] Gregor-Svetec, D. Intelligent Packaging. *Nanomaterials for Food Packaging: Materials, Processing Technologies, and Safety Issues*, **2018**, 203–247. https://doi.org/10.1016/B978-0-323-51271-8.00008-5.

[47] Tran, T. S.; Dutta, N. K.; Choudhury, N. R. Graphene Inks for Printed Flexible Electronics: Graphene Dispersions, Ink Formulations, Printing Techniques and Applications. *Advances in Colloid and Interface Science*, **2018**, *261*, 41–61. https://doi.org/10.1016/J.CIS.2018.09.003.

[48] Qamar, M. Z.; Khalid, Z.; Shahid, R.; Tsoi, W. C.; Mishra, Y. K.; Kyaw, A. K. K.; Saeed, M. A. Advancement in Indoor Energy Harvesting through Flexible Perovskite Photovoltaics for Self-Powered IoT Applications. *Nano Energy*, **2024**, *129*, 109994. https://doi.org/10.1016/J.NANOEN.2024.109994.

[49] Agresti, A.; Di Giacomo, F.; Pescetelli, S.; Di Carlo, A. Scalable Deposition Techniques for Large-Area Perovskite Photovoltaic Technology: A Multi-Perspective Review. *Nano Energy*, **2024**, *122*, 109317. https://doi.org/10.1016/J.NANOEN.2024.109317.

[50] Harifi, T. Textiles for Flexible Solar Cells and 3D Printable Materials. *Nanosensors and Nanodevices for Smart Multifunctional Textiles*, **2021**, 125–134. https://doi.org/10.1016/B978-0-12-820777-2.00008-X.

[51] Mishra, A.; Singh, P. K.; Chauhan, N.; Roy, S.; Tiwari, A.; Gupta, S.; Tiwari, A.; Patra, S.; Das, T. R.; Mishra, P.; et al. Emergence of Integrated Biosensing-Enabled Digital Healthcare Devices. *Sensors and Diagnostics*, **2024**, *3* (5), 718–744. https://doi.org/10.1039/D4SD00017J.

[52] Chowdhury, S.; Yadaiah, N.; Prakash, C.; Ramakrishna, S.; Dixit, S.; Gupta, L. R.; Buddhi, D. Laser Powder Bed Fusion: A State-of-the-Art Review of the Technology, Materials, Properties & Defects, and Numerical Modelling. *Journal of Materials Research and Technology*, **2022**, *20*, 2109–2172. https://doi.org/10.1016/J.JMRT.2022.07.121.

[53] Mansour, M.; Saeed Darweesh, M.; Soltan, A. Wearable Devices for Glucose Monitoring: A Review of State-of-the-Art Technologies and Emerging Trends. *Alexandria Engineering Journal*, **2024**, *89*, 224–243. https://doi.org/10.1016/J.AEJ.2024.01.021.

[54] Aftab, S.; Li, X.; Kabir, F.; Akman, E.; Aslam, M.; Pallavolu, M. R.; Koyyada, G.; Assiri, M. A.; Rajpar, A. H. Lighting the Future: Perovskite Nanorods and Their Advances across Applications. *Nano Energy*, **2024**, *124*, 109504. https://doi.org/10.1016/J.NANOEN.2024.109504.

[55] Carlé, J. E.; Helgesen, M.; Hagemann, O.; Hösel, M.; Heckler, I. M.; Bundgaard, E.; Gevorgyan, S. A.; Søndergaard, R. R.; Jørgensen, M.; García-Valverde, R.; et al. Overcoming the Scaling Lag for Polymer Solar Cells. *Joule*, **2017**, *1* (2), 274–289. https://doi.org/10.1016/J.JOULE.2017.08.002.

[56] Leschok, M.; Cheibas, I.; Piccioni, V.; Seshadri, B.; Schlüter, A.; Gramazio, F.; Kohler, M.; Dillenburger, B. 3D Printing Facades: Design, Fabrication, and Assessment Methods. *Autom Constr*, **2023**, *152*, 104918. https://doi.org/10.1016/J.AUTCON.2023.104918.

[57] Muthukutty, B.; Kumar, P. S.; Vivekanandan, A. K.; Sivakumar, M.; Lee, S.; Lee, D. Progress and Perspective in Harnessing MXene–Carbon-Based Composites (0–3D): Synthesis, Performance, and Applications. *Chemosphere*, **2024**, *355*, 141838. https://doi.org/10.1016/J.CHEMOSPHERE.2024.141838.

[58] Singh, A.; Dhau, J.; Kumar, R.; Badru, R.; Singh, P.; Kumar Mishra, Y.; Kaushik, A. Tailored Carbon Materials (TCM) for Enhancing Photocatalytic Degradation of Polyaromatic Hydrocarbons. *Progress in Materials Science*, **2024**, *144*, 101289. https://doi.org/10.1016/J.PMATSCI.2024.101289.

[59] Qin, H.; Ding, S.; Ashour, A.; Zheng, Q.; Han, B. Revolutionizing Infrastructure: The Evolving Landscape of Electricity-Based Multifunctional Concrete from Concept to Practice. *Progress in Materials Science*, **2024**, *145*, 101310. https://doi.org/10.1016/J.PMATSCI.2024.101310.

[60] Kumar, S.; Taunk, M. 2D MXenes for Flexible Device Applications. *Materials Today Physics*, **2024**, *46*, 101483. https://doi.org/10.1016/J.MTPHYS.2024.101483.

Index